CONTENTS

CHAPTER 1

EXERCISES 1.1, page 10

1. The statement is false because -3 is greater than -20. (See the number line that follows).

    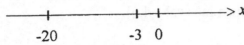

    $$-20 \qquad -3 \quad 0$$

3. The statement is false because 2/3 [which is equal to (4/6)] is less than 5/6.

    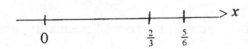

    $$0 \qquad\qquad \tfrac{2}{3} \quad \tfrac{5}{6}$$

5. The interval (3,6) is shown on the number line that follows. Note that this is an open interval indicated by (and).

    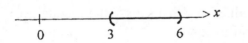

    $$0 \qquad 3 \qquad 6$$

7. The interval [-1,4) is shown on the number line that follows. Note that this is a half-open interval indicated by [(closed) and) (open).

    $$-1 \qquad\qquad 4$$

9. The infinite interval (0,∞) is shown on the number line that follows.

    $$0$$

11. First, $2x + 4 < 8$ (Add -4 to each side of the inequality.)
    Next, $2x < 4$ (Multiply each side of the inequality by 1/2)
    and $x < 2$.
    We write this in interval notation as $(-\infty,2)$.

13. We are given the inequality $-4x \geq 20$.
Then $x \leq -5$.   (Multiply both sides of the inequality by -1/4 and reverse the sign of the inequality.)
We write this in interval notation as $(-\infty, -5]$.

15. We are given the inequality $-6 < x - 2 < 4$.
First $-6 + 2 < x < 4 + 2$ (Add +2 to each member of the inequality.)
and $-4 < x < 6$,
so the solution set is the open interval $(-4,6)$.

17. We want to find the values of x that satisfy the inequalities
$$x + 1 > 4 \text{ or } x + 2 < -1.$$
Adding -1 to both sides of the first inequality, we obtain
$$x + 1 - 1 > 4 - 1,$$
or $x > 3$.
Similarly, adding -2 to both sides of the second inequality, we obtain
$$x + 2 - 2 < -1 - 2,$$
or $x < -3$.
Therefore, the solution set is $(-\infty, -3) \cup (3, \infty)$.

19. We want to find the values of x that satisfy the inequalities
$$x + 3 > 1 \text{ and } x - 2 < 1.$$
Adding -3 to both sides of the first inequality, we obtain
$$x + 3 - 3 > 1 - 3,$$
or $x > -2$.
Similarly, adding 2 to each side of the second inequality, we obtain
$$x < 3,$$
and the solution set is $(-2,3)$.

21. $|-6 + 2| = 4$.

23. $\dfrac{|-12 + 4|}{|16 - 12|} = \dfrac{|-8|}{|4|} = 2$.

25. $\sqrt{3}|-2| + 3|-\sqrt{3}| = \sqrt{3}(2) + 3\sqrt{3} = 5\sqrt{3}$.

27. $|\pi - 1| + 2 = \pi - 1 + 2 = \pi + 1$.

29. $\left|\sqrt{2}-1\right|+\left|3-\sqrt{2}\right| = \sqrt{2}-1+3-\sqrt{2} = 2.$

31. False. If $a > b$, then $-a < -b$, $-a + b < -b + b$, and $b - a < 0$.

33. False. Let $a = -2$ and $b = -3$. Then $a^2 = 4$ and $b^2 = 9$, and $4 < 9$. Note that we only need to provide a counterexample to show that the statement is not always true.

35. True. There are three possible cases.

   Case 1 If $a > 0$, $b > 0$, then $a^3 > b^3$, since $a^3 - b^3 = (a - b)(a^2 + ab + b^2) > 0$.

   Case 2 If $a > 0$, $b < 0$, then $a^3 > 0$ and $b^3 < 0$ and it follows that $a^3 > b^3$.

   Case 3 If $a < 0$ and $b < 0$, then $a^3 - b^3 = (a - b)(a^2 + ab + b^2) > 0$, and we see that $a^3 > b^3$. (Note that $(a - b) > 0$ and $ab > 0$.)

37. False. Take $a = -2$, then $\left|-a\right| = \left|-(-2)\right| = \left|2\right| = 2 \neq a$.

39. True. If $a - 4 < 0$, then $\left|a-4\right| = 4-a = \left|4-a\right|$. If $a - 4 > 0$, then
$$\left|4-a\right| = a-4 = \left|a-4\right|.$$

41. False. Take $a = 3$, $b = -1$. Then $\left|a+b\right| = \left|3-1\right| = 2 \neq \left|a\right| + \left|b\right| = 3+1 = 4.$

43. $27^{2/3} = (3^3)^{2/3} = 3^2 = 9.$

45. $\left(\dfrac{1}{\sqrt{3}}\right)^0 = 1.$ Recall that any number raised to the zero power is 1.

47. $\left[\left(\dfrac{1}{8}\right)^{1/3}\right]^{-2} = \left(\dfrac{1}{2}\right)^{-2} = (2^2) = 4.$

49. $\left(\dfrac{7^{-5} \cdot 7^2}{7^{-2}}\right)^{-1} = (7^{-5+2+2})^{-1} = (7^{-1})^{-1} = 7^1 = 7.$

51. $(125^{2/3})^{-1/2} = 125^{(2/3)(-1/2)} = 125^{-1/3} = \dfrac{1}{125^{1/3}} = \dfrac{1}{5}.$

53. $\dfrac{\sqrt{32}}{\sqrt{8}} = \sqrt{\dfrac{32}{8}} = \sqrt{4} = 2.$

55. $\dfrac{16^{5/8}16^{1/2}}{16^{7/8}} = 16^{(5/8+1/2-7/8)} = 16^{1/4} = 2.$

57. $16^{1/4} \cdot 8^{-1/3} = 2 \cdot \left(\dfrac{1}{8}\right)^{1/3} = 2 \cdot \dfrac{1}{2} = 1.$

59. True.

61. False. $x^3 \times 2x^2 = 2x^{3+2} = 2x^5 \neq 2x^6.$

63. False. $\dfrac{2^{4x}}{1^{3x}} = \dfrac{2^{4x}}{1} = 2^{4x}.$

65. False. $\dfrac{1}{4^{-3}} = 4^3 = 64.$

67. False. $(1.2^{1/2})^{-1/2} = (1.2)^{-1/4} \neq 1.$

69. $(xy)^{-2} = \dfrac{1}{(xy)^2}.$

71. $\dfrac{x^{-1/3}}{x^{1/2}} = x^{(-1/3)-(1/2)} = x^{-5/6} = \dfrac{1}{x^{5/6}}.$

73. $12^0(s+t)^{-3} = 1 \cdot \dfrac{1}{(s+t)^3} = \dfrac{1}{(s+t)^3}.$

75. $\dfrac{x^{7/3}}{x^{-2}} = x^{(7/3)+2} = x^{(7/3)+(6/3)} = x^{13/3}.$

77. $(x^2 y^{-3})(x^{-5}y^3) = (x^{2-5}y^{-3+3}) = x^{-3}y^0 = x^{-3} = \dfrac{1}{x^3}.$

79. $\dfrac{x^{3/4}}{x^{-1/4}} = x^{(3/4)-(-1/4)} = x^{4/4} = x.$

81. $\left(\dfrac{x^3}{-27y^{-6}}\right)^{-2/3} = x^{3(-2/3)}\left(-\dfrac{1}{27}\right)^{-2/3} y^{6(-2/3)} = x^{-2}\left(-\dfrac{1}{3}\right)^{-2} y^{-4} = \dfrac{9}{x^2 y^4}.$

83. $\left(\dfrac{x^{-3}}{y^{-2}}\right)^2 \left(\dfrac{y}{x}\right)^4 = \dfrac{x^{-3(2)}y^4}{y^{-2(2)}x^4} = \left(\dfrac{y^{4+4}}{x^{4+6}}\right) = \dfrac{y^8}{x^{10}}.$

85. $\sqrt[3]{x^{-2}} \cdot \sqrt{4x^5} = x^{-2/3} \cdot 4^{1/2} \cdot x^{5/2} = x^{-(2/3)+(5/2)} \cdot 2 = 2x^{11/6}.$

87. $-\sqrt[4]{16x^4y^8} = -(16^{1/4} \cdot x^{4/4} \cdot y^{8/4}) = -2xy^2.$

89. $\sqrt[6]{64x^8y^3} = (64)^{1/6} \cdot x^{8/6}y^{3/6} = 2x^{4/3}y^{1/2}.$

91. $2^{3/2} = (2)(2^{1/2}) = 2(1.414) = 2.828.$

93. $9^{3/4} = (3^2)^{3/4} = 3^{6/4} = 3^{3/2} = 3 \cdot 3^{1/2} = 3(1.732) = 5.196.$

95. $10^{3/2} = 10^{1/2} \cdot 10 = (3.162)(10) = 31.62.$

97. $10^{2.5} = 10^2 \cdot 10^{1/2} = 100(3.162) = 316.2.$

99. $\dfrac{3}{2\sqrt{x}} \cdot \dfrac{\sqrt{x}}{\sqrt{x}} = \dfrac{3\sqrt{x}}{2x}.$

101. $\dfrac{2y}{\sqrt{3y}} \cdot \dfrac{\sqrt{3y}}{\sqrt{3y}} = \dfrac{2y\sqrt{3y}}{3y} = \dfrac{2}{3}\sqrt{3y}.$

103. $\dfrac{1}{\sqrt[3]{x}} \cdot \dfrac{\sqrt[3]{x^2}}{\sqrt[3]{x^2}} = \dfrac{\sqrt[3]{x^2}}{\sqrt[3]{x^3}} = \dfrac{\sqrt[3]{x^2}}{x}.$

105. $\dfrac{2\sqrt{x}}{3} \cdot \dfrac{\sqrt{x}}{\sqrt{x}} = \dfrac{2x}{3\sqrt{x}}.$

107. $\sqrt{\dfrac{2y}{x}} = \dfrac{\sqrt{2y}}{\sqrt{x}} \cdot \dfrac{\sqrt{2y}}{\sqrt{2y}} = \dfrac{2y}{\sqrt{2xy}}.$

109. $\dfrac{\sqrt[3]{x^2z}}{y} \cdot \dfrac{\sqrt[3]{xz^2}}{\sqrt[3]{xz^2}} = \dfrac{\sqrt[3]{x^3z^3}}{y\sqrt[3]{xz^2}} = \dfrac{xz}{y\sqrt[3]{xz^2}}.$

111. If the car is driven in the city, then it can be expected to cover
$$(18.1)(20) = 362 \qquad \text{(miles/gal} \cdot \text{gal)}$$
or 362 miles on a full tank. If the car is driven on the highway, then it can be expected to cover
$$(18.1)(27) = 488.7 \qquad \text{(miles/gal} \cdot \text{gal)}$$
or 488.7 miles on a full tank. Thus, the driving range of the car may be described by the interval [362, 488.7].

113.
$$6(P - 2500) \le 4(P + 2400)$$
$$6P - 15000 \le 4P + 9600$$
$$2P \le 24600, \text{ or } P \le 12300.$$
Therefore, the maximum profit is $12,300.

115. Let x represent the salesman's monthly sales in dollars. Then
$$0.15(x - 12000) \geq 3000$$
$$15(x - 12000) \geq 300000$$
$$15x - 180000 \geq 300000$$
$$15x \geq 480000$$
$$x \geq 32000.$$
We conclude that the salesman must earn at least $32,000 to reach his goal.

117. The rod is acceptable if $0.49 < x < 0.51$ or $-0.01 < x - 0.5 < 0.01$. This gives the required inequality $|x - 0.5| < 0.01$.

119. We want to solve the inequality
$$-6x^2 + 30x - 10 \geq 14. \qquad \text{(Remember } x \text{ is expressed in thousands.)}$$
Adding -14 to both sides of this inequality, we have
$$-6x^2 + 30x - 10 - 14 \geq 14 - 14,$$
or $\qquad -6x^2 + 30x - 24 \geq 0.$
Dividing both sides of the inequality by -6 (which reverses the sign of the inequality), we have
$$x^2 - 5x + 4 \leq 0.$$
Factoring this last expression, we have
$$(x - 4)(x - 1) \leq 0.$$
From the following sign diagram,

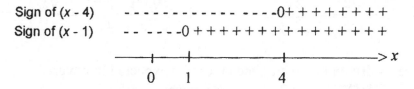

we see that x must lie between 1 and 4. (The inequality is only satisfied when the two factors have opposite signs.) Since x is expressed in thousands of units, we see that the manufacturer must produce between 1000 and 4000 units of the commodity.

121. False. Take $a = 1$, $b = 2$, and $c = 3$. Then $a < b$, but
$$a - c = 1 - 3 = -2 \not> 2 - 3 = -1 = b - c.$$

123. True. $|a - b| = |a + (-b)| \leq |a| + |-b| = |a| + |b|.$

EXERCISES 1.2, page 22

1. $(7x^2 - 2x + 5) + (2x^2 + 5x - 4) = 7x^2 - 2x + 5 + 2x^2 + 5x - 4$
$$= 9x^2 + 3x + 1.$$

3. $(5y^2 - 2y + 1) - (y^2 - 3y - 7) = 5y^2 - 2y + 1 - y^2 + 3y + 7$
$$= 4y^2 + y + 8.$$

5. $x - \{2x - [-x - (1 - x)]\} = x - \{2x - [-x - 1 + x]\}$
$$= x - \{2x + 1\}$$
$$= x - 2x - 1$$
$$= -x - 1.$$

7. $(\dfrac{1}{3} - 1 + e) - (-\dfrac{1}{3} - 1 + e^{-1}) = \dfrac{1}{3} - 1 + e + \dfrac{1}{3} + 1 - \dfrac{1}{e}$
$$= \dfrac{2}{3} + e - \dfrac{1}{e}$$
$$= \dfrac{3e^2 + 2e - 3}{3e}.$$

9. $3\sqrt{8} + 8 - 2\sqrt{y} + \dfrac{1}{2}\sqrt{x} - \dfrac{3}{4}\sqrt{y} = 3\sqrt{4 \cdot 2} + 8 + \dfrac{1}{2}\sqrt{x} - \dfrac{11}{4}\sqrt{y}$
$$= 6\sqrt{2} + 8 + \dfrac{1}{2}\sqrt{x} - \dfrac{11}{4}\sqrt{y}.$$

11. $(x + 8)(x - 2) = x(x - 2) + 8(x - 2) = x^2 - 2x + 8x - 16 = x^2 + 6x - 16.$

13. $(a + 5)^2 = (a + 5)(a + 5) = a(a + 5) + 5(a + 5) = a^2 + 5a + 5a + 25$
$$= a^2 + 10a + 25.$$

15. $(x + 2y)^2 = (x + 2y)(x + 2y) = x(x + 2y) + 2y(x + 2y)$
$$= x^2 + 2xy + 2yx + 4y^2 = x^2 + 4xy + 4y^2.$$

17. $(2x + y)(2x - y) = 2x(2x - y) + y(2x - y) = 4x^2 - 2xy + 2xy - y^2$
$$= 4x^2 - y^2.$$

19. $(x^2 - 1)(2x) - x^2(2x) = 2x^3 - 2x - 2x^3 = -2x.$

21. $2\left(t + \sqrt{t}\right)^2 - 2t^2 = 2(t + \sqrt{t})(t + \sqrt{t}) - 2t^2$
$$= 2(t^2 + 2t\sqrt{t} + t) - 2t^2$$
$$= 2t^2 + 4t\sqrt{t} + 2t - 2t^2$$
$$= 4t\sqrt{t} + 2t = 2t(2\sqrt{t} + 1).$$

23. $4x^5 - 12x^4 - 6x^3 = 2x^3(2x^2 - 6x - 3).$

25. $7a^4 - 42a^2b^2 + 49a^3b = 7a^2(a^2 - 6b^2 + 7ab).$

27. $e^{-x} - xe^{-x} = e^{-x}(1 - x).$

29. $2x^{-5/2} - \frac{3}{2}x^{-3/2} = \frac{1}{2}x^{-5/2}(4 - 3x).$

31. $6ac + 3bc - 4ad - 2bd = 3c(2a + b) - 2d(2a + b) = (2a + b)(3c - 2d).$

33. $4a^2 - b^2 = (2a + b)(2a - b).$ (Difference of two squares)

35. $10 - 14x - 12x^2 = -2(6x^2 + 7x - 5) = -2(3x + 5)(2x - 1).$

37. $3x^2 - 6x - 24 = 3(x^2 - 2x - 8) = 3(x - 4)(x + 2).$

39. $12x^2 - 2x - 30 = 2(6x^2 - x - 15) = 2(3x - 5)(2x + 3).$

41. $9x^2 - 16y^2 = (3x)^2 - (4y)^2 = (3x - 4y)(3x + 4y).$

43. $x^6 + 125 = (x^2)^3 + (5)^3 = (x^2 + 5)(x^4 - 5x^2 + 25).$

45. $(x^2 + y^2)x - xy(2y) = x^3 + xy^2 - 2xy^2 = x^3 - xy^2.$

47. $2(x - 1)(2x + 2)^3[4(x - 1) + (2x + 2)]$
$$= 2(x - 1)(2x + 2)^3[4x - 4 + 2x + 2]$$
$$= 2(x - 1)(2x + 2)^3[6x - 2]$$
$$= 4(x - 1)(3x - 1)(2x + 2)^3.$$

49. $4(x - 1)^2(2x + 2)^3(2) + (2x + 2)^4(2)(x - 1)$
$$= 2(x - 1)(2x + 2)^3[4(x - 1) + (2x + 2)] = 2(x - 1)(2x + 2)^3(6x - 2)$$
$$= 4(x - 1)(3x - 1)(2x + 2)^3.$$

51. $(x^2 + 2)^2[5(x^2 + 2)^2 - 3](2x) = (x^2 + 2)^2[5(x^4 + 4x^2 + 4) - 3](2x)$
$$= (2x)(x^2 + 2)^2(5x^4 + 20x^2 + 17).$$

53. $x^2 + x - 12 = 0$, or $(x + 4)(x - 3) = 0$, so that $x = -4$ or $x = 3$. We conclude that the roots are $x = -4$ and $x = 3$.

55. $4t^2 + 2t - 2 = (2t - 1)(2t + 2) = 0$. Thus, $t = 1/2$ and $t = -1$ are the roots.

57. $\frac{1}{4}x^2 - x + 1 = (\frac{1}{2}x - 1)(\frac{1}{2}x - 1) = 0$. Thus $\frac{1}{2}x = 1$, and $x = 2$ is a double root of the equation.

59. Here we use the quadratic formula to solve the equation $4x^2 + 5x - 6 = 0$. Then, $a = 4$, $b = 5$, and $c = -6$. Therefore,

$$x = \frac{-b \pm \sqrt{b^2 - 4ac}}{2a} = \frac{-(5) \pm \sqrt{(5)^2 - 4(4)(-6)}}{2(4)} = \frac{-5 \pm \sqrt{121}}{8}$$
$$= \frac{-5 \pm 11}{8}.$$

Thus, $x = -\frac{16}{8} = -2$ and $x = \frac{6}{8} = \frac{3}{4}$ are the roots of the equation.

61. We use the quadratic formula to solve the equation $8x^2 - 8x - 3 = 0$. Here $a = 8$, $b = -8$, and $c = -3$. Therefore,

$$x = \frac{-b \pm \sqrt{b^2 - 4ac}}{2a} = \frac{-(-8) \pm \sqrt{(-8)^2 - 4(8)(-3)}}{2(8)} = \frac{8 \pm \sqrt{160}}{16}$$
$$= \frac{8 \pm 4\sqrt{10}}{16} = \frac{2 \pm \sqrt{10}}{4}.$$

Thus, $x = \frac{1}{2} + \frac{1}{4}\sqrt{10}$ and $x = \frac{1}{2} - \frac{1}{4}\sqrt{10}$ are the roots of the equation.

63. We use the quadratic formula to solve $2x^2 + 4x - 3 = 0$. Here, $a = 2$, $b = 4$, and $c = -3$. Therefore

$$x = \frac{-b \pm \sqrt{b^2 - 4ac}}{2a} = \frac{-(4) \pm \sqrt{(4)^2 - 4(2)(-3)}}{2(2)} = \frac{-4 \pm \sqrt{40}}{4}$$

$$= \frac{-4 \pm 2\sqrt{10}}{4} = \frac{-2 \pm \sqrt{10}}{2}.$$

Thus, $x = -1 + \frac{1}{2}\sqrt{10}$ and $x = -1 - \frac{1}{2}\sqrt{10}$ are the roots of the equation.

65. $\dfrac{x^2 + x - 2}{x^2 - 4} = \dfrac{(x+2)(x-1)}{(x+2)(x-2)} = \dfrac{x-1}{x-2}.$

67. $\dfrac{12t^2 + 12t + 3}{4t^2 - 1} = \dfrac{3(4t^2 + 4t + 1)}{4t^2 - 1} = \dfrac{3(2t+1)(2t+1)}{(2t+1)(2t-1)} = \dfrac{3(2t+1)}{2t-1}.$

69. $\dfrac{(4x-1)(3) - (3x+1)(4)}{(4x-1)^2} = \dfrac{12x - 3 - 12x - 4}{(4x-1)^2} = -\dfrac{7}{(4x-1)^2}.$

71. $\dfrac{2a^2 - 2b^2}{b-a} \cdot \dfrac{4a + 4b}{a^2 + 2ab + b^2} = \dfrac{2(a+b)(a-b)4(a+b)}{-(a-b)(a+b)(a+b)} = -8.$

73. $\dfrac{3x^2 + 2x - 1}{2x + 6} \div \dfrac{x^2 - 1}{x^2 + 2x - 3} = \dfrac{(3x-1)(x+1)}{2(x+3)} \cdot \dfrac{(x+3)(x-1)}{(x+1)(x-1)} = \dfrac{3x-1}{2}.$

75. $\dfrac{58}{3(3t+2)} + \dfrac{1}{3} = \dfrac{58 + 3t + 2}{3(3t+2)} = \dfrac{3t + 60}{3(3t+2)} = \dfrac{t + 20}{3t + 2}.$

77. $\dfrac{2x}{2x-1} - \dfrac{3x}{2x+5} = \dfrac{2x(2x+5) - 3x(2x-1)}{(2x-1)(2x+5)} = \dfrac{4x^2 + 10x - 6x^2 + 3x}{(2x-1)(2x+5)}$

$$= \dfrac{-2x^2 + 13x}{(2x-1)(2x+5)} = -\dfrac{x(2x-13)}{(2x-1)(2x+5)}.$$

79. $\dfrac{4}{x^2 - 9} - \dfrac{5}{x^2 - 6x + 9} = \dfrac{4}{(x+3)(x-3)} - \dfrac{5}{(x-3)^2}$

$$= \dfrac{4(x-3) - 5(x+3)}{(x-3)^2(x+3)} = -\dfrac{x + 27}{(x-3)^2(x+3)}.$$

81. $\dfrac{1+\dfrac{1}{x}}{1-\dfrac{1}{x}}=\dfrac{\dfrac{x+1}{x}}{\dfrac{x-1}{x}}=\dfrac{x+1}{x}\cdot\dfrac{x}{x-1}=\dfrac{x+1}{x-1}.$

83. $\dfrac{4x^2}{2\sqrt{2x^2+7}}+\sqrt{2x^2+7}=\dfrac{4x^2+2\sqrt{2x^2+7}\sqrt{2x^2+7}}{2\sqrt{2x^2+7}}=\dfrac{4x^2+4x^2+14}{2\sqrt{2x^2+7}}$

$$=\dfrac{4x^2+7}{\sqrt{2x^2+7}}.$$

85. $\dfrac{2x(x+1)^{-1/2}-(x+1)^{1/2}}{x^2}=\dfrac{(x+1)^{-1/2}(2x-x-1)}{x^2}=\dfrac{(x+1)^{-1/2}(x-1)}{x^2}$

$$=\dfrac{x-1}{x^2\sqrt{x+1}}.$$

87. $\dfrac{(2x+1)^{1/2}-(x+2)(2x+1)^{-1/2}}{2x+1}=\dfrac{(2x+1)^{-1/2}(2x+1-x-2)}{2x+1}$

$$=\dfrac{(2x+1)^{-1/2}(x-1)}{2x+1}=\dfrac{x-1}{(2x+1)^{3/2}}.$$

89. $\dfrac{1}{\sqrt{3}-1}\cdot\dfrac{\sqrt{3}+1}{\sqrt{3}+1}=\dfrac{\sqrt{3}+1}{3-1}=\dfrac{\sqrt{3}+1}{2}.$

91. $\dfrac{1}{\sqrt{x}-\sqrt{y}}\cdot\dfrac{\sqrt{x}+\sqrt{y}}{\sqrt{x}+\sqrt{y}}=\dfrac{\sqrt{x}+\sqrt{y}}{x-y}.$

93. $\dfrac{\sqrt{a}+\sqrt{b}}{\sqrt{a}-\sqrt{b}}\cdot\dfrac{\sqrt{a}+\sqrt{b}}{\sqrt{a}+\sqrt{b}}=\dfrac{(\sqrt{a}+\sqrt{b})^2}{a-b}.$

95. $\dfrac{\sqrt{x}}{3}\cdot\dfrac{\sqrt{x}}{\sqrt{x}}=\dfrac{x}{3\sqrt{x}}.$

97. $\dfrac{1-\sqrt{3}}{3}\cdot\dfrac{1+\sqrt{3}}{1+\sqrt{3}}=\dfrac{1^2-(\sqrt{3})^2}{3(1+\sqrt{3})}=-\dfrac{2}{3(1+\sqrt{3})}.$

99. $\dfrac{1+\sqrt{x+2}}{\sqrt{x+2}}\cdot\dfrac{1-\sqrt{x+2}}{1-\sqrt{x+2}}=\dfrac{1-(x+2)}{\sqrt{x+2}(1-\sqrt{x+2})}=-\dfrac{x+1}{\sqrt{x+2}(1-\sqrt{x+2})}.$

101. True. The two real roots are $\dfrac{-b \pm \sqrt{b^2 - 4ac}}{2a}$.

103. False. Take $a = 2$, $b = 3$, and $c = 4$. Then

$$\frac{a}{b+c} = \frac{2}{3+4} = \frac{2}{7}. \text{ But } \frac{a}{b} + \frac{a}{c} = \frac{2}{3} + \frac{3}{4} = \frac{8+9}{12} = \frac{17}{12}.$$

EXERCISES 1.3, page 29

1. The coordinates of A are (3,3) and it is located in Quadrant I.

3. The coordinates of C are (2,-2) and it is located in Quadrant IV.

5. The coordinates of E are (-4,-6) and it is located in Quadrant III.

7. A   9. E, F, and G.   11. F

For Exercises 13-20, refer to the following figure.

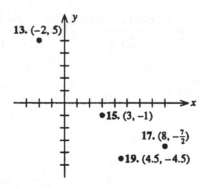

21. Using the distance formula, we find that $\sqrt{(4-1)^2 + (7-3)^2} = \sqrt{3^2 + 4^2} = \sqrt{25} = 5$.

23. Using the distance formula, we find that
$$\sqrt{(4-(-1))^2 + (9-3)^2} = \sqrt{5^2 + 6^2} = \sqrt{25+36} = \sqrt{61}.$$

25. The coordinates of the points have the form $(x, -6)$. Since the points are 10 units away from the origin, we have

$$(x - 0)^2 + (-6 - 0)^2 = 10^2$$
$$x^2 = 64,$$
or $x = \pm 8$. Therefore, the required points are $(-8, -6)$ and $(8, -6)$.

27. The points are shown in the diagram that follows.

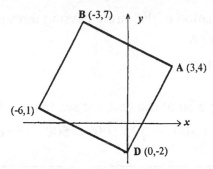

To show that the four sides are equal, we compute the following:

$$d(A,B) = \sqrt{(-3-3)^2 + (7-4)^2} = \sqrt{(-6)^2 + 3^2} = \sqrt{45}$$
$$d(B,C) = \sqrt{[(-6-(-3)]^2 + (1-7)^2} = \sqrt{(-3)^2 + (-6)^2} = \sqrt{45}$$
$$d(C,D) = \sqrt{[0-(-6)]^2 + [(-2)-1]^2} = \sqrt{(6)^2 + (-3)^2} = \sqrt{45}$$
$$d(A,D) = \sqrt{(0-3)^2 + (-2-4)^2} = \sqrt{(3)^2 + (-6)^2} = \sqrt{45}.$$

Next, to show that $\triangle ABC$ is a right triangle, we show that it satisfies the Pythagorean Theorem. Thus,

$$d(A,C) = \sqrt{(-6-3)^2 + (1-4)^2} = \sqrt{(-9)^2 + (-3)^2} = \sqrt{90} = 3\sqrt{10}$$

zand $[d(A,B)]^2 + [d(B,C)]^2 = 90 = [d(A,C)]^2$. Similarly, $d(B,D) = \sqrt{90} = 3\sqrt{10}$, so $\triangle BAD$ is a right triangle as well. It follows that $\angle B$ and $\angle D$ are right angles, and we conclude that $ADCB$ is a square

29. The equation of the circle with radius 5 and center $(2,-3)$ is given by
$$(x-2)^2 + [y-(-3)]^2 = 5^2$$
or $\qquad (x-2)^2 + (y+3)^2 = 25.$

31. The equation of the circle with radius 5 and center $(0, 0)$ is given by

$$(x-0)^2 + (y-0)^2 = 5^2$$
or $\qquad x^2 + y^2 = 25$

33. The distance between the points (5,2) and (2,-3) is given by
$$d = \sqrt{(5-2)^2 + (2-(-3))^2} = \sqrt{3^2 + 5^2} = \sqrt{34}.$$
Therefore $r = \sqrt{34}$ and the equation of the circle passing through (5,2) and (2,-3) is
$$(x-2)^2 + [y-(-3)]^2 = 34$$
or $\qquad (x-2)^2 + (y+3)^2 = 34.$

35. Referring to the diagram on page 30 of the text, we see that the distance from A to B is given by $d(A,B) = \sqrt{400^2 + 300^2} = \sqrt{250,000} = 500$. The distance from B to C is given by
$$d(B,C) = \sqrt{(-800-400)^2 + (800-300)^2} = \sqrt{(-1200)^2 + (500)^2}$$
$$= \sqrt{1,690,000} = 1300.$$
The distance from C to D is given by
$$d(C,D) = \sqrt{[-800-(-800)]^2 + (800-0)^2} = \sqrt{0 + 800^2} = 800 \ .$$
The distance from D to A is given by
$$d(D,A) = \sqrt{[(-800)-0]^2 + (0-0)} = \sqrt{640000} = 800.$$
Therefore, the total distance covered on the tour, is
$$d(A,B) + d(B,C) + d(C,D) + d(D,A) = 500 + 1300 + 800 + 800$$
$$= 3400, \quad \text{or } 3400 \text{ miles.}$$

37. Referring to the following diagram,

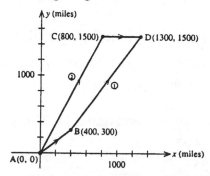

we see that the distance he would cover if he took Route (1) is given by

$$d(A,B)+d(B,D)=\sqrt{400^2+300^2}+\sqrt{(1300-400)^2+(1500-300)^2}$$
$$=\sqrt{250{,}000}+\sqrt{2{,}250{,}000}=500+1500=2000,$$

or 2000 miles. On the other hand, the distance he would cover if he took Route (2) is given by

$$d(A,C)+d(C,D)=\sqrt{800^2+1500^2}+\sqrt{(1300-800)^2}$$
$$=\sqrt{2{,}890{,}000}+\sqrt{250{,}000}=1700+500=2200,$$

or 2200 miles. Comparing these results, we see that he should take Route (1).

39. Calculations to determine VHF requirements:
$$d=\sqrt{25^2+35^2}\ =\sqrt{625+1225}=\sqrt{1850}\ \approx 43.01.$$
Models B through D satisfy this requirement.

Calculations to determine UHF requirements:
$$d=\sqrt{20^2+32^2}=\sqrt{400+1024}=\sqrt{1424}=37.74$$
Models C through D satisfy this requirement. Therefore, Model C will allow him to receive both channels at the least cost.

41. a. Let the position of ship A and ship B after t hours be $A(0, y)$ and $B(0, y)$, respectively. Then $x = 30t$ and $y = 20t$. Therefore, the distance between the two ships is
$$D=\sqrt{(30t)^2+(20t)^2}=\sqrt{900t^2+400t^2}=10\sqrt{13}t.$$

b. The required distance is obtained by letting $t = 2$ giving $D=10\sqrt{13}(2)$ or approximately 72.11 miles.

43. True. Plot the points.

45. False. The distance between $P_1(a,b)$ and $P_3(kc,kd)$ is
$$d=\sqrt{(kc-a)^2+(kd-b)^2}$$
$$\neq |k|D=|k|\sqrt{(c-a)^2+(d-b)^2}$$
$$=\sqrt{k^2(c-a)^2+k^2(d-b)^2}=\sqrt{[k(c-a)]^2+[k(d-b)]^2}$$

47. Referring to the figure in the text, we see that the distance between the two points is given by the length of the hypotenuse of the right triangle. That is,

$$d = \sqrt{(x_2 - x_1)^2 + (y_2 - y_1)^2}$$

EXERCISES 1.4, page 41

1. e   3. a   5. f

7. Referring to the figure shown in the text, we see that $m = \dfrac{2-0}{0-(-4)} = \dfrac{1}{2}$.

9. This is a vertical line, and hence its slope is undefined.

11. $m = \dfrac{y_2 - y_1}{x_2 - x_1} = \dfrac{8-3}{5-4} = 5$.

13. $m = \dfrac{y_2 - y_1}{x_2 - x_1} = \dfrac{8-3}{4-(-2)} = \dfrac{5}{6}$.

15. $m = \dfrac{y_2 - y_1}{x_2 - x_1} = \dfrac{d-b}{c-a}$.

17. Since the equation is in the slope-intercept form, we read off the slope $m = 4$.
    a. If x increases by 1 unit, then y increases by 4 units.
    b. If x decreases by 2 units, y decreases by $4(-2) = -8$ units.

19. The slope of the line through A and B is $\dfrac{-10-(-2)}{-3-1} = \dfrac{-8}{-4} = 2$.

    The slope of the line through C and D is $\dfrac{1-5}{-1-1} = \dfrac{-4}{-2} = 2$.

    Since the slopes of these two lines are equal, the lines are parallel.

21. The slope of the line through A and B is $\dfrac{2-5}{4-(-2)} = -\dfrac{3}{6} = -\dfrac{1}{2}$.

    The slope of the line through C and D is $\dfrac{6-(-2)}{3-(-1)} = \dfrac{8}{4} = 2$. Since the slopes of these

    two lines are the negative reciprocals of each other, the lines are perpendicular.

23. The slope of the line through the point $(1, a)$ and $(4, -2)$ is $m_1 = \dfrac{-2-a}{4-1}$ and the

    slope of the line through $(2,8)$ and $(-7, a+4)$ is $m_2 = \dfrac{a+4-8}{-7-2}$. Since these two lines are parallel, m_1 is equal to m_2. Therefore,

$$\frac{-2-a}{3} = \frac{a-4}{-9}$$
$$-9(-2-a) = 3(a-4)$$
$$18 + 9a = 3a - 12$$
$$6a = -30 \qquad \text{and} \quad a = -5$$

25. An equation of a horizontal line is of the form $y = b$. In this case $b = -3$, so $y = -3$ is an equation of the line.

27. We use the point-slope form of an equation of a line with the point $(3, -4)$ and slope $m = 2$. Thus $\qquad y - y_1 = m(x - x_1),$
and
$$y - (-4) = 2(x - 3)$$
$$y + 4 = 2x - 6$$
$$y = 2x - 10.$$

29. Since the slope $m = 0$, we know that the line is a horizontal line of the form $y = b$. Since the line passes through $(-3, 2)$, we see that $b = 2$, and an equation of the line is $y = 2$.

31. We first compute the slope of the line joining the points $(2, 4)$ and $(3, 7)$. Thus,
$$m = \frac{7 - 4}{3 - 2} = 3.$$
Using the point-slope form of an equation of a line with the point $(2, 4)$ and slope $m = 3$, we find
$$y - 4 = 3(x - 2)$$
$$y = 3x - 2.$$

33. We first compute the slope of the line joining the points $(1, 2)$ and $(-3, -2)$. Thus,
$$m = \frac{-2 - 2}{-3 - 1} = \frac{-4}{-4} = 1.$$
Using the point-slope form of an equation of a line with the point $(1, 2)$ and slope $m = 1$, we find
$$y - 2 = x - 1$$
$$y = x + 1.$$

35. We use the slope-intercept form of an equation of a line: $y = mx + b$. Since $m = 3$, and $b = 4$, the equation is $y = 3x + 4$.

37. We use the slope-intercept form of an equation of a line: $y = mx + b$. Since $m = 0$, and $b = 5$, the equation is $y = 5$.

39. We first write the given equation in the slope-intercept form:
$$x - 2y = 0$$
$$-2y = -x$$
$$y = \tfrac{1}{2}x \; .$$
From this equation, we see that $m = 1/2$ and $b = 0$.

41. We write the equation in slope-intercept form:
$$2x - 3y - 9 = 0$$
$$-3y = -2x + 9$$
$$y = \tfrac{2}{3}x - 3.$$
From this equation, we see that $m = 2/3$ and $b = -3$.

43. We write the equation in slope-intercept form:
$$2x + 4y = 14$$
$$4y = -2x + 14$$
$$y = -\tfrac{2}{4}x + \tfrac{14}{4}$$
$$= -\tfrac{1}{2}x + \tfrac{7}{2}.$$
From this equation, we see that $m = -1/2$ and $b = 7/2$.

45. We first write the equation $2x - 4y - 8 = 0$ in slope-intercept form:
$$2x - 4y - 8 = 0$$
$$4y = 2x - 8$$
$$y = \tfrac{1}{2}x - 2$$
Now the required line is parallel to this line, and hence has the same slope. Using the point-slope equation of a line with $m = 1/2$ and the point $(-2,2)$, we have
$$y - 2 = \tfrac{1}{2}[x - (-2)]$$
$$y = \tfrac{1}{2}x + 3.$$

47. A line parallel to the x-axis has slope 0 and is of the form $y = b$. Since the line is 6 units below the axis, it passes through $(0,-6)$ and its equation is $y = -6$.

49. We use the point-slope form of an equation of a line to obtain
$$y - b = 0(x - a) \quad \text{or} \quad y = b.$$

51. Since the required line is parallel to the line joining $(-3,2)$ and $(6,8)$, it has slope
$$m = \frac{8-2}{6-(-3)} = \frac{6}{9} = \frac{2}{3}.$$
We also know that the required line passes through $(-5,-4)$. Using the point-slope form of an equation of a line, we find
$$y - (-4) = \frac{2}{3}(x - (-5))$$
or $\quad y = \frac{2}{3}x + \frac{10}{3} - 4; \quad$ that is $\quad y = \frac{2}{3}x - \frac{2}{3}$.

53. Since the point $(-3,5)$ lies on the line $kx + 3y + 9 = 0$, it satisfies the equation. Substituting $x = -3$ and $y = 5$ into the equation gives
$$-3k + 15 + 9 = 0, \quad \text{or} \quad k = 8.$$

55. $3x - 2y + 6 = 0$

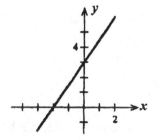

57. $x + 2y - 4 = 0$

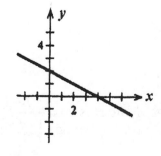

59. $y + 5 = 0$

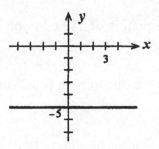

61. Since the line passes through the points $(a, 0)$ and $(0, b)$, its slope is
$m = \dfrac{b-0}{0-a} = -\dfrac{b}{a}$. Then, using the point-slope form of an equation of a line with the
point $(a, 0)$ we have
$$y - 0 = -\tfrac{b}{a}(x - a)$$
$$y = -\tfrac{b}{a}x + b$$
which may be written in the form $\tfrac{b}{a}x + y = b$.

Multiplying this last equation by $1/b$, we have $\dfrac{x}{a} + \dfrac{y}{b} = 1$.

63. Using the equation $\dfrac{x}{a} + \dfrac{y}{b} = 1$ with $a = -2$ and $b = -4$, we have $-\dfrac{x}{2} - \dfrac{y}{4} = 1$.
Then
$$-4x - 2y = 8$$
$$2y = -8 - 4x$$
$$y = -2x - 4.$$

65. Using the equation $\dfrac{x}{a} + \dfrac{y}{b} = 1$ with $a = 4$ and $b = -1/2$, we have
$$\dfrac{x}{4} + \dfrac{y}{-\frac{1}{2}} = 1$$
$$-\tfrac{1}{4}x + 2y = -1$$
$$2y = \tfrac{1}{4}x - 1$$
$$y = \tfrac{1}{8}x - \tfrac{1}{2}.$$

67. The slope of the line passing through A and B is $m = \dfrac{7-1}{1-(-2)} = \dfrac{6}{3} = 2$,

and the slope of the line passing through B and C is $m = \dfrac{13-7}{4-1} = \dfrac{6}{3} = 2$.

Since the slopes are equal, the points lie on the same line.

69. a.

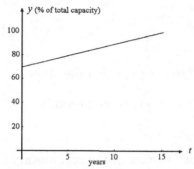

b. The slope is 1.9467 and the y-intercept is 70.082.

c. The output is increasing at the rate of 1.9467%/yr; the output at the beginning of 1990 was 70.082%.

d. We solve the equation $1.9467t + 70.082 = 100$ giving $t = 15.37$. We conclude that the plants will be generating at maximum capacity shortly after 2005.

71. a. $y = 0.55x$

b. Solving the equation $1100 = 0.55x$ for x, we have $x = \dfrac{1100}{0.55} = 2000$.

73. Using the points $(0, 0.68)$ and $(10, 0.80)$, we see that the slope of the required line is

$$m = \frac{0.80 - 0.68}{10 - 0} = \frac{0.12}{10} = .012.$$

Next, using the point-slope form of the equation of a line, we have

$$y - 0.68 = 0.012(t - 0)$$

or $\qquad y = 0.012t + 0.68.$

Therefore, when $t = 14$, we have

$$y = 0.012(14) + 0.68$$

$$= .848$$

or 84.8%. That is, in 2004 women's wages are expected to be 84.8% of men's wages.

75. a. – b.

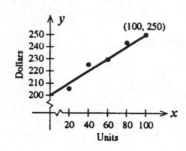

c. Using the points (0,200) and (100,250), we see that the slope of the required line
is $m = \dfrac{250-200}{100} = \dfrac{1}{2}$. Therefore, the required equation is

$$y - 200 = \tfrac{1}{2}x \quad \text{or} \quad y = \tfrac{1}{2}x + 200.$$

d. The approximate cost for producing 54 units of the commodity is
$\tfrac{1}{2}(54) + 200$, or $227.

77. True. The slope of the line is given by $-\dfrac{2}{4} = -\dfrac{1}{2}$.

79. False. Let the slope of L_1 be $m_1 > 0$. Then the slope of L_2 is $m_2 = -\dfrac{1}{m_1} < 0$.

81. True. Set $y = 0$ and we have $Ax + C = 0$ or $x = -C/A$ and this is where the line
cuts the x-axis.

83. Writing each equation in the slope-intercept form, we have

$$y = -\frac{a_1}{b_1}x - \frac{c_1}{b_1} \ \ (b_1 \neq 0) \quad \text{and} \quad y = -\frac{a_2}{b_2}x - \frac{c_2}{b_2} \ \ (b_2 \neq 0)$$

Since two lines are parallel if and only if their slopes are equal, we see that the
lines are parallel if and only if $-\dfrac{a_1}{b_1} = -\dfrac{a_2}{b_2}$, or $a_1 b_2 - b_1 a_2 = 0$.

CHAPTER 1, REVIEW EXERCISES, page 47

1. Adding x to both sides yields $3 \le 3x + 9$ or $3x \ge -6$, and $x \ge -2$.
   We conclude that the solution set is $[-2, \infty)$.

3. The inequalities imply $x > 5$ or $x < -4$. So the solution set is

$(-\infty, -4) \cup (5, \infty)$.

5. $|-5+7| + |-2| = |2| + |-2| = 2 + 2 = 4$.   7. $|2\pi - 6| - \pi = 2\pi - 6 - \pi = \pi - 6$.

9. $\left(\dfrac{9}{4}\right)^{3/2} = \dfrac{9^{3/2}}{4^{3/2}} = \dfrac{27}{8}$.   11. $(3 \cdot 4)^{-2} = 12^{-2} = \dfrac{1}{12^2} = \dfrac{1}{144}$.

13. $\dfrac{(3 \cdot 2^{-3})(4 \cdot 3^5)}{2 \cdot 9^3} = \dfrac{3 \cdot 2^{-3} \cdot 2^2 \cdot 3^5}{2 \cdot (3^2)^3} = \dfrac{2^{-1} \cdot 3^6}{2 \cdot 3^6} = \dfrac{1}{4}$.

15. $\dfrac{4(x^2 + y)^3}{x^2 + y} = 4(x^2 + y)^2$.

17. $\dfrac{\sqrt[4]{16x^5 yz}}{\sqrt[4]{81xyz^5}} = \dfrac{(2^4 x^5 yz)^{1/4}}{(3^4 xyz^5)^{1/4}} = \dfrac{2x^{5/4} y^{1/4} z^{1/4}}{3x^{1/4} y^{1/4} z^{5/4}} = \dfrac{2x}{3z}$.

19. $\left(\dfrac{3xy^2}{4x^3 y}\right)^{-2}\left(\dfrac{3xy^3}{2x^2}\right)^3 = \left(\dfrac{3y}{4x^2}\right)^{-2}\left(\dfrac{3y^3}{2x}\right)^3 = \left(\dfrac{4x^2}{3y}\right)^2\left(\dfrac{3y^3}{2x}\right)^3 = \dfrac{(16x^4)(27y^9)}{(9y^2)(8x^3)} = 6xy^7$.

21. $-2\pi^2 r^3 + 100\pi r^2 = -2\pi r^2(\pi r - 50)$.

23. $16 - x^2 = 4^2 - x^2 = (4 - x)(4 + x)$.

25. $8x^2 + 2x - 3 = (4x + 3)(2x - 1) = 0$ and $x = -3/4$ and $x = 1/2$ are the roots of the equation.

27. $-x^3 - 2x^2 + 3x = -x(x^2 + 2x - 3) = -x(x + 3)(x - 1) = 0$ and the roots of the equation are $x = 0$, $x = -3$, and $x = 1$.

29. Here we use the quadratic formula to solve the equation $x^2 - 2x - 5$. Then $a = 1$, $b = -2$, and $c = -5$. Thus,
$$x = \dfrac{-b \pm \sqrt{b^2 - 4ac}}{2a} = \dfrac{-(-2) \pm \sqrt{(-2)^2 - 4(1)(-5)}}{2(1)} = \dfrac{2 \pm \sqrt{24}}{2} = 1 \pm \sqrt{6}.$$

1 Preliminaries

31. $\dfrac{(t+6)(60)-(60t+180)}{(t+6)^2} = \dfrac{60t+360-60t-180}{(t+6)^2} = \dfrac{180}{(t+6)^2}.$

33. $\dfrac{2}{3}\left(\dfrac{4x}{2x^2-1}\right)+3\left(\dfrac{3}{3x-1}\right) = \dfrac{8x}{3(2x^2-1)}+\dfrac{9}{3x-1} = \dfrac{8x(3x-1)+27(2x^2-1)}{3(2x^2-1)(3x-1)}$

$$= \dfrac{78x^2-8x-27}{3(2x^2-1)(3x-1)}.$$

35. $\dfrac{\sqrt{x}-1}{x-1} = \dfrac{\sqrt{x}-1}{x-1}\cdot\dfrac{\sqrt{x}+1}{\sqrt{x}+1} = \dfrac{(\sqrt{x})^2-1}{(x-1)(\sqrt{x}+1)} = \dfrac{x-1}{(x-1)(\sqrt{x}+1)} = \dfrac{1}{\sqrt{x}+1}.$

37. The distance is
$$d = \sqrt{[1-(-2)]^2+[-7-(-3)]^2} = \sqrt{3^2+(-4)^2} = \sqrt{9+16} = \sqrt{25} = 5.$$

39. An equation is $x = -2$.

41. The slope of L is $m = \dfrac{\frac{7}{2}-4}{3-(-2)} = -\dfrac{1}{10}$ and an equation of L is
$$y-4 = -\tfrac{1}{10}\left[x-(-2)\right] = -\tfrac{1}{10}x-\tfrac{1}{5},$$
or $\qquad y = -\tfrac{1}{10}x+\tfrac{19}{5}$
The general form of this equation is $x + 10y - 38 = 0$.

43. Writing the given equation in the form $y = \tfrac{5}{2}x-3$, we see that the slope of the given line is 5/2. So a required equation is
$$y-4 = \tfrac{5}{2}(x+2) \quad \text{or} \quad y = \tfrac{5}{2}x+9$$
The general form of this equation is $5x - 2y + 18 = 0$.

45. Rewriting the given equation in the slope-intercept form, we have $4y = -3x + 8$
or $\qquad y = -\tfrac{3}{4}x+2$
and conclude that the slope of the required line is $-3/4$. Using the point-slope form of the equation of a line with the point $(2,3)$ and slope $-3/4$, we obtain
$$y-3 = -\tfrac{3}{4}(x-2)$$
$$y = -\tfrac{3}{4}x+\tfrac{6}{4}+3$$
$$= -\tfrac{3}{4}x+\tfrac{9}{2}.$$

The general form of this equation is $3x + 4y - 18 = 0$.

47. The slope of the line passing through (-2,-4) and (1,5) is
$$m = \frac{5-(-4)}{1-(-2)} = \frac{9}{3} = 3.$$
So the required line is
$$y-(-2) = 3[x-(-3)]$$
$$y+2 = 3x+9$$
or $\quad y = 3x+7$

49. Setting $x = 0$ gives $y = -6$ as the y-intercept. Setting $y = 0$ gives $x = 8$ as the x-intercept. The graph of the equation $3x - 4y = 24$ is follows.

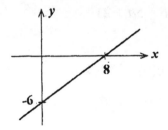

51. $2(1.5C + 80) \leq 2(2.5C - 20) \Rightarrow 1.5C + 80 \leq 2.5C - 20$, so $C \geq 100$ and the minimum cost is \$100.

CHAPTER 2

1. $f(x) = 5x + 6$. Therefore

$f(3) = 5(3) + 6 = 21$

$f(-3) = 5(-3) + 6 = -9$

$f(a) = 5(a) + 6 = 5a + 6$

$f(-a) = 5(-a) + 6 = -5a + 6$

$f(a + 3) = 5(a + 3) + 6 = 5a + 15 + 6 = 5a + 21.$

3. $g(x) = 3x^2 - 6x - 3;$ $\quad g(0) = 3(0) - 6(0) - 3 = -3$

$g(-1) = 3(-1)^2 - 6(-1) - 3 = 3 + 6 - 3 = 6$

$g(a) = 3(a)^2 - 6(a) - 3 = 3a^2 - 6a - 3$

$g(-a) = 3(-a)^2 - 6(-a) - 3 = 3a^2 + 6a - 3$

$g(x + 1) = 3(x + 1)^2 - 6(x + 1) - 3 = 3(x^2 + 2x + 1) - 6x - 6 - 3$

$\qquad = 3x^2 + 6x + 3 - 6x - 9 = 3x^2 - 6.$

5. $f(a + h) = 2(a + h) + 5 = 2a + 2h + 5$

$f(-a) = 2(-a) + 5 = -2a + 5$

$f(a^2) = 2(a^2) + 5 = 2a^2 + 5$

$f(a - 2h) = 2(a - 2h) + 5 = 2a - 4h + 5$

$f(2a - h) = 2(2a - h) + 5 = 4a - 2h + 5$

7. $s(t) = \dfrac{2t}{t^2 - 1}$. Therefore, $s(4) = \dfrac{2(4)}{(4)^2 - 1} = \dfrac{8}{15}.$ $\quad s(0) = \dfrac{2(0)}{0^2 - 1} = 0$

$s(a) = \dfrac{2(a)}{a^2 - 1} = \dfrac{2a}{a^2 - 1}$

$$s(2+a) = \frac{2(2+a)}{(2+a)^2 - 1} = \frac{2(2+a)}{a^2 + 4a + 4 - 1} = \frac{2(2+a)}{a^2 + 4a + 3}$$

$$s(t+1) = \frac{2(t+1)}{(t+1)^2 - 1} = \frac{2(t+1)}{t^2 + 2t + 1 - 1} = \frac{2(t+1)}{t(t+2)}.$$

9. $f(t) = \dfrac{2t^2}{\sqrt{t-1}}$. Therefore, $f(2) = \dfrac{2(2^2)}{\sqrt{2-1}} = 8$. $f(a) = \dfrac{2a^2}{\sqrt{a-1}}$

$$f(x+1) = \frac{2(x+1)^2}{\sqrt{(x+1)-1}} = \frac{2(x+1)^2}{\sqrt{x}} ; f(x-1) = \frac{2(x-1)^2}{\sqrt{(x-1)-1}} = \frac{2(x-1)^2}{\sqrt{x-2}}.$$

11. Since $x = -2 \leq 0$, we see that $f(-2) = (-2)^2 + 1 = 4 + 1 = 5$

Since $x = 0 \leq 0$, we see that $f(0) = (0)^2 + 1 = 1$

Since $x = 1 > 0$, we see that $f(1) = \sqrt{1} = 1$.

13. Since $x = -1 < 1$, $f(-1) = -\frac{1}{2}(-1)^2 + 3 = \frac{5}{2}$.

Since $x = 0 < 1$, $f(0) = -\frac{1}{2}(0)^2 + 3 = 3$.

Since $x = 1 \geq 1$, $f(1) = 2(1^2) + 1 = 3$.

Since $x = 2 \geq 1, f(2) = 2(2^2) + 1 = 9$.

15. a. $f(0) = -2$;
   b. (i) $f(x) = 3$ when $x \approx 2$ (ii) $f(x) = 0$ when $x = 1$
   c. $[0,6]$ d. $[-2, 6]$

17. $g(2) = \sqrt{2^2 - 1} = \sqrt{3}$ and the point $(2, \sqrt{3})$ lies on the graph of g.

19. $f(-2) = \dfrac{|-2 - 1|}{-2 + 1} = \dfrac{|-3|}{-1} = -3$ and the point $(-2,-3)$ does lie on the graph of f.

21. Since $f(x)$ is a real number for any value of x, the domain of f is $(-\infty, \infty)$.

23. $f(x)$ is not defined at $x = 0$ and so the domain of f is $(-\infty, 0) \cup (0, \infty)$.

25. $f(x)$ is a real number for all values of x. Note that $x^2 + 1 \geq 1$ for all x. Therefore, the domain of f is $(-\infty, \infty)$.

27. Since the square root of a number is defined for all real numbers greater than or equal to zero, we have
$$5 - x \geq 0, \text{ or } \qquad -x \geq -5$$
and $\quad x \leq 5$. (Recall that multiplying by -1 reverses the sign of an inequality.)
Therefore, the domain of g is $(-\infty, 5]$.

29. The denominator of f is zero when $x^2 - 1 = 0$ or $x = \pm 1$.
Therefore, the domain of f is $(-\infty, -1) \cup (-1, 1) \cup (1, \infty)$.

31. f is defined when $x + 3 \geq 0$, that is, when $x \geq -3$. Therefore, the domain of f is $[-3, \infty)$.

33. The numerator is defined when $1 - x \geq 0$, $\quad -x \geq -1 \quad$ or $\quad x \leq 1$.
Furthermore, the denominator is zero when $x = \pm 2$. Therefore, the domain is the set of all real numbers in $(-\infty, -2) \cup (-2, 1]$.

35. a. The domain of f is the set of all real numbers.
b. $f(x) = x^2 - x - 6$. Therefore,
$f(-3) = (-3)^2 - (-3) - 6 = 9 + 3 - 6 = 6$;

$f(-2) = (-2)^2 - (-2) - 6 = 4 + 2 - 6 = 0$.

$f(-1) = (-1)^2 - (-1) - 6 = 1 + 1 - 6 = -4$;

$f(0) = (0)^2 - (0) - 6 = -6$.

$f\left(\tfrac{1}{2}\right) = \left(\tfrac{1}{2}\right)^2 - \left(\tfrac{1}{2}\right) - 6 = \tfrac{1}{4} - \tfrac{2}{4} - \tfrac{24}{4} = -\tfrac{25}{4}$;

$f(1) = (1)^2 - 1 - 6 = -6$.

$f(2) = (2)^2 - 2 - 6 = 4 - 2 - 6 = -4$; $\quad f(3) = (3)^2 - 3 - 6 = 9 - 3 - 6 = 0$.

c.

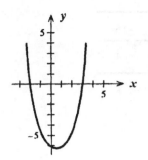

37. $f(x) = 2x^2 + 1$

x	-3	-2	-1	0	1	2	3
$f(x)$	19	9	3	1	3	9	19

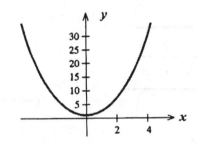

$(-\infty, \infty)$; $[1, \infty)$

39. $f(x) = 2 + \sqrt{x}$

x	0	1	2	4	9	16
$f(x)$	2	3	3.41	4	5	6

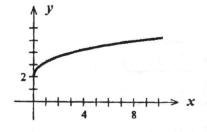

$[0, \infty)$; $[2, \infty)$

41. $f(x) = \sqrt{1-x}$

x	0	-1	-3	-8	-15
$f(x)$	1	1.4	2	3	4

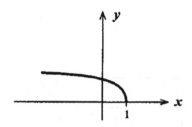

$(-\infty, 1]; [0, \infty)$

43. $f(x) = |x| - 1$

x	-3	-2	-1	0	1	2	3
$f(x)$	2	1	0	-1	0	1	2

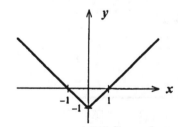

$(-\infty, \infty); [-1, \infty)$

45. $f(x) = \begin{cases} x & \text{if } x < 0 \\ 2x+1 & \text{if } x \geq 0 \end{cases}$

x	-3	-2	-1	0	1	2	3
$f(x)$	-3	-2	-1	1	3	5	7

$(-\infty, \infty); (-\infty, 0) \cup [1, \infty)$

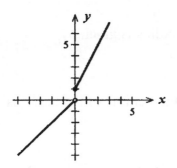

47. If $x \leq 1$, the graph of f is the half-line $y = -x + 1$. For $x > 1$, use the table

x	2	3	4
$f(x)$	3	8	15

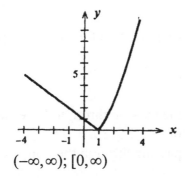

$(-\infty, \infty); [0, \infty)$

49. Each vertical line cuts the given graph at exactly one point, and so the graph represents y as a function of x.

51. Since there is a vertical line that intersects the graph at three points, the graph does not represent y as a function of x.

53. Each vertical line intersects the graph of f at exactly one point, and so the graph represents y as a function of x.

55. Each vertical line intersects the graph of f at exactly one point, and so the graph represents y as a function of x.

57. The circumference of a circle with a 5-inch radius is given by
$$C(5) = 2\pi(5) = 10\pi, \text{ or } 10\pi \text{ inches.}$$

59. $\frac{4}{3}(\pi)(2r)^3 = \frac{4}{3}\pi 8r^3 = 8(\frac{4}{3}\pi r^3)$. Therefore, the volume of the tumor is increased by a factor of 8.

61. a. From $t = 0$ to $t = 5$, the graph for cassettes lies above that for CDs so from 1985 to 1990, sales of prerecorded cassettes were greater than that of CDs.
b. Sales of prerecorded CDs were greater than that of prerecorded cassettes from 1990 on.
c. The graphs intersect at the point with coordinates $x = 5$ and $y \approx 3.5$, and this tells us that the sales of the two formats were the same in 1990 with the level of sales at approximately $3.5 billion.

63. a. The slope of the straight line passing through the points (0, 0.58) and (20, 0.95) is
$$m = \frac{0.95 - 0.58}{20 - 0} = 0.0185,$$
and so an equation of the straight line passing through these two points is
$$y - 0.58 = 0.0185(t - 0) \text{ or } y = 0.0185t + 0.58$$
Next, the slope of the straight line passing through the points (20, 0.95) and (30, 1.1) is
$$m = \frac{1.1 - 0.95}{30 - 20} = 0.015$$
and so an equation of the straight line passing through the two points is
$$y - 0.95 = 0.015(t - 20) \text{ or } y = 0.015t + 0.65.$$
Therefore, the rule for f is
$$f(t) = \begin{cases} 0.0185t + 0.58 & 0 \le t \le 20 \\ 0.015t + 0.65 & 20 < t \le 30 \end{cases}$$
b. The ratios were changing at the rates of 0.0185/yr and 0.015/yr from 1960 through 1980, and from 1980 through 1990, respectively.
c. The ratio was 1 when $t \approx 23.3$. This shows that the number of bachelor's

degrees earned by women equaled the number earned by men for the first time around 1983.

65. a. $T(x) = 0.06x$ b. $T(200) = 0.06(200) = 12$, or $12.00.

$T(5.65) = 0.06(5.65) = 0.34$, or $0.34.

67. The child should receive $D(4) = \frac{2}{25}(500)(4) = 160$, or 160 mg.

69. a. The daily cost of leasing from Ace is $C_1(x) = 30 + 0.45x$, while the daily cost of leasing from Acme is $C_2(x) = 25 + 0.50x$, where x is the number of miles driven.
    b.

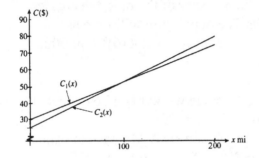

   c. The costs will be the same when $C_1(x) = C_2(x)$, that is, when
$$30 + 0.45x = 25 + 0.50x \quad \text{or} \quad -0.05x = -5, \quad \text{or} \quad x = 100.$$
   Since $C_1(70) = 30 + 0.45(70) = 61.5$ (Ace)
   and $C_2(70) = 25 + 0.50(70) = 60$,   (Acme)
   and the customer plans to drive less than 70 miles, she should rent from Acme.

71. Here $V = -20{,}000n + 1{,}000{,}000$.
   The book value in 1999 will be $V = -20{,}000(15) + 1{,}000{,}000$, or $700,000.
   The book value in 2003 will be $V = -20{,}000(19) + 1{,}000{,}000$, or $620,000.
   The book value in 2007 will be $V = -20{,}000(23) + 1{,}000{,}000$, or $540,000.

73. a. We require that $0.04 - r^2 \geq 0$ and $r \geq 0$. This is true if $0 \leq r \leq 0.2$. Therefore, the domain of v is $[0, 0.2]$.
    b. Compute $v(0) = 1000[0.04 - (0)^2] = 1000(0.04) = 40$.

$$v(0.1) = 1000[0.04 - (0.1)^2] = 1000(0.04 - .01)$$
$$= 1000(0.03) = 30.$$
$$v(0.2) = 1000[0.04 - (0.2)^2] = 1000(0.04 - 0.04) = 0.$$

c.

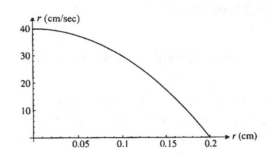

d. As the distance r increases, the velocity of the blood decreases.
or 990 during the next 9 months.The population will increase by
$$P(16) - P(0) = [50,000 + 30(16)^{3/2} + 20(16)] - 50,000,$$
or 2240 during the next 16 months.

75. Between 8 A.M. and 9 A.M., the average worker can be expected to assemble
$$N(1) - N(0) = (-1 + 6 + 15) - 0 = 20,$$
or 20 walkie-talkies. Between 9 A.M. and 10 A.M., we can expect
$$N(2) - N(1) = [-2^3 + 6(2^2) + 15(2)] - (-1 + 6 + 15)$$
$$= 46 - 20 = 26,$$
or 26 walkie-talkies can be assembled by the average worker.

77. The percentage at age 65 that are expected to have Alzheimer's disease is given by
$$P(0) = 0.0726(0)^2 + 0.7902(0) + 4.9623 = 4.9623, \quad \text{or } 4.96\%.$$

The percentage at age 90 that are expected to have Alzheimer's disease is given by
$$P(5) = 0.0726(25)^2 + 0.7902(25) + 4.9623 = 70.09, \text{ or } 70.09\%.$$

79. a. The amount of solids discharged in 1989 ($t = 0$) was 130 tons/day; in 1992
($t = 3$), it was 100 tons/day; and in 1996 ($t = 7$), it was
$$f(7) = 1.25(7)^2 - 26.25(7) + 162.5 = 40, \text{ or } 40 \text{ tons/day.}$$

b.

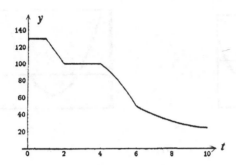

81. True, by definition the of a function .

83. False. Let $f(x) = x^2$, then take $a = 1$, and $b = 2$. Then $f(a) = f(1) = 1$ and
$f(b) = f(2) = 4$ and $f(a) + f(b) = 1 + 4 \neq f(a+b) = f(3) = 9$.

USING TECHNOLOGY EXERCISES 2.1, page 67

1.

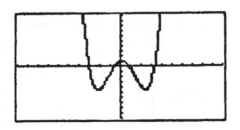

3.

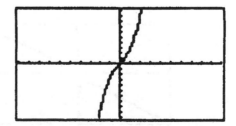

5.

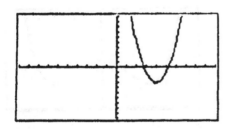

7.

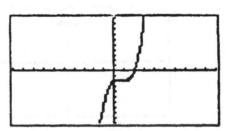

9. a.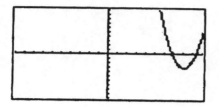

b.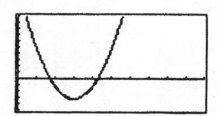

11. a.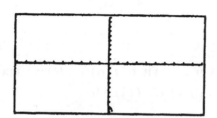

b.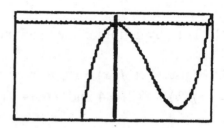

13. a.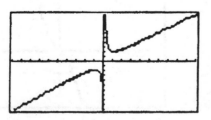

b.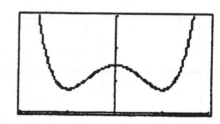

15. a.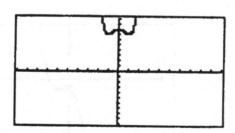

b.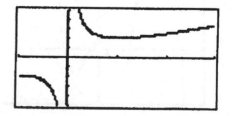

17. a. b.

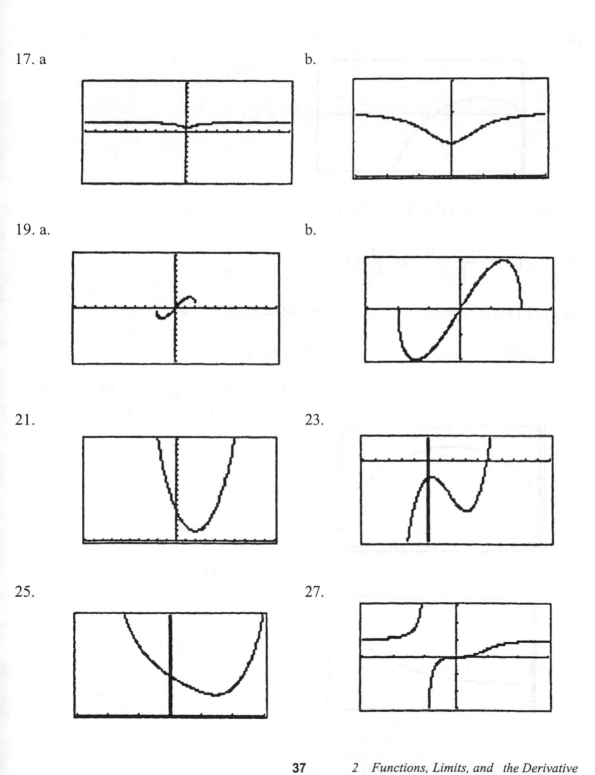

19. a. b.

21. 23.

25. 27.

29.

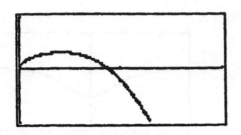

31. $18; f(-1) = -3(-1)^3 + 5(-1)^2 - 2(-1) + 8 = 3 + 5 + 2 + 8 = 18.$

33. $2; f(1) = \dfrac{(1)^4 - 3(1)^2}{1-2} = \dfrac{1-3}{-1} = 2.$

35. $f(2.145) \approx 18.5505$

37. $f(1.28) \approx 17.3850$

39. $f(2.41) \approx 4.1616$

41. $f(0.62) \approx 1.7214$

43. a.

b. $f(2) \approx 9.4066$, or approximately 9.4066%/yr
$f(4) \approx 8.7062$, or approximately 8.7062 %/yr.

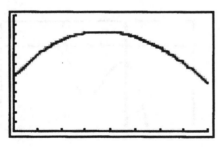

45. a.

b. $f(6) = 44.7;$

$f(8) = 52.7;$

$f(11) = 129.2.$

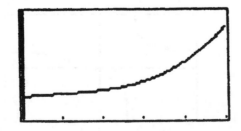

EXERCISES 2.2, page 75

1. $(f+g)(x) = f(x) + g(x) = (x^3 + 5) + (x^2 - 2) = x^3 + x^2 + 3.$

3. $fg(x) = f(x)g(x) = (x^3 + 5)(x^2 - 2) = x^5 - 2x^3 + 5x^2 - 10.$

5. $\dfrac{f}{g}(x) = \dfrac{f(x)}{g(x)} = \dfrac{x^3 + 5}{x^2 - 2}.$

7. $\dfrac{fg}{h}(x) = \dfrac{f(x)g(x)}{h(x)} = \dfrac{(x^3 + 5)(x^2 - 2)}{2x + 4} = \dfrac{x^5 - 2x^3 + 5x^2 - 10}{2x + 4}$

9. $(f + g)(x) = x - 1 + \sqrt{x + 1}.$

11. $(f\,g)(x) = (x - 1)\sqrt{x + 1}$

13. $\dfrac{g}{h}(x) = \dfrac{g(x)}{h(x)} = \dfrac{\sqrt{x + 1}}{2x^3 - 1}.$

15. $\dfrac{fg}{h}(x) = \dfrac{(x - 1)(\sqrt{x + 1})}{2x^3 - 1}$

17. $\dfrac{f - h}{g}(x) = \dfrac{x - 1 - (2x^3 - 1)}{\sqrt{x + 1}} = \dfrac{x - 2x^3}{\sqrt{x + 1}}.$

19. $(f + g)(x) = x^2 + 5 + \sqrt{x} - 2 = x^2 + \sqrt{x} + 3.$
    $(f - g)(x) = x^2 + 5 - (\sqrt{x} - 2) = x^2 - \sqrt{x} + 7.$
    $(f\,g)(x) = (x^2 + 5)(\sqrt{x} - 2); \; (\dfrac{f}{g})(x) = \dfrac{x^2 + 5}{\sqrt{x} - 2}.$

21. $(f + g)(x) = \sqrt{x + 3} + \dfrac{1}{x - 1} = \dfrac{(x - 1)\sqrt{x + 3} + 1}{x - 1}.$
    $(f - g)(x) = \sqrt{x + 3} - \dfrac{1}{x - 1} = \dfrac{(x - 1)\sqrt{x + 3} - 1}{x - 1}.$

$$(f\,g)(x) = \sqrt{x+3}\left(\frac{1}{x-1}\right) = \frac{\sqrt{x+3}}{x-1}. \qquad (\frac{f}{g}) = \sqrt{x+3}(x-1).$$

23. $(f+g)(x) = \dfrac{x+1}{x-1} + \dfrac{x+2}{x-2} = \dfrac{(x+1)(x-2)+(x+2)(x-1)}{(x-1)(x-2)}$

$$= \frac{x^2 - x - 2 + x^2 + x - 2}{(x-1)(x-2)} = \frac{2x^2 - 4}{(x-1)(x-2)} = \frac{2(x^2-2)}{(x-1)(x-2)}.$$

$(f-g)(x) = \dfrac{x+1}{x-1} - \dfrac{x+2}{x-2} = \dfrac{(x+1)(x-2)-(x+2)(x-1)}{(x-1)(x-2)}$

$$= \frac{x^2 - x - 2 - x^2 - x + 2}{(x-1)(x-2)} = \frac{-2x}{(x-1)(x-2)}.$$

$(f\,g)(x) = \dfrac{(x+1)(x+2)}{(x-1)(x-2)}; \quad (\dfrac{f}{g}) = \dfrac{(x+1)(x-2)}{(x-1)(x+2)}.$

25. $(f \circ g)(x) = f(g(x)) = f(x^2) = (x^2)^2 + x^2 + 1 = x^4 + x^2 + 1.$

   $(g \circ f)(x) = g(f(x)) = g(x^2 + x + 1) = (x^2 + x + 1)^2.$

27. $(f \circ g)(x) = f(g(x)) = f(x^2 - 1) = \sqrt{x^2 - 1} + 1.$

   $(g \circ f)(x) = g(f(x)) = g(\sqrt{x} + 1) = (\sqrt{x} + 1)^2 - 1 = x + 2\sqrt{x} + 1 - 1 = x + 2\sqrt{x}.$

29. $(f \circ g)(x) = f(g(x)) = f\left(\dfrac{1}{x}\right) = \dfrac{1}{x} \div \left(\dfrac{1}{x^2} + 1\right) = \dfrac{1}{x} \cdot \dfrac{x^2}{x^2+1} = \dfrac{x}{x^2+1}.$

   $(g \circ f)(x) = g(f(x)) = g\left(\dfrac{x}{x^2+1}\right) = \dfrac{x^2+1}{x}.$

31. $h(2) = g[f(2)]$. But $f(2) = 4 + 2 + 1 = 7$, so $h(2) = g(7) = 49$.

33. $h(2) = g[f(2)]$. But $f(2) = \dfrac{1}{2(2)+1} = \dfrac{1}{5}$, so $h(2) = g(\dfrac{1}{5}) = \dfrac{1}{\sqrt{5}} = \dfrac{\sqrt{5}}{5}.$

35. $f(x) = 2x^3 + x^2 + 1, g(x) = x^5.$ 37. $f(x) = x^2 - 1, g(x) = \sqrt{x}.$

39. $f(x) = x^2 - 1, g(x) = \dfrac{1}{x}.$ 41. $f(x) = 3x^2 + 2, g(x) = \dfrac{1}{x^{3/2}}.$

43. $f(a+h) - f(a) = [3(a+h) + 4] - (3a + 4) = 3a + 3h + 4 - 3a - 4 = 3h.$

45. $f(a+h) - f(a) = 4 - (a+h)^2 - (4 - a^2)$
$$= 4 - a^2 - 2ah - h^2 - 4 + a^2 = -2ah - h^2 = -h(2a + h).$$

47. $\dfrac{f(a+h) - f(a)}{h} = \dfrac{[(a+h)^2 + 1] - (a^2 + 1)}{h} = \dfrac{a^2 + 2ah + h^2 + 1 - a^2 - 1}{h} = \dfrac{2ah + h^2}{h}$

$$= \dfrac{h(2a+h)}{h} = 2a + h$$

49. $\dfrac{f(a+h) - f(a)}{h} = \dfrac{[(a+h)^3 - (a+h)] - (a^3 - a)}{h}$

$$= \dfrac{a^3 + 3a^2h + 3ah^2 + h^3 - a - h - a^3 + a}{h}$$

$$= \dfrac{3a^2h + 3ah^2 + h^3 - h}{h} = 3a^2 + 3ah + h^2 - 1$$

51. $\dfrac{f(a+h) - f(a)}{h} = \dfrac{\dfrac{1}{a+h} - \dfrac{1}{a}}{h} = \dfrac{\dfrac{a - (a+h)}{a(a+h)}}{h}$

$$= -\dfrac{1}{a(a+h)}.$$

53. $F(t)$ represents the total revenue for the two restaurants at time t.

55. $f(t)g(t)$ represents the (dollar) value of Nancy's holdings at time t.

57. $g \circ f$ is the function giving the amount of carbon monoxide pollution at time t.

59. $C(x) = 0.6x + 12{,}100$.

61. a. $P(x) = R(x) - C(x)$
$$= -0.1x^2 + 500x - (0.000003x^3 - 0.03x^2 + 200x + 100{,}000)$$
$$= -0.000003x^3 - 0.07x^2 + 300x - 100{,}000.$$
  b. $P(1500) = -0.000003(1500)^3 - 0.07(1500)^2 + 300(1500) - 100{,}000$
$$= 182{,}375 \quad \text{or } \$182{,}375.$$

63. a. The gap is
$$G(t) - C(t) = (3.5t^2 + 26.7t + 436.2) - (24.3t + 365)$$
$$= 3.5t^2 + 2.4t + 71.2.$$
  b. At the beginning of 1983, the gap was
$$G(0) = 3.5(0)^2 + 2.4(0) + 71.2 = 71.2, \text{ or } 71{,}200.$$
  At the beginning of 1986, the gap was
$$G(3) = 3.5(3)^2 + 2.4(3) + 71.2 = 109.9, \text{ or } 109{,}900.$$

65. a. The occupancy rate at the beginning of January is
$$r(0) = \frac{10}{81}(0)^3 - \frac{10}{3}(0)^2 + \frac{200}{9}(0) + 55 = 55, \text{ or } 55 \text{ percent.}$$
$$r(5) = \frac{10}{81}(5)^3 - \frac{10}{3}(5)^2 + \frac{200}{9}(5) + 55 = 98.2, \text{ or } 98.2 \text{ percent.}$$
  b. The monthly revenue at the beginning of January is
$$R(55) = -\frac{3}{5000}(55)^3 + \frac{9}{50}(55)^2 = 444.68, \text{ or } \$444{,}700.$$
  The monthly revenue at the beginning of June is
$$R(98.2) = -\frac{3}{5000}(98.2)^3 + \frac{9}{50}(98.2)^2 = 1167.6, \text{ or } \$1{,}167{,}600.$$

67. True. $(f + g)(x) = f(x) + g(x) = g(x) + f(x) = (g + f)(x)$.

69. False. Take $f(x) = \sqrt{x}$ and $g(x) = x + 1$. Then $(g \circ f)(x) = \sqrt{x} + 1$, but $(f \circ g)(x) = \sqrt{x + 1}$.

EXERCISES 2.3, page 87

1. Yes. $2x + 3y = 6$ and so $y = -\frac{2}{3}x + 2$.

3. Yes. $2y = x + 4$ and so $y = \frac{1}{2}x + 2$.

5. Yes. $4y = 2x + 9$ and so $y = \frac{1}{2}x + \frac{9}{4}$.

7. No, because of the term x^2.

9. f is a polynomial function in x of degree 6.

11. Expanding $G(x) = 2(x^2 - 3)^3$, we have
    $$G(x) = 2x^6 - 18x^4 + 54x^2 - 54,$$
    and we conclude that G is a polynomial function in x of degree 6.

13. f is neither a polynomial nor a rational function.

15. $f(0) = 2$ gives $g(0) = m(0) + b = b = 2$. Next, $f(3) = -1$ gives
    $f(3) = m(3) + b = -1$.
    Substituting $b = 2$ in this last equation, we have $3m + 2 = -1$, and $3m = -3$, or
    $m = -1$. So $m = -1$ and $b = 2$.

17. a. $C(x) = 8x + 40,000$ b. $R(x) = 12x$
    c. $P(x) = R(x) - C(x) = 12x - (8x + 40,000) = 4x - 40,000$.
    d. $P(8000) = 4(8000) - 40,000 = -8000$, or a loss of $8000.
        $P(12,000) = 4(12,000) - 40,000 = 8000$, or a profit of $8000.

19. The individual's disposable income is
    $$D = (1 - 0.28)40,000 = 28,800, \text{ or } \$28,800.$$

21. The child should receive
    $$D(4) = \left(\frac{4+1}{24}\right)(500) = 104.17, \text{ or } 104 \text{ mg.}$$

23. When 1000 units are produced,
$$R(1000) = -0.1(1000)^2 + 500(1000) = 400,000, \text{ or } \$400,000.$$

25. a.

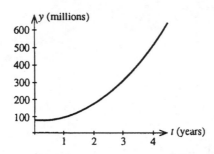

b. $f(4) = 38.57(4^2) - 24.29(4) + 79.14 = 599.1,$ or $599,100,000.$

27. a. The given data implies that $R(40) = 50$, that is,
$$\frac{100(40)}{b + 40} = 50$$
$$50(b + 40) = 4000, \text{ or } b = 40.$$

Therefore, the required response function is $R(x) = \dfrac{100x}{40 + x}.$

b. The response will be $R(60) = \dfrac{100(60)}{40 + 60} = 60$, or approximately 60 percent.

29. a. We are given that $T = aN + b$ where a and b are constants to be determined. The given conditions imply
$$70 = 120a + b$$
and $\quad 80 = 160a + b$

Subtracting the first equation from the second gives
$$10 = 40a, \text{ or } a = \tfrac{1}{4}.$$

Substituting this value of a into the first equation gives
$$70 = 120(\tfrac{1}{4}) + b, \text{ or } b = 40.$$

Therefore, $T = \tfrac{1}{4} N + 40.$

b. Solving the equation in (a) for N, we find
$$\tfrac{1}{4} N = T - 40,$$
or $\quad N = f(t) = 4T - 160.$

When $T = 102$, we find $N = 4(102) - 160 = 248$, or 248 times per minute.

31. Using the formula given in Problem 28, we have

$$V(2) = 100,000 - \frac{(100,000 - 30,000)}{5}(2) = 100,000 - \frac{70,000}{5}(2)$$

$$= 72,000, \text{ or } \$72,000.$$

33. $f(t) = 0.1714t^2 + 0.6657t + 0.7143$
   a. $f(0) = 0.7143$ or $714,300$.
   b. $f(5) = 0.1714(25) + 0.6657(5) + 0.7143 = 8.3278$, or 8.33 million.

35. a.

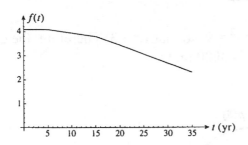

   b. At the beginning of 2005, the ratio will be $f(10) = -0.03(10) + 4.25 = 3.95$
      At the beginning of 2020, the ratio will be $f(25) = -0.075(25) + 4.925 = 3.05$.
   c. The ratio is constant from 1995 to 2000.
   d. The decline of the ratio is greatest from 2010 through 2030. It is

$$\frac{f(35) - f(15)}{35 - 15} = \frac{2.3 - 3.8}{20} = -0.075.$$

37. $h(t) = f(t) - g(t) = \dfrac{110}{\frac{1}{2}t + 1} - 26(\frac{1}{4}t^2 - 1)^2 - 52.$

$h(0) = f(0) - g(0) = \dfrac{110}{\frac{1}{2}(0) + 1} - 26\left[\frac{1}{4}(0)^2 - 1\right]^2 - 52 = 110 - 26 - 52 = 32$, or \$32.

$h(1) = f(1) - g(1) = \dfrac{110}{\frac{1}{2}(1) + 1} - 26\left[\frac{1}{4}(1)^2 - 1\right]^2 - 52 = 6.71$, or \$6.71.

$$h(2) = f(2) - g(2) = \frac{110}{\frac{1}{2}(2)+1} - 26\left[\frac{1}{4}(2)^2 - 1\right]^2 - 52 = 3, \text{ or } \$3.$$

We conclude that the price gap was narrowing.

39. a.

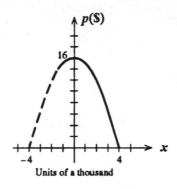

b. If $p = 7$, we have $7 = -x^2 + 16$, or $x^2 = 9$, so that $x = \pm3$. Therefore, the quantity demanded when the unit price is $7 is 3000 units.

41. a.

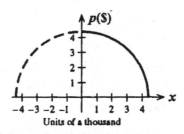

b. If $p = 3$, then $3 = \sqrt{18 - x^2}$, and $9 = 18 - x^2$, so that $x^2 = 9$ and $x = \pm 3$. Therefore, the quantity demanded when the unit price is $3 is 3000 units.

43. a.

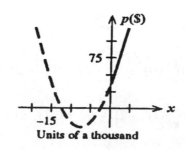

b. If $x = 2$, then $p = 2^2 + 16(2) + 40 = 76$, or \$76.

45. a.

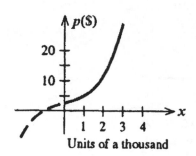

Units of a thousand

b. $p = 2^3 + 2(2) + 3 = 15$, or \$15.

47. The slope of L_2 is greater than that of L_1. This tells us that for each increase of a dollar in the price of a clock radio, more model A clock radios will be made available in the market place than model B clock radios.

49. Substituting $x = 6$ and $p = 8$ into the given equation gives
$$8 = \sqrt{-36a + b}, \quad \text{or } -36a + b = 64.$$
Next, substituting $x = 8$ and $p = 6$ into the equation gives
$$6 = \sqrt{-64a + b}, \quad \text{or } -64a + b = 36.$$
Solving the system
$$-36a + b = 64$$
$$-64a + b = 36$$
for a and b, we find $a = 1$ and $b = 100$. Therefore the demand equation is
$p = \sqrt{-x^2 + 100}$. When the unit price is set at \$7.50, we have $7.5 = \sqrt{-x^2 + 100}$, or
$56.25 = -x^2 + 100$ from which we deduce that $x = \pm 6.614$. So, the quantity demanded is 6614 units.

51. Substituting $x = 10,000$ and $p = 20$ into the given equation yields
$$20 = a\sqrt{10,000} + b = 100a + b.$$
Next, substituting $x = 62,500$ and $p = 35$ into the equation yields
$$35 = a\sqrt{62,500} + b = 250a + b.$$
Subtracting the first equation from the second yields
$$15 = 150a, \quad \text{or } a = \tfrac{1}{10}.$$

Substituting this value of a into the first equation gives $b = 10$. Therefore, the required equation is $p = \frac{1}{10}\sqrt{x} + 10$. The graph of the supply function follows.

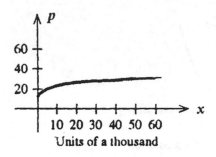

Units of a thousand

Substituting $x = 40,000$ into the supply equation yields
$$p = \tfrac{1}{10}\sqrt{40,000} + 10 = 30, \quad \text{or } \$30.$$

53. We solve the equation $-2x^2 + 80 = 15x + 30$, or $2x^2 + 15x - 50 = 0$ for x. Thus, $(2x - 5)(x + 10) = 0$, or $x = 5/2$ or $x = -10$. Rejecting the negative root, we have $x = 5/2$. The corresponding value of p is $p = -2(\tfrac{5}{2})^2 + 80 = 67.5$. We conclude that the equilibrium quantity is 2500 and the equilibrium price is $67.50.

55. Solving both equations for x, we have $x = -(11/3)p + 22$ and $x = 2p^2 + p - 10$. Equating these two equations, we have
$$-\tfrac{11}{3}p + 22 = 2p^2 + p - 10,$$
or $\qquad -11p + 66 = 6p^2 + 3p - 30$
and $\qquad 6p^2 + 14p - 96 = 0$.
Dividing this last equation by 2 and then factoring, we have
$$(3p + 16)(p - 3) = 0,$$
or $p = 3$. The corresponding value of x is $2(3)^2 + 3 - 10 = 11$. We conclude that the equilibrium quantity is 11,000 and the equilibrium price is $3.

57. Equating the two equations, we have
$$0.1x^2 + 2x + 20 = -0.1x^2 - x + 40$$
$$0.2x^2 + 3x - 20 = 0$$
$$2x^2 + 30x - 200 = 0$$
$$x^2 + 15x - 100 = 0$$
$$(x + 20)(x - 5) = 0,$$
and $x = -20$ or 5. Substituting $x = 5$ into the first equation gives

$$p = -0.1(25) - 5 + 40 = 32.5.$$

Therefore, the equilibrium quantity is 500 tents (x is measured in hundreds) and the equilibrium price is $32.50.

59. a.

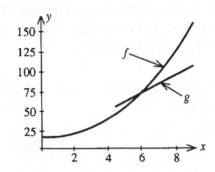

b. $5x^2 + 5x + 30 = 33x + 30$

$\quad\quad 5x^2 - 28x = 0$

$\quad\quad x(5x - 28) = 0$

$\quad\quad x = 0$ or $x = \dfrac{28}{5} = 5.6$, or 5.6 mph.

$\quad g(x) = 11(5.6) + 10 = 71.6$, or 71.6 mL/lb/min

c. The oxygen consumption of the walker is greater than that of the runner.

61. The area of Juanita's garden is 250 sq ft. Therefore $xy = 250$ and $y = \dfrac{250}{x}$.

The amount of fencing needed is given by $2x + 2y$.

Therefore, $f = 2x + 2\left(\dfrac{250}{x}\right) = 2x + \dfrac{500}{x}$. The domain of f is $x > 0$.

63. Since the volume of the box is given by

$\quad\quad V = $ (area of the base) $\times$ the height of the box

$\quad\quad\quad = x^2 y = 20,$

we have $y = \dfrac{20}{x^2}$. Next, the amount of material used in constructing the box is given by the area of the base of the box, plus the area of the 4 sides, plus the area of the top of the box, or $x^2 + 4xy + x^2$. Then, the cost of constructing the box is given by $f(x) = 0.30x^2 + 0.40x \cdot \dfrac{20}{x^2} + .20x^2 = 0.5x^2 + \dfrac{8}{x}$.

65. The average yield of the apple orchard is 36 bushels/tree when the density is 22 trees/acre. Let $x =$ the unit increase in tree density beyond 22. Then the yield of the apple orchard in bushels/acre is given by $(22 + x)(36 - 2x)$.

67. a. Let x denote the number of bottles sold beyond 10,000 bottles. Then
$$P(x) = (10,000 + x)(5 - 0.0002x)$$
$$= -0.0002x^2 + 3x + 50,000$$
b. He can expect a profit
$$P(6000) = -0.0002(6000^2) + 3(6000) + 50,000 = 60,800$$
or $60,800.

69. a. $f(r) = \pi r^2$ b. $g(t) = 2t$
c. $h(t) = (f \circ g)(t) = f(g(t)) = \pi[g(t)]^2 = 4\pi t^2$.
d. $h(30) = 4\pi(30^2) = 3600\pi$, or $3600\,\pi$ sq ft.

71. False. $f(x) = 3x^{3/4} + x^{1/2} + 1$ is not a polynomial function. The powers in x must be nonnegative integers.

73. False. $f(x) = x^{1/2}$ is not defined for negative values of x or $x = 0$.

USING TECHNOLOGY EXERCISES 2.3, page 93

1. (-3.0414, 0.1503); (3.0414, 7.4497)   3. (-2.3371, 2.4117); (6.0514, -2.5015)

5. (-1.0219, -6.3461); (1.2414, -1.5931), and (5.7805, 7.9391)

7. a.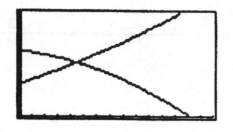

b. 438 wall clocks; $40.92

9. a. $y = 0.1375t^2 + 0.675t + 3.1$

b.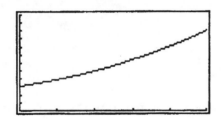

c. 3.1; 3.9; 5; 6.4; 8; 9.9

11. a. $y = -0.02028t^3 + 0.31393t^2 + 0.40873t + 0.66024$

b.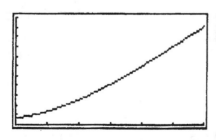

c. 0.66; 1.36; 2.57; 4.16; 6.02; 8.02; 10.03

13. a. $f(t) = 0.0012t^3 - 0.053t^2 + 0.497t + 2.55$ $\quad$ $(0 \le t \le 20)$

b.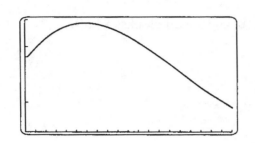

c.

t	0	7	10	20
$f(t)$	2.55	3.84	3.42	0.89

15. a. $y = 0.05833t^3 - 0.325t^2 + 1.8881t + 5.07143$

     b.

                                          c. 6.7; 8.0; 9.4; 11.2; 13.7

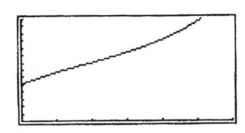

17. a. $y = 0.0125t^4 - 0.01389t^3 + 0.55417t^2 + 0.53294t + 4.95238$ $(0 \leq t \leq 5)$

     b.   c. 5.0; 6.0; 8.3; 12.2; 18.3; 27.5

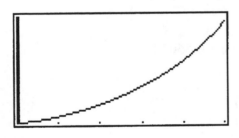

EXERCISES 2.4, page 111

1. $\lim_{x \to -2} f(x) = 3$.   3. $\lim_{x \to 3} f(x) = 3$.   5. $\lim_{x \to -2} f(x) = 3$.

7. The limit does not exist. If we consider any value of x to the right of $x = -2$, $f(x) \leq 2$. If we consider values of x to the left of $x = -2$, $f(x) \geq -2$. Since $f(x)$ does

not approach any one number as x approaches $x = -2$, we conclude that the limit does not exist.

9. $\lim_{x \to 2} (x^2 + 1) = 5.$

x	1.9	1.99	1.999	2.001	2.01	2.1
$f(x)$	4.61	4.9601	4.9960	5.004	5.0401	5.41

11.

x	-0.1	-0.01	-0.001	0.001	0.01	0.1
$f(x)$	-1	-1	-1	1	1	1

The limit does not exist.

13.

x	0.9	0.99	0.999	1.001	1.01	1.1
$f(x)$	100	10,000	1,000,000	1,000,000	10,000	100

The limit does not exist.

15.

x	0.9	0.99	0.999	1.001	1.01	1.1
$f(x)$	2.9	2.99	2.999	3.001	3.01	3.1

$$\lim_{x \to 1} \frac{x^2 + x - 2}{x - 1} = 3.$$

17.

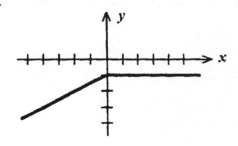

$$\lim_{x \to 0} f(x) = -1$$

19.

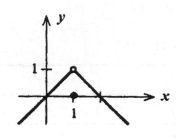

$$\lim_{x \to 1} f(x) = 1$$

21.

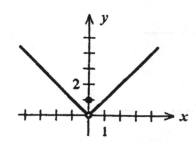

$$\lim_{x \to 0} f(x) = 0$$

23. $\lim_{x \to 2} 3 = 3$ **25.** $\lim_{x \to 3} x = 3$ **27.** $\lim_{x \to 1}(1 - 2x^2) = 1 - 2(1)^2 = -1$

29. $\lim_{x \to 1}(2x^3 - 3x^2 + x + 2) = 2(1)^3 - 3(1)^2 + 1 + 2 = 2.$

31. $\lim_{s \to 0}(2s^2 - 1)(2s + 4) = (-1)(4) = -4.$ **33.** $\lim_{x \to 2} \dfrac{2x + 1}{x + 2} = \dfrac{2(2) + 1}{2 + 2} = \dfrac{5}{4}.$

35. $\lim_{x \to 2} \sqrt{x + 2} = \sqrt{2 + 2} = 2.$

37. $\lim_{x \to -3} \sqrt{2x^4 + x^2} = \sqrt{2(-3)^4 + (-3)^2} = \sqrt{162 + 9} = \sqrt{171} = 3\sqrt{19}.$

39. $\lim\limits_{x \to -1} \dfrac{\sqrt{x^2+8}}{2x+4} = \dfrac{\sqrt{(-1)^2+8}}{2(-1)+4} = \dfrac{\sqrt{9}}{2} = \dfrac{3}{2}$.

41. $\lim\limits_{x \to a}[f(x) - g(x)] = \lim\limits_{x \to a} f(x) - \lim\limits_{x \to a} g(x) = 3 - 4 = -1$.

43. $\lim\limits_{x \to a}[2f(x) - 3g(x)] = \lim\limits_{x \to a} 2f(x) - \lim\limits_{x \to a} 3g(x) = 2(3) - 3(4) = -6$.

45. $\lim\limits_{x \to a}\sqrt{g(x)} = \lim\limits_{x \to a}\sqrt{4} = 2$.

47. $\lim\limits_{x \to a}\dfrac{2f(x) - g(x)}{f(x)g(x)} = \dfrac{2(3) - (4)}{(3)(4)} = \dfrac{2}{12} = \dfrac{1}{6}$.

49. $\lim\limits_{x \to 1}\dfrac{x^2 - 1}{x - 1} = \lim\limits_{x \to 1}\dfrac{(x-1)(x+1)}{x - 1} = \lim\limits_{x \to 1}(x + 1) = 1 + 1 = 2$.

51. $\lim\limits_{x \to 0}\dfrac{x^2 - x}{x} = \lim\limits_{x \to 0}\dfrac{x(x-1)}{x} = \lim\limits_{x \to 0}(x - 1) = 0 - 1 = -1$.

53. $\lim\limits_{x \to -5}\dfrac{x^2 - 25}{x + 5} = \lim\limits_{x \to -5}\dfrac{(x+5)(x-5)}{x + 5} = \lim\limits_{x \to -5}(x - 5) = -10$.

55. $\lim\limits_{x \to 1}\dfrac{x}{x - 1}$ does not exist.

57. $\lim\limits_{x \to -2}\dfrac{x^2 - x - 6}{x^2 + x - 2} = \lim\limits_{x \to -2}\dfrac{(x-3)(x+2)}{(x+2)(x-1)} = \lim\limits_{x \to -2}\dfrac{x-3}{x-1} = \dfrac{-2-3}{-2-1} = \dfrac{5}{3}$.

59. $\lim\limits_{x \to 1}\dfrac{\sqrt{x} - 1}{x - 1} = \lim\limits_{x \to 1}\dfrac{\sqrt{x} - 1}{x - 1} \cdot \dfrac{\sqrt{x} + 1}{\sqrt{x} + 1} = \lim\limits_{x \to 1}\dfrac{x - 1}{(x-1)(\sqrt{x} + 1)} = \lim\limits_{x \to 1}\dfrac{1}{\sqrt{x} + 1} = \dfrac{1}{2}$.

61. $\lim\limits_{x\to 1}\dfrac{x-1}{x^3+x^2-2x}=\lim\limits_{x\to 1}\dfrac{x-1}{x(x-1)(x+2)}=\lim\limits_{x\to 1}\dfrac{1}{x(x+2)}=\dfrac{1}{3}.$

63. $\lim\limits_{x\to\infty}f(x)=\infty$ (does not exist) and $\lim\limits_{x\to-\infty}f(x)=\infty$ (does not exist).

65. $\lim\limits_{x\to\infty}f(x)=0$ and $\lim\limits_{x\to-\infty}f(x)=0.$

67. $\lim\limits_{x\to\infty}f(x)=-\infty$ (does not exist) and $\lim\limits_{x\to-\infty}f(x)=-\infty$ (does not exist).

69.

x	1	10	100	1000
$f(x)$	0.5	0.009901	0.0001	0.000001

x	-1	-10	-100	-1000
$f(x)$	0.5	0.009901	0.0001	0.000001

$\lim\limits_{x\to\infty}f(x)=0$ and $\lim\limits_{x\to-\infty}f(x)=0$

71.

x	1	5	10	100	1000
$f(x)$	12	360	2910	2.99×10^6	2.999×10^9

x	-1	-5	-10	-100	-1000
$f(x)$	6	-390	-3090	-3.01×10^6	-3.0×10^9

$\lim\limits_{x\to\infty}f(x)=\infty$ (does not exist) and $\lim\limits_{x\to-\infty}f(x)=-\infty$ (does not exist).

73. $\lim\limits_{x\to\infty}\dfrac{3x+2}{x-5}=\lim\limits_{x\to\infty}\dfrac{3+\dfrac{2}{x}}{1-\dfrac{5}{x}}=\dfrac{3}{1}=3.$

75. $\lim\limits_{x\to-\infty}\dfrac{3x^3+x^2+1}{x^3+1}=\lim\limits_{x\to-\infty}\dfrac{3+\dfrac{1}{x}+\dfrac{1}{x^3}}{1+\dfrac{1}{x^3}}=3.$

77. $\lim\limits_{x\to-\infty}\dfrac{x^4+1}{x^3-1}=\lim\limits_{x\to-\infty}\dfrac{x+\dfrac{1}{x^3}}{1-\dfrac{1}{x^3}}=-\infty$; that is, the limit does not exist.

79. $\lim\limits_{x\to\infty}\dfrac{x^5-x^3+x-1}{x^6+2x^2+1}=\lim\limits_{x\to\infty}\dfrac{\dfrac{1}{x}-\dfrac{1}{x^3}+\dfrac{1}{x^5}-\dfrac{1}{x^6}}{1+\dfrac{2}{x^4}+\dfrac{1}{x^6}}=0.$

81. a. The cost of removing 50 percent of the pollutant is

$$C(50)=\frac{0.5(50)}{100-50}=0.5\,,\text{ or }\$500{,}000.$$

Similarly, we find that the cost of removing 60, 70, 80, 90, and 95 percent of the pollutants is \$750,000; \$1,166,667; \$2,000,000, \$4,500,000, and \$9,500,000, respectively.

b. $\lim\limits_{x\to100}\dfrac{0.5x}{100-x}=\infty,$

which means that the cost of removing the pollutant increases astronomically if we wish to remove almost all of the pollutant.

83. $\lim\limits_{x\to\infty}\overline{C}(x)=\lim\limits_{x\to\infty}2.2+\dfrac{2500}{x}=2.2,$ or \$2.20 per video disc.

In the long-run, the average cost of producing x video discs will approach \$2.20/disc.

85. a. $T(1)=\dfrac{120}{1+4}=24,$ or \$24 million. $\qquad T(2)=\dfrac{120(4)}{8}=60,$ \$60 million.

$T(3)=\dfrac{120(9)}{13}=83.1,$ or \$83.1 million.

b. In the long run, the movie will gross

$$\lim_{x \to \infty} \frac{120x^2}{x^2 + 4} = \lim_{x \to \infty} \frac{120}{1 + \dfrac{4}{x^2}} = 120, \text{ or } \$120 \text{ million.}$$

87. a. The average cost of driving 5000 miles per year is

$$C(5) = \frac{2010}{5^{2.2}} + 17.80 = 76.07,$$

or 76.1 cents per mile. Similarly, we see that the average cost of driving 10,000 miles per year; 15,000 miles per year; 20,000 miles per year; and 25,000 miles per year is 30.5, 23; 20.6, and 19.5 cents per mile, respectively.

b.

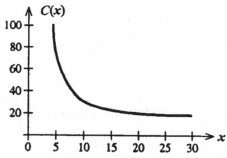

c. It approaches 17.80 cents per mile.

89. False. Let $f(x) = \begin{cases} -1 & \text{if } x < 0 \\ 1 & \text{if } x > 0 \end{cases}$. Then $\lim_{x \to 0} f(x) = 1$, but $f(1)$ is not defined.

91. True. Division by zero is not permitted.

93. True. Each limit in the sum exists. Therefore,

$$\lim_{x \to 2}\left(\frac{x}{x+1} + \frac{3}{x-1} \right) = \lim_{x \to 2} \frac{x}{x+1} + \lim_{x \to 2} \frac{3}{x-1} = \frac{2}{3} + \frac{3}{1} = \frac{11}{3}.$$

95. $\lim_{x \to \infty} \frac{ax}{x+b} = \lim_{x \to \infty} \frac{a}{1 + \frac{b}{x}} = a.$ As the amount of substrate becomes very large, the initial

speed approaches the constant a moles per liter per second.

97. Consider the functions $f(x) = \begin{cases} -1 & \text{if } x < 0 \\ 1 & \text{if } x \geq 0 \end{cases}$ and $g(x) = \begin{cases} 1 & \text{if } x < 0 \\ -1 & \text{if } x \geq 0 \end{cases}.$

Then $\lim_{x \to 0} f(x)$ and $\lim_{x \to 0} g(x)$ do not exist, but $\lim_{x \to 0} [f(x)g(x)] = \lim_{x \to 0}(-1) = -1$.

This example does not contradict Theorem 1 because the hypothesis of Theorem 1 says that if $\lim_{x \to 0} f(x)$ and $\lim_{x \to 0} g(x)$ both exist, then the limit of the product of f and g also exists. It does not say that if the former do not exist, then the latter might not exist.

USING TECHNOLOGY EXERCISES 2.4, page 116

1. 5   3. 3   5. $\dfrac{2}{3}$ 7. $\dfrac{1}{2}$

9. e^2, or 7.38906

11. From the graph we see that $f(x)$ does not approach any finite number as x approaches 3.

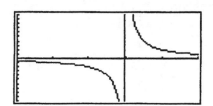

13. a.

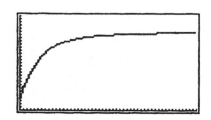

b. $\lim_{t \to \infty} \dfrac{25t^2 + 125t + 200}{t^2 + 5t + 40} = 25$, so in the long run the population will approach 25,000.

EXERCISES 2.5, page 127

1. $\lim_{x \to 2^-} f(x) = 3$, $\lim_{x \to 2^+} f(x) = 2$, $\lim_{x \to 2} f(x)$ does not exist.

3. $\lim_{x \to -1^-} f(x) = \infty$, $\lim_{x \to -1^+} f(x) = 2$. Therefore $\lim_{x \to -1} f(x)$ does not exist.

5. $\lim_{x \to 1^-} f(x) = 0$, $\lim_{x \to 1^+} f(x) = 2$, $\lim_{x \to 1} f(x)$ does not exist.

7. $\lim_{x \to 0^-} f(x) = -2$, $\lim_{x \to 0^+} f(x) = 2$, $\lim_{x \to 0} f(x)$ does not exist.

9. True   11. True   13. False   15. True   17. False   19. True

21. $\lim_{x \to 1^+} (2x + 4) = 6$.

23. $\lim_{x \to 2^-} \dfrac{x-3}{x+2} = \dfrac{2-3}{2+2} = -\dfrac{1}{4}$.

25. $\lim_{x \to 0^+} \dfrac{1}{x}$ does not exist because $1/x \to \infty$ as $x \to 0$ from the right..

27. $\lim_{x \to 0^+} \dfrac{x-1}{x^2 + 1} = \dfrac{-1}{1} = -1$.

29. $\lim_{x \to 0^+} \sqrt{x} = \sqrt{\lim_{x \to 0^+} x} = 0$.

31. $\lim_{x \to -2^+} (2x + \sqrt{2+x}) = \lim_{x \to -2^+} 2x + \lim_{x \to -2^+} \sqrt{2+x} = -4 + 0 = -4$.

33. $\lim_{x \to 1^-} \dfrac{1+x}{1-x} = \infty$, that is, the limit does not exist.

35. $\lim_{x \to 2^-} \dfrac{x^2 - 4}{x - 2} = \lim_{x \to 2^-} \dfrac{(x+2)(x-2)}{x-2} = \lim_{x \to 2^-} (x+2) = 4$.

37. $\lim_{x \to 3^+} \dfrac{x^2 - 9}{x + 3} = \dfrac{9-9}{3+3} = 0$.

39. $\lim_{x \to 0^+} f(x) = \lim_{x \to 0^+} x^2 = 0$, $\lim_{x \to 0^-} f(x) = \lim_{x \to 0^-} 2x = 0$

41. $\lim_{x \to 1^+} f(x) = \lim_{x \to 1^+} \sqrt{x+3} = \sqrt{4} = 2$.

$\lim_{x \to 1^-} f(x) = \lim_{x \to 1^-} (2 + \sqrt{x}) = 2 + \sqrt{x} = 2 + \sqrt{1} = 3$.

43. The function is discontinuous at $x = 0$. Conditions 2 and 3 are violated.

45. The function is continuous everywhere.

47. The function is discontinuous at $x = 0$. Condition 3 is violated.

49. The function is discontinuous at $x = 0$. Condition 3 is violated.

51. f is continuous for all values of x.

53. f is continuous for all values of x. Note that $x^2 + 1 \geq 1 > 0$.

55. f is discontinuous at $x = 1/2$, where the denominator is 0.

57. Observe that $x^2 + x - 2 = (x + 2)(x - 1) = 0$ if $x = -2$ or $x = 1$. So, f is discontinuous at these values of x.

59. f is continuous everywhere since all three conditions are satisfied.

61. f is continuous everywhere because all three conditions are satisfied.

63. f is continuous everywhere since all three conditions are satisfied. Observe that
$$\lim_{x \to 1} f(x) = \lim_{x \to 1} \frac{x^2 - 1}{x - 1} = \lim_{x \to 1} \frac{(x - 1)(x + 1)}{x - 1} = \lim_{x \to 1}(x + 1) = 2 = f(1).$$

65. f is continuous everywhere since all three conditions are satisfied.

67. Since the denominator $x^2 - 1 = (x - 1)(x + 1) = 0$ if $x = -1$ or 1, we see that f is discontinuous at these points.

69. Since $x^2 - 3x + 2 = (x - 2)(x - 1) = 0$ if $x = 1$ or 2, we see that the denominator is zero at these points and so f is discontinuous at these points.

71. The function f is discontinuous at $x = 1, 2, 3, ..., 11$ because the limit of f does not exist at these points.

73. Having made steady progress up to $x = x_1$, Michael's progress came to a standstill. Then at $x = x_2$ a sudden break-through occurs and he then continues to successfully

complete the solution to the problem.

75. Conditions 2 and 3 are not satisfied at each of these points.

77. The graph of f follows.

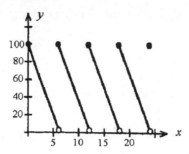

f is discontinuous at $x = 6, 12, 18, 24$.

79.

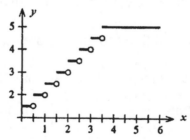

f is discontinuous at $x = \frac{1}{2}, 1, 1\frac{1}{2}, \ldots, 4$.

81. a. $\lim\limits_{t \to 0^+} S(t) = \lim\limits_{t \to 0^+} \dfrac{a}{t} + b = \infty$. As the time taken to excite the tissue is made smaller

and smaller, the strength of the electric current gets stronger and stronger.

b. $\lim\limits_{t \to \infty} \dfrac{a}{t} + b = b$. As the time taken to excite the tissue is made larger and larger,

the strength of the electric current gets smaller and smaller and approaches b.

83. We require that $f(1) = 1 + 2 = 3 = \lim\limits_{x \to 1^+} kx^2 = k$, or $k = 3$.

85. a. Yes, because if $f + g$ were continuous at a, then $g = (f + g) - f$
would be continuous (the difference of two continuous functions is continuous),
and this would imply that g is continuous, a contradiction.

b. No. Consider the functions f and g defined by

$$f(x) = \begin{cases} -1 & \text{if } x < 0 \\ 1 & \text{if } x \geq 0 \end{cases} \quad \text{and} \quad g(x) = \begin{cases} 1 & \text{if } x < 0 \\ -1 & \text{if } x \geq 0 \end{cases}.$$

Both f and g are discontinuous at $x = 0$, but $f + g$ is continuous everywhere.

87. a. f is a polynomial of degree 2 and is therefore continuous everywhere, and in particular in [1,3].
b. $f(1) = 3$ and $f(3) = -1$ and so f must have at least one zero in (1,3).

89. f is a polynomial and is therefore continuous on [-1,1].
$$f(-1) = (-1)^3 - 2(-1)^2 + 3(-1) + 2 = -1 - 2 - 3 + 2 = -4.$$
$$f(1) = 1 - 2 + 3 + 2 = 4.$$
Since $f(-1)$ and $f(1)$ have opposite signs, we see that f has at least one zero in (-1,1).

91. $f(0) = 6$ and $f(3) = 3$ and f is continuous on [0,3]. So the Intermediate Value Theorem guarantees that there is at least one value of x for which $f(x) = 4$. Solving $f(x) = x^2 - 4x + 6 = 4$, we find $x^2 - 4x + 2 = 0$. Using the quadratic formula, we find that $x = 2 \pm \sqrt{2}$. Since $2 \pm \sqrt{2}$ does not lie in [0,3], we see that $x = 2 - \sqrt{2} \approx 0.59$.

93. $x^5 + 2x - 7 = 0$

Step	Root of f(x) = 0 lies in
1	(1,2)
2	(1,1.5)
3	(1.25,1.5)
4	(1.25,1.375)
5	(1.3125,1.375)
6	(1.3125,1.34375)
7	(1.328125,1.34375)
8	(1.3359375,1.34375)
9	(1.33984375,1.34375)

We see that the required root is approximately 1.34.

95. a. $h(t) = 4 + 64(0) - 16(0) = 4$, and $h(2) = 4 + 64(2) - 16(4) = 68$.

b. The function h is continuous on $[0,2]$. Furthermore, the number 32 lies between 4 and 68. Therefore, the Intermediate Value Theorem guarantees that there is at least one value of t such that $h(t) = 32$, that is, Joan must see the ball at least once during the time the ball is in the air.

c. We solve

$$h(t) = 4 + 64t - 16t^2 = 32$$

or

$$16t^2 - 64t + 28 = 0$$
$$4t^2 - 16t + 7 = 0$$
$$(2t - 1)(2t - 7) = 0$$

giving $t = \frac{1}{2}$ or $t = \frac{7}{2}$. Joan sees the ball on its way up half a second after it was thrown and again $3\frac{1}{2}$ seconds later when it is on its way down.

97. False. Take

$$f(x) = \begin{cases} -1 & \text{if } x < 2 \\ 4 & \text{if } x = 2 \\ 1 & \text{if } x > 2 \end{cases}$$

Then $f(2) = 4$ but $\lim\limits_{x \to 2}$ does not exist.

99. False. Consider the function $f(x) = x^2 - 1$ on the interval $[-2, 2]$. Here, $f(-2) = f(2) = 3$, but f has zeros at $x = -1$ and $x = 1$.

101. False. Let $f(x) = \begin{cases} x & \text{if } x \neq 0 \\ 1 & \text{if } x = 0 \end{cases}$. Then $\lim\limits_{x \to 0^+} f(x) = \lim\limits_{x \to 0^-} f(x)$, but $f(0) = 1$.

103. False. Take $f(x) = \begin{cases} \frac{1}{x} & \text{if } x \neq 0 \\ 0 & \text{if } x = 0 \end{cases}$. Then f is continuous for all $x \neq 0$ but

$\lim\limits_{x \to 0} f(x)$ does not exist.

105. True. Since the number 2 lies between $f(-2) = 3$ and $f(3) = 1$, the intermediate value theorem guarantees that there exists at least one number $-2 \leq c \leq 3$ such that $f(c) = 2$.

107. a. Both $g(x) = x$ and $h(x) = \sqrt{1 - x^2}$ are continuous on $[-1,1]$ and so

$f(x) = x - \sqrt{1-x^2}$ is continuous on $[-1,1]$.

b. $f(-1) = -1$ and $f(1) = 1$ and so f has at least one zero in $(-1,1)$.

c. Solving $f(x) = 0$, we have $x = \sqrt{1-x^2}$, $x^2 = 1-x^2$, $2x^2 = 1$, or $x = \frac{\pm\sqrt{2}}{2}$.

109.   a. (i). Repeated use of property 3 shows that $g(x) = x^n = x \cdot x \cdots x$ (n times) is a continuous function since $f(x) = x$ is continuous by Property 1.

(ii). Properties 1 and 5 combine to show that $c \cdot x^n$ is continuous using the results of (a).

(iii). Each of the terms of $p(x) = a_0 x^n + a_1 x^{n-1} + \cdots + a_n$ is continuous and so Property 4 implies that p is continuous.

b. Property 6 now shows that $R(x) = \dfrac{p(x)}{q(x)}$ is continuous if $q(a) \neq 0$ since p

and q are continuous at $x = a$.

USING TECHNOLOGY EXERCISES 2.5, page 134

1. $x = 0,1$ 3. $x = 2$ 5. $x = 0, \frac{1}{2}$ 7. $x = -\frac{1}{2}, 2$ 9. $x = -2, 1$

11.

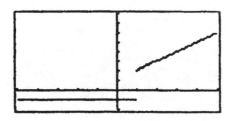

13

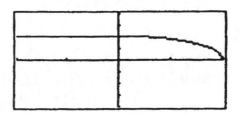

15.

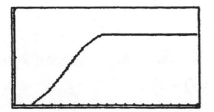

EXERCISES 2.6, page 148

1. The rate of change of the average infant's weight when $t = 3$ is (7.5)/5, or 1.5 lb/month. The rate of change of the average infant's weight when $t = 18$ is (3.5)/6, or approximately 0.6 lb/month. The average rate of change over the infant's first year of life is (22.5 − 7.5)/(12), or 1.25 lb/month.

3. The rate of change of the percentage of households watching television at 4 P.M. is (12.3)/4, or approximately 3.1 percent per hour. The rate at 11 P.M. is (−42.3)/2 = −21.15; that is, it is dropping off at the rate of 21.15 percent per hour.

5. a. Car A is travelling faster than Car B at t_1 because the slope of the tangent line to the graph of f is greater than the slope of the tangent line to the graph of g at t_1.

   b. Their speed is the same because the slope of the tangent lines are the same at t_2.

   c. Car B is travelling faster than Car A.

   d. They have both covered the same distance and are once again side by side at t_3.

7. a. P_2 is decreasing faster at t_1 because the slope of the tangent line to the graph of g at t_1 is greater than the slope of the tangent line to the graph of f at t_1.
   b. P_1 is decreasing faster than P_2 at t_2.
   c. Bactericide B is more effective in the short run, but bactericide A is more effective in the long run.

9. $f(x) = 13$
   Step 1 $f(x + h) = 13$
   Step 2 $f(x + h) - f(x) = 13 - 13 = 0$
   Step 3 $\dfrac{f(x+h) - f(x)}{h} = \dfrac{0}{h} = 0$
   Step 4 $f'(x) = \lim\limits_{h \to 0} \dfrac{f(x+h) - f(x)}{h} = \lim\limits_{h \to 0} 0 = 0$

11. $f(x) = 2x + 7$
    Step 1 $f(x + h) = 2(x + h) + 7$
    Step 2 $f(x + h) - f(x) = 2(x + h) + 7 - (2x + 7) = 2h$
    Step 3 $\dfrac{f(x+h) - f(x)}{h} = \dfrac{2h}{h} = 2$

Step 4 $\quad f'(x) = \lim\limits_{h \to 0} \dfrac{f(x+h) - f(x)}{h} = \lim\limits_{h \to 0} 2 = 2$

13. $f(x) = 3x^2$

    Step 1 $\quad f(x+h) = 3(x+h)^2 = 3x^2 + 6xh + 3h^2$

    Step 2 $\quad f(x+h) - f(x) = (3x^2 + 6xh + 3h^2) - 3x^2 = 6xh + 3h^2 = h(6x + 3h)$

    Step 3 $\quad \dfrac{f(x+h) - f(x)}{h} = \dfrac{h(6x + 3h)}{h} = 6x + 3h$

    Step 4 $\quad f'(x) = \lim\limits_{h \to 0} \dfrac{f(x+h) - f(x)}{h} = \lim\limits_{h \to 0} (6x + 3h) = 6x.$

15. $f(x) = -x^2 + 3x$

    Step 1 $\quad f(x+h) = -(x+h)^2 + 3(x+h) = -x^2 - 2xh - h^2 + 3x + 3h$

    Step 2 $\quad f(x+h) - f(x) = (-x^2 - 2xh - h^2 + 3x + 3h) - (-x^2 + 3x)$

$$= -2xh - h^2 + 3h = h(-2x - h + 3)$$

    Step 3 $\quad \dfrac{f(x+h) - f(x)}{h} = \dfrac{h(-2x - h + 3)}{h} = -2x - h + 3$

    Step 4 $\quad f'(x) = \lim\limits_{h \to 0} \dfrac{f(x+h) - f(x)}{h} = \lim\limits_{h \to 0} (-2x - h + 3) = -2x + 3.$

17. $f(x) = 2x + 7$. Using the four-step process,

    Step 1 $\quad f(x+h) = 2(x+h) + 7 = 2x + 2h + 7$

    Step 2 $\quad f(x+h) - f(x) = 2x + 2h + 7 - 2x - 7 = 2h$

    Step 3 $\quad \dfrac{f(x+h) - f(x)}{h} = \dfrac{2h}{h} = 2$

    Step 4 $\quad f'(x) = \lim\limits_{h \to 0} \dfrac{f(x+h) - f(x)}{h} = \lim\limits_{h \to 0} 2 = 2$

    we find that $f'(x) = 2$. In particular, the slope at $x = 2$ is also 2. Therefore, a required equation is $y - 11 = 2(x - 2)$ or $y = 2x + 7$.

19. $f(x) = 3x^2$. We first compute $f'(x) = 6x$ (see Problem 13). Since the slope of the

tangent line is $f'(1) = 6$, we use the point-slope form of the equation of a line and find that a required equation is $y - 3 = 6(x - 1)$, or $y = 6x - 3$.

21. $f(x) = -1/x$. We first compute $f'(x)$ using the four-step process.

Step 1 $f(x + h) = -\dfrac{1}{x + h}$

Step 2 $f(x + h) - f(x) = -\dfrac{1}{x + h} + \dfrac{1}{x} = \dfrac{-x + (x + h)}{x(x + h)} = \dfrac{h}{x(x + h)}$

Step 3 $\dfrac{f(x + h) - f(x)}{h} = \dfrac{\frac{h}{x(x + h)}}{h} = \dfrac{1}{x(x + h)}$

Step 4 $f'(x) = \lim\limits_{h \to 0} \dfrac{f(x + h) - f(x)}{h} = \lim\limits_{h \to 0} \dfrac{1}{x(x + h)} = \dfrac{1}{x^2}.$

The slope of the tangent line is $f'(3) = 1/9$. Therefore, a required equation is
$$y - (-\tfrac{1}{3}) = \tfrac{1}{9}(x - 3) \quad \text{or} \quad y = \tfrac{1}{9}x - \tfrac{2}{3}.$$

23. a. $f(x) = 2x^2 + 1$. We use the four-step process.

Step 1 $f(x + h) = 2(x + h)^2 + 1 = 2x^2 + 4xh + 2h^2 + 1$

Step 2 $f(x + h) - f(x) = (2x^2 + 4xh + 2h^2 + 1) - (2x^2 + 1) = 4xh + 2h^2$
$$= h(4x + 2h)$$

Step 3 $\dfrac{f(x + h) - f(x)}{h} = \dfrac{h(4x + 2h)}{h} = 4x + 2h$

Step 4 $f'(x) = \lim\limits_{h \to 0} \dfrac{f(x + h) - f(x)}{h} = \lim\limits_{h \to 0} (4x + 2h) = 4x$

b. The slope of the tangent line is $f'(1) = 4(1) = 4$. Therefore, an equation is
$y - 3 = 4(x - 1)$ or $y = 4x - 1$.

c.

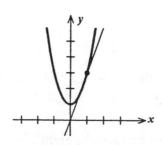

25. a. $f(x) = x^2 - 2x + 1$. We use the four-step process:

Step 1 $f(x+h) = (x+h)^2 - 2(x+h) + 1 = x^2 + 2xh + h^2 - 2x - 2h + 1$

Step 2 $f(x+h) - f(x) = (x^2 + 2xh + h^2 - 2x - 2h + 1) - (x^2 - 2x + 1)]$
$$= 2xh + h^2 - 2h = h(2x + h - 2)$$

Step 3 $\dfrac{f(x+h) - f(x)}{h} = \dfrac{h(2x+h-2)}{h} = 2x + h - 2$

Step 4 $f'(x) = \lim_{h \to 0} \dfrac{f(x+h) - f(x)}{h} = \lim_{h \to 0} (2x + h - 2) = 2x - 2.$

b. At a point on the graph of f where the tangent line to the curve is horizontal, $f'(x) = 0$. Then $2x - 2 = 0$, or $x = 1$. Since $f(1) = 1 - 2 + 1 = 0$, we see that the required point is $(1,0)$.

c.

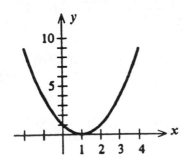

d. It is changing at the rate of 0 units per unit change in x.

27. a.
$$f(x) = x^2 + x$$

$$\frac{f(3) - f(2)}{3 - 2} = \frac{(3^2 + 3) - (2^2 + 2)}{1} = 6$$

$$\frac{f(2.5) - f(2)}{2.5 - 2} = \frac{(2.5^2 + 2.5) - (2^2 + 2)}{0.5} = 5.5$$

$$\frac{f(2.1) - f(2)}{2.1 - 2} = \frac{(2.1^2 + 2.1) - (2^2 + 2)}{0.1} = 5.1$$

b. We first compute $f'(x)$ using the four-step process.

Step 1 $f(x+h) = (x+h)^2 + (x+h) = x^2 + 2xh + h^2 + x + h$

Step 2 $f(x+h) - f(x) = (x^2 + 2xh + h^2 + x + h) - (x^2 + x)]$

$$= 2xh + h^2 + h = h(2x + h + 1)$$

Step 3 $\quad \dfrac{f(x+h) - f(x)}{h} = \dfrac{h(2x+h+1)}{h} = 2x + h + 1$

Step 4 $\quad f'(x) = \lim_{h \to 0} \dfrac{f(x+h) - f(x)}{h} = \lim_{h \to 0} (2x + h + 1) = 2x + 1.$

The instantaneous rate of change of y at $x = 2$ is $f'(2) = 5$ or 5 units per unit change in x.

c. The results in (a) suggest that the average rates of change of f at $x = 2$ approach 5 as the interval $[2, 2+h]$ gets smaller and smaller ($h = 1, 0.5,$ and 0.1). This number is the instantaneous rate of change of f at $x = 2$ as computed in (b).

29. a. $f(t) = 2t^2 + 48t$. The average velocity of the car over $[20,21]$ is

$$\dfrac{f(21) - f(20)}{21 - 20} = \dfrac{[2(21)^2 + 48(21)] - [2(20)^2 + 48(20)]}{1} = 130 \text{ ft / sec}$$

Its average velocity over $[20,20.1]$ is

$$\dfrac{f(20.1) - f(20)}{20.1 - 20} = \dfrac{[2(20.1)^2 + 48(20.1)] - [2(20)^2 + 48(20)]}{0.1} = 128.2 \text{ ft / sec}$$

Its average velocity over $[20,20.01]$

$$\dfrac{f(20.01) - f(20)}{20.01 - 20} = \dfrac{[2(20.01)^2 + 48(20.01)] - [2(20)^2 + 48(20)]}{0.01} = 128.02 \text{ ft / sec}$$

b. We first compute $f'(t)$ using the four-step process.

Step 1 $\quad f(t + h) = 2(t + h)^2 + 48(t + h) = 2t^2 + 4th + 2h^2 + 48t + 48h$

Step 2 $\quad f(t + h) - f(t) = (2t^2 + 4th + 2h^2 + 48t + 48h) - (2t^2 + 48t)]$

$$= 4th + 2h^2 + 48h = h(4t + 2h + 48).$$

Step 3 $\quad \dfrac{f(t+h) - f(t)}{h} = \dfrac{h(4t + 2h + 48)}{h} = 4t + 2h + 48$

Step 4 $\quad f'(t) = \lim_{t \to 0} \dfrac{f(t+h) - f(t)}{h} = \lim_{t \to 0} 4t + 2h + 48 = 4t + 48$

The instantaneous velocity of the car at $t = 20$ is $f'(20) = 4(20) + 48$, or 128 ft/sec.

c. Our results shows that the average velocities do approach the instantaneous velocity as the intervals over which they are computed decreases.

31. a. We solve the equation $16t^2 = 400$ obtaining $t = 5$ which is the time it takes the screw driver to reach the ground.

b. The average velocity over the time [0,5] is

$$\frac{f(5)-f(0)}{5-0} = \frac{16(25)-0}{5} = 80, \text{ or } 80 \text{ ft/sec. } \quad [\text{Let } s = f(t) = 16t^2.]$$

c. The velocity of the screwdriver at time t is

$$v(t) = \lim_{h \to 0} \frac{f(t+h)-f(t)}{h} = \lim_{h \to 0} \frac{16(t+h)^2 - 16t^2}{h}$$

$$= \lim_{h \to 0} \frac{16t^2 + 32th + 16h^2 - 16t^2}{h} = \lim_{h \to 0} \frac{(32t+16h)h}{h} = 32t.$$

In particular, the velocity of the screwdriver when it hits the ground (at $t = 5$) is
   $v(5) = 32(5) = 160$, or 160 ft/sec.

33. a. $V = \dfrac{1}{p}$. The average rate of change of V is

$$\frac{f(3)-f(2)}{3-2} = \frac{\frac{1}{3}-\frac{1}{2}}{1} = -\frac{1}{6}, \qquad [\text{Write } V = f(p) = \frac{1}{p}.]$$

or a decrease of $\frac{1}{6}$ liter/atmosphere.

b.

$$V'(t) = \lim_{h \to 0} \frac{\dfrac{f(p+h)-f(p)}{h}}{h} = \lim_{h \to 0} \frac{\dfrac{1}{p+h}-\dfrac{1}{p}}{h}$$

$$= \lim_{h \to 0} \frac{p-(p+h)}{hp(p+h)} = \lim_{h \to 0} -\frac{1}{p(p+h)} = -\frac{1}{p^2}.$$

In particular, the rate of change of V when $p = 2$ is

$$V'(2) = -\frac{1}{2^2}, \text{ or a decrease of } \frac{1}{4} \text{ liter/atmosphere}$$

35. a. Using the four-step process, we find that

$$P'(x) = \lim_{h \to 0} \frac{P(x+h)-P(x)}{h}$$

$$= \lim_{h \to 0} \frac{-\frac{1}{3}(x^2+2xh+h^2)+7x+7h+30-(-\frac{1}{3}x^2+7x+30)}{h}$$

$$P'(x) = \lim_{h \to 0} \frac{P(x+h)-P(x)}{h}$$

$$= \lim_{h \to 0} \frac{-\frac{2}{3}xh - \frac{1}{3}h + 7h}{h} = \lim_{h \to 0}(-\frac{2}{3}x - \frac{1}{3}h + 7) = -\frac{2}{3}x + 7.$$

b. $P'(10) = -\frac{2}{3}(10) + 7 \approx 0.333$, or \$333 per quarter.

  $P'(30) = -\frac{2}{3}(30) + 7 \approx -13$, or a decrease of \$13,000 per quarter.

37. $N(t) = t^2 + 2t + 50$. We first compute $N'(t)$ using the four–step process.

Step 1 $N(t + h) = (t + h)^2 + 2(t + h) + 50$
  $= t^2 + 2th + h^2 + 2t + 2h + 50$

Step 2 $N(t + h) - N(t)$

$$= (t^2 + 2th + h^2 + 2t + 2h + 50) - (t^2 + 2t + 50)$$

$$= 2th + h^2 + 2h = h(2t + h + 2).$$

Step 3 $\dfrac{N(t + h) - N(t)}{h} = 2t + h + 2.$

Step 4 $N'(t) = \lim_{h \to 0}(2t + h + 2) = 2t + 2.$

The rate of change of the country's GNP two years from now will be $N'(2) = 6$, or \$6 billion/yr. The rate of change four years from now will be $N'(4) = 10$, or \$10 billion/yr.

39. $\dfrac{f(a + h) - f(a)}{h}$ gives the average rate of change of the seal population over the time interval $[a, a + h]$.

  $\lim_{h \to 0} \dfrac{f(a + h) - f(a)}{h}$ gives the instantaneous rate of change of the seal population at $x = a$.

41. $\dfrac{f(a + h) - f(a)}{h}$ gives the average rate of change of the country's industrial production over the time interval $[a, a + h]$.

  $\lim_{h \to 0} \dfrac{f(a + h) - f(a)}{h}$ gives the instantaneous rate of change of the country's industrial production at $x = a$.

43. $\dfrac{f(a+h)-f(a)}{h}$ gives the average rate of change of the atmospheric pressure over

the altitudes $[a, a+h]$.

$\displaystyle\lim_{h\to 0}\dfrac{f(a+h)-f(a)}{h}$ gives the instantaneous rate of change of the atmospheric

pressure at $x = a$.

45. a. f has a limit at $x = a$.
    b. f is not continuous at $x = a$ because $f(a)$ is not defined.
    c. f is not differentiable at $x = a$ because it is not continuous there.

47. a. f has a limit at $x = a$.   b. f is continuous at $x = a$.
    c. f is not differentiable at $x = a$ because f has a kink at the point $x = a$.

49. a. f does not have a limit at $x = a$ because it is unbounded in the neighborhood of a.
    b. f is not continuous at $x = a$.
    c. f is not differentiable at $x = a$ because it is not continuous there.

51. Our computations yield the following results:
    32.1, 30.939, 30.814, 30.8014, 30.8001, 30.8000.
    The motorcycle's instantaneous velocity at $t = 2$ is approximately 30.8 ft/sec.

53. False. Let $f(x)=|x|$. Then f is continuous at $x = 0$, but is not differentiable there.

55. Observe that the graph of f has a kink at $x = -1$. We have
    $$\dfrac{f(-1+h)-f(-1)}{h} = 1 \text{ if } h > 0, \text{ and } -1 \text{ if } h < 0,$$

so that $\displaystyle\lim_{h\to 0}\dfrac{f(-1+h)-f(-1)}{h}$ does not exist.

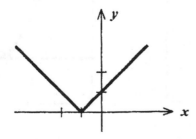

57. For continuity, we require that $f(1) = 1 = \lim_{x \to 1^+}(ax + b) = a + b$, or $a + b = 1$.

In order that the derivative exist at $x = 1$, we require that $\lim_{x \to 1^-} 2x = \lim_{x \to 1^+} a$, or $2 = a$.

Therefore, $b = -1$ and so $f(x) = \begin{cases} x^2 & \text{if } x \leq 1 \\ 2x - 1 & \text{if } x > 1 \end{cases}$. The graph of f follows.

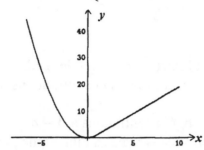

59. We have $f(x) = x$ if $x > 0$ and $f(x) = -x$ if $x < 0$. Therefore, when $x > 0$

$$f'(x) = \lim_{h \to 0} \frac{f(x+h) - f(x)}{h} = \lim_{h \to 0} \frac{x + h - x}{h} = \lim_{h \to 0} \frac{h}{h} = 1,$$

and when $x < 0$

$$f'(x) = \lim_{h \to 0} \frac{f(x+h) - f(x)}{h} = \lim_{h \to 0} \frac{-x - h - (-x)}{h} = \lim_{h \to 0} \frac{-h}{h} = -1.$$

Since the right–hand limit does not equal the left–hand limit, we conclude that $\lim_{h \to 0} f(x)$ does not exist.

USING TECHNOLOGY EXERCISES 2.6, page 153

1. a. $y = 4x - 3$
    b.

3. a. $y = -7x - 8$
    b.

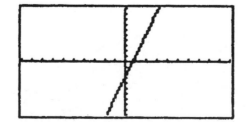

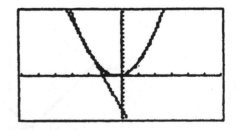

5. a. $y = 9x - 11$ b.

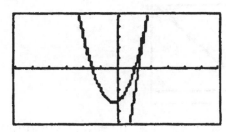

7. a. $y = 2$
    b.

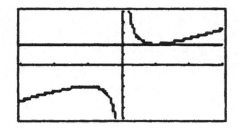

9. a. $y = \frac{1}{4}x + 1$
    b.

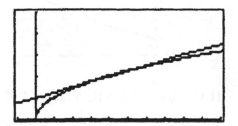

11. a. 4   b. $y = 4x - 1$
    c.

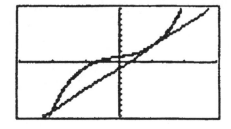

13. a. 20   b. $y = 20x - 35$
    c.

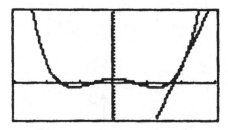

15. a. $\frac{3}{4}$ b. $y = \frac{3}{4}x - 1$
    c.

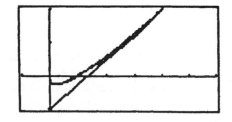

17. a. $-\frac{1}{4}$ b. $y = -\frac{1}{4}x + \frac{3}{4}$
    c.

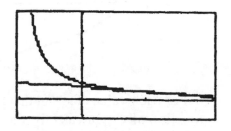

19. a. 4.02   b. $y = 4.02x - 3.57$

    c.

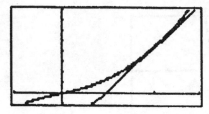

21. a.   b. 41.22 cents/mile   c. 1.22 cents/mile

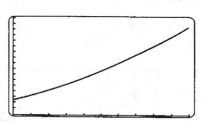

CHAPTER 2 REVIEW, page 158

1.   a. $9 - x \geq 0$ gives $x \leq 9$ and the domain is $(-\infty, 9]$.
   b. $2x^2 - x - 3 = (2x - 3)(x + 1)$, and $x = 3/2$ or $x = -1$.
   Since the denominator of the given expression is zero at these points, we see that
   the domain of f cannot include these points and so the domain of f is
   $(-\infty, -1) \cup (-1, \frac{3}{2}) \cup (\frac{3}{2}, \infty)$.

3.   a.

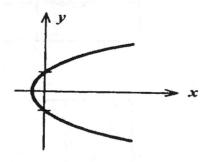

    b. For each value of $x > 0$, there are two values of y. We conclude that y is not a
   function of x. Equivalently, the function fails the vertical line test.
   c. Yes. For each value of y, there is only 1 value of x.

5. a. $f(x)g(x) = \dfrac{2x+3}{x}$

    b. $\dfrac{f(x)}{g(x)} = \dfrac{1}{x(2x+3)}$

    c. $f(g(x)) = \dfrac{1}{2x+3}$.

    d. $g(f(x)) = 2\left(\dfrac{1}{x}\right) + 3 = \dfrac{2}{x} + 3.$

7. $\lim\limits_{x\to1}(x^2+1) = [(1)^2 + 1] = 1+1 = 2.$

9. $\lim\limits_{x\to3}\dfrac{x-3}{x+4} = \dfrac{3-3}{3+4} = 0.$

11. $\lim\limits_{x\to-2}\dfrac{x^2-2x-3}{x^2+5x+6}$ does not exist. (The denominator is 0 at $x=-2$.)

13. $\lim\limits_{x\to3}\dfrac{4x-3}{\sqrt{x+1}} = \dfrac{12-3}{\sqrt{4}} = \dfrac{9}{2}.$

15. $\lim\limits_{x\to1^-}\dfrac{\sqrt{x}-1}{x-1} = \lim\limits_{x\to1^-}\dfrac{(\sqrt{x}-1)(\sqrt{x}+1)}{(x-1)(\sqrt{x}+1)} = \lim\limits_{x\to1^-}\dfrac{x-1}{(x-1)(\sqrt{x}+1)} = \lim\limits_{x\to1^-}\dfrac{1}{\sqrt{x}+1} = \dfrac{1}{2}.$

17. $\lim\limits_{x\to-\infty}\dfrac{x+1}{x} = \lim\limits_{x\to-\infty}\left(1+\dfrac{1}{x}\right) = 1.$

19. $\lim\limits_{x\to-\infty}\dfrac{x^2}{x+1} = \lim\limits_{x\to-\infty} x\cdot\dfrac{1}{1+\dfrac{1}{x}} = -\infty$, so the limit does not exist.

21. $\lim\limits_{x\to2^+} f(x) = \lim\limits_{x\to2^+}(x+2) = 4;$

    $\lim\limits_{x\to2^-} f(x) = \lim\limits_{x\to2^-}(4-x) = 2.$

    Therefore, $\lim\limits_{x\to2} f(x)$ does not exist.

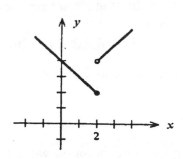

23. Since the denominator
    $$4x^2 - 2x - 2 = 2(2x^2 - x - 1) = 2(2x+1)(x-1) = 0$$
    if $x = -1/2$ or 1, we see that f is discontinuous at these points.

25. The function is discontinuous at $x = 0$.

27. $f(x) = 3x + 5$. Using the four-step process, we find
   Step 1 $f(x + h) = 3(x + h) + 5 = 3x + 3h + 5$
   Step 2 $f(x + h) - f(x) = 3x + 3h + 5 - 3x - 5 = 3h$
   Step 3 $\dfrac{f(x + h) - f(x)}{h} = \dfrac{3h}{h} = 3.$
   Step 4 $f'(x) = \lim\limits_{h \to 0} \dfrac{f(x + h) - f(x)}{h} = \lim\limits_{h \to 0}(3) = 3.$

29. $f(x) = 3x + 5$. We use the four-step process to obtain
   Step 1 $f(x + h) = \frac{3}{2}(x + h) + 5 = \frac{3}{2}x + \frac{3}{2}h + 5.$

   Step 2 $f(x + h) - f(x) = \frac{3}{2}x + \frac{3}{2}h + 5 - \frac{3}{2}x - 5 = \frac{3}{2}h.$

   Step 3 $\dfrac{f(x + h) - f(x)}{h} = \dfrac{3}{2}.$

   Step 4 $f'(x) = \lim\limits_{h \to 0} \dfrac{f(x + h) - f(x)}{h} = \lim\limits_{h \to 0}\dfrac{3}{2} = \dfrac{3}{2}.$
   Therefore, the slope of the tangent line to the graph of the function f at the point
   (-2,2) is 3/2. To find the equation of the tangent line to the curve at the point
   (-2,2), we use the point–slope form of the equation of a line obtaining
   $$y - 2 = \tfrac{3}{2}[x - (-2)] \quad \text{or} \quad y = \tfrac{3}{2}x + 5.$$

31. a. f is continuous at $x = a$ because the three conditions for continuity are satisfied at
   $x = a$; that is,
       i. $f(x)$ is defined   ii. $\lim\limits_{x \to a} f(x)$ exists   iii. $\lim\limits_{x \to a} f(x) = f(a)$
   b. f is not differentiable at $x = a$ because the graph of f has a kink at $x = a$.

33. a. The line passes through (0, 2.4) and (5, 7.4) and has slope $m = \dfrac{7.4 - 2.4}{5 - 0} = 1.$

   Letting y denote the sales, we see that an equation of the line is
   $$y - 2.4 = 1(t - 0), \text{ or } y = t + 2.4.$$
   We can also write this in the form $S(t) = t + 2.4.$
   b. The sales in 2002 were $S(3) = 3 + 2.4 = 5.4$, or \$5.4 million.

35. Substituting the first equation into the second yields
$$3x - 2(\tfrac{3}{4}x + 6) + 3 = 0 \quad \text{or} \quad \tfrac{3}{2}x - 12 + 3 = 0$$
or $x = 6$. Substituting this value of x into the first equation then gives $y = 21/2$, so the point of intersection is $(6, \tfrac{21}{2})$.

37. We solve the system
$$3x + p - 40 = 0$$
$$2x - p + 10 = 0.$$
Adding these two equations, we obtain $5x - 30 = 0$, or $x = 6$. So,
$$p = 2x + 10 = 12 + 10 = 22.$$
Therefore, the equilibrium quantity is 6000 and the equilibrium price is $22.

39. $R(30) = -\tfrac{1}{2}(30)^2 + 30(30) = 450$, or $45,000.

41. $T = f(n) = 4n\sqrt{n-4}$.
$$f(4) = 0, \quad f(5) = 20\sqrt{1} = 20, \quad f(6) = 24\sqrt{2} \approx 33.9, \quad f(7) = 28\sqrt{3} \approx 48.5,$$
$$f(8) = 32\sqrt{4} = 64, \quad f(9) = 36\sqrt{5} \approx 80.5, \quad f(10) = 40\sqrt{6} \approx 98,$$
$$f(11) = 44\sqrt{7} \approx 116 \quad \text{and} \quad f(12) = 48\sqrt{8} \approx 135.8.$$
The graph of f follows:

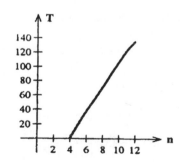

43.

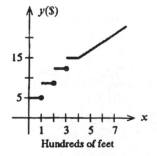

2 Functions, Limits, and the Derivative

CHAPTER 3

EXERCISES 3.1, page 168

1. $f'(x) = \dfrac{d}{dx}(-3) = 0.$

3. $f'(x) = \dfrac{d}{dx}(x^5) = 5x^4.$

5. $f'(x) = \dfrac{d}{dx}(x^{2.1}) = 2.1x^{1.1}.$

7. $f'(x) = \dfrac{d}{dx}(3x^2) = 6x.$

9. $f'(r) = \dfrac{d}{dr}(\pi r^2) = 2\pi r.$

11. $f'(x) = \dfrac{d}{dx}(9x^{1/3}) = \dfrac{1}{3}(9)x^{(1/3-1)} = 3x^{-2/3}.$

13. $f'(x) = \dfrac{d}{dx}(3\sqrt{x}) = \dfrac{d}{dx}(3x^{1/2}) = \dfrac{1}{2}(3)x^{-1/2} = \dfrac{3}{2}x^{-1/2} = \dfrac{3}{2\sqrt{x}}.$

15. $f'(x) = \dfrac{d}{dx}(7x^{-12}) = (-12)(7)x^{(-12-1)} = -84x^{-13}.$

17. $f'(x) = \dfrac{d}{dx}(5x^2 - 3x + 7) = 10x - 3.$

19. $f'(x) = \dfrac{d}{dx}(-x^3 + 2x^2 - 6) = -3x^2 + 4x.$

21. $f'(x) = \dfrac{d}{dx}(0.03x^2 - 0.4x + 10) = 0.06x - 0.4.$

23.   If $f(x) = \dfrac{x^3 - 4x^2 + 3}{x} = x^2 - 4x + \dfrac{3}{x},$

     then $f'(x) = \dfrac{d}{dx}(x^2 - 4x + 3x^{-1}) = 2x - 4 - \dfrac{3}{x^2}.$

25. $f'(x) = \dfrac{d}{dx}\left(4x^4 - 3x^{5/2} + 2\right) = 16x^3 - \tfrac{15}{2}x^{3/2}$.

27. $f'(x) = \dfrac{d}{dx}\left(3x^{-1} + 4x^{-2}\right) = -3x^{-2} - 8x^{-3}$.

29. $f'(t) = \dfrac{d}{dt}\left(4t^{-4} - 3t^{-3} + 2t^{-1}\right) = -16t^{-5} + 9t^{-4} - 2t^{-2}$.

31. $f'(x) = \dfrac{d}{dx}\left(2x - 5x^{1/2}\right) = 2 - \dfrac{5}{2}x^{-1/2} = 2 - \dfrac{5}{2\sqrt{x}}$.

33. $f'(x) = \dfrac{d}{dx}\left(2x^{-2} - 3x^{-1/3}\right) = -4x^{-3} + x^{-4/3} = -\dfrac{4}{x^3} + \dfrac{1}{x^{4/3}}$.

35. a. $f'(x) = \dfrac{d}{dx}\left(2x^3 - 4x\right) = 6x^2 - 4$. $f'(-2) = 6(-2)^2 - 4 = 20$.

   b. $f'(0) = 6(0) - 4 = -4$.   c. $f'(2) = 6(2)^2 - 4 = 20$.

37. The given limit is $f'(1)$ where $f(x) = x^3$. Since $f'(x) = 3x^2$, we have
$$\lim_{h \to 0} \dfrac{(1+h)^3 - 1}{h} = f'(1) = 3$$

39. Let $f(x) = 3x^2 - x$. Then $\displaystyle\lim_{h \to 0} \dfrac{3(2+h)^2 - (2+h) - 10}{h} = \lim_{h \to 0} \dfrac{f(2+h) - f(2)}{h}$

   because $f(2 + h) - f(2) = 3(2 + h)^2 - (2 + h) - [3(4) - 2]$
   $$= 3(2 + h)^2 - (2 + h) - 10.$$
   But the last limit is $f'(2)$. Since $f'(x) = 6x - 1$, we have $f'(2) = 11$.

   Therefore, $\displaystyle\lim_{h \to 0} \dfrac{3(2+h)^2 - (2+h) - 10}{h} = 11$.

41. $f(x) = 2x^2 - 3x + 4$. The slope of the tangent line at any point $(x, f(x))$ on the graph of f is $f'(x) = 4x - 3$. In particular, the slope of the tangent line at the point $(2,6)$ is $f'(2) = 4(2) - 3 = 5$. An equation of the required tangent line is
   $$y - 6 = 5(x - 2) \qquad \text{or} \qquad y = 5x - 4.$$

43. $f(x) = x^4 - 3x^3 + 2x^2 - x + 1$. $f'(x) = 4x^3 - 9x^2 + 4x - 1$.
The slope is $f'(1) = 4 - 9 + 4 - 1 = -2$. An equation
of the tangent line is $y - 0 = -2(x - 1)$ or $y = -2x + 2$.

45. a. $f'(x) = 3x^2$. At a point where the tangent line is horizontal,
$f'(x) = 0$, or $3x^2 = 0$ giving $x = 0$. Therefore, the point is (0,0).

b.

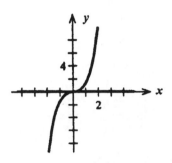

47. a. $f(x) = x^3 + 1$. The slope of the tangent line at any point $(x, f(x))$ on the graph
of f is $f'(x) = 3x^2$. At the point(s) where the slope is 12, we have
$3x^2 = 12$, or $x = \pm 2$. The required points are (-2,-7) and (2,9).
b. The tangent line at (-2,-7) has equation
            $y - (-7) = 12[x - (-2)]$,   or $y = 12x + 17$,
and the tangent line at (2,9) has equation
            $y - 9 = 12(x - 2)$,   or $y = 12x - 15$.
c.

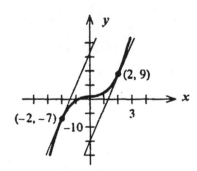

49. If $f(x) = \frac{1}{4}x^4 - \frac{1}{3}x^3 - x^2$, then $f'(x) = x^3 - x^2 - 2x$.
a. $f'(x) = x^3 - x^2 - 2x = -2x$
            $x^3 - x^2 = 0$

$$x^2(x-1) = 0 \qquad \text{and} \qquad x = 0 \text{ or } x = 1.$$
$$f(1) = \tfrac{1}{4}(1)^4 - \tfrac{1}{3}(1)^3 - (1)^2 = -\tfrac{13}{12}.$$
$$f(0) = \tfrac{1}{4}(0)^4 - \tfrac{1}{3}(0)^3 - (0)^2 = 0.$$

We conclude that the corresponding points on the graph are $(1, -\tfrac{13}{12})$ and $(0,0)$.

b.
$$f'(x) = x^3 - x^2 - 2x = 0$$
$$x(x^2 - x - 2) = 0$$
$$x(x-2)(x+1) = 0 \qquad \text{and} \qquad x = 0, 2, \text{ or } -1.$$
$$f(0) = 0$$
$$f(2) = \frac{1}{4}(2)^4 - \frac{1}{3}(2)^3 - (2)^2 = 4 - \frac{8}{3} - 4 = -\frac{8}{3}.$$
$$f(-1) = \frac{1}{4}(-1)^4 - \frac{1}{3}(-1)^3 - (-1)^2 = \frac{1}{4} + \frac{1}{3} - 1 = -\frac{5}{12}.$$

We conclude that the corresponding points are $(0,0)$, $(2, -\tfrac{8}{3})$ and $(-1, -\tfrac{5}{12})$.

c.
$$f'(x) = x^3 - x^2 - 2x = 10x$$
$$x^3 - x^2 - 12x = 0$$
$$x(x^2 - x - 12) = 0$$
$$x(x-4)(x+3) = 0$$

and $x = 0, 4$, or -3.
$$f(0) = 0$$
$$f(4) = \tfrac{1}{4}(4)^4 - \tfrac{1}{3}(4)^3 - (4)^2 = 48 - \tfrac{64}{3} = \tfrac{80}{3}.$$
$$f(-3) = \tfrac{1}{4}(-3)^4 - \tfrac{1}{3}(-3)^3 - (-3)^2 = \tfrac{81}{4} + 9 - 9 = \tfrac{81}{4}.$$

We conclude that the corresponding points are $(0,0)$, $(4, \tfrac{80}{3})$ and $(-3, \tfrac{81}{4})$.

51. $V(r) = \tfrac{4}{3}\pi r^3$. $V'(r) = 4\pi r^2$.

a. $V'(\tfrac{2}{3}) = 4\pi(\tfrac{4}{9}) = \tfrac{16}{9}\pi$ cm^3/cm. b. $V'(\tfrac{5}{4}) = 4\pi(\tfrac{25}{16}) = \tfrac{25}{4}\pi$ cm^3/cm.

53. $\dfrac{dA}{dx} = 26.5\dfrac{d}{dx}(x^{-0.45}) = 26.5(-0.45)x^{-1.45} = -\dfrac{11.925}{x^{1.45}}.$

Therefore, $\dfrac{dA}{dx}\Big|_{x=0.25} = -\dfrac{11.925}{(0.25)^{1.45}} \approx -89.01$ and $\dfrac{dA}{dx}\Big|_{x=2} = -\dfrac{11.925}{(2)^{1.45}} \approx -4.36$

Our computations reveal that if you make 0.25 stops per mile, your average speed will decrease at the rate of approximately 89.01 mph per stop per mile. If you make 2 stops per mile, your average speed will decrease at the rate of approximately 4.36 mph per stop per mile.

3 *Differentiation*

55. $I'(t) = -0.6t^2 + 6t$.

a. In 1999, it was changing at a rate of $I'(5) = -0.6(25) + 6(5)$, or 15 points/yr. In 2001, it was $I'(7) = -0.6(49) + 6(7)$, or 12.6 pts/yr. In 2004, it was $I'(10) = -0.6(100) + 6(10)$, or 0 pts/yr.

b. The average rate of increase of the CPI over the period from 1999 to 2004 was

$$\frac{I(10) - I(5)}{5} = \frac{[-0.2(1000) + 3(100) + 100] - [-0.2(125) + 3(25) + 100]}{5}$$

$$= \frac{200 - 150}{5} = 10, \text{ or } 10 \text{ pts/yr.}$$

57. $P(t) = 50,000 + 30t^{3/2} + 20t$. The rate at which the population will be increasing at any time t is $P'(t) = 45t^{1/2} + 20$. Nine months from now the population will be increasing at the rate of $P'(9) = 45(3) + 20$, or 155 people/month. Sixteen months from now the population will be increasing at the rate of
$$P'(16) = 45(4) + 20, \text{ or } 200 \text{ people/month.}$$

59. $N(t) = 2t^3 + 3t^2 - 4t + 1000.$ $N'(t) = 6t^2 + 6t - 4$.

$N'(2) = 6(4) + 6(2) - 4 = 32$, or 32 turtles/yr.

$N'(8) = 6(64) + 6(8) - 4 = 428$, or 428 turtles/yr.

The population ten years after implementation of the conservation measures will be $N(10) = 2(10^3) + 3(10^2) - 4(10) + 1000$, or 3260 turtles.

61. a. $f(t) = 120t - 15t^2$. $v = f'(t) = 120 - 30t$ b. $v(0) = 120$ ft/sec

c. Setting $v = 0$ gives $120 - 30t = 0$, or $t = 4$. Therefore, the stopping distance is
$$f(4) = 120(4) - 15(16) \text{ or } 240 \text{ ft.}$$

63. a. The number of temporary workers at the beginning of 1994 ($t = 3$) was
$$N(3) = 0.025(3^2) + 0.255(3) + 1.505 = 2.495 \text{ million.}$$

b. $N(t) = 0.025t^2 + 0.255t + 1.505$ $N'(t) = 0.05t + 0.255$.

So, at the beginning of 1994 ($t = 3$), the number of temporary workers was growing at the rate of $N'(3) = 0.05(3) + 0.255 = 0.405$, or 405,000 per year.

65. a. $f'(x) = \frac{d}{dx}\left[0.0001x^{5/4} + 10\right] = \frac{5}{4}(0.0001x^{1/4}) = 0.000125x^{1/4}$

b. $f'(10,000) = 0.000125(10,000)^{1/4} = 0.00125$, or $0.00125/radio.

67.　a. $f(t) = 20t - 40\sqrt{t} + 50$. $f'(t) = 20 - 40\left(\dfrac{1}{2}\right)t^{-1/2} = 20\left(1 - \dfrac{1}{\sqrt{t}}\right)$.

　　b. $f(0) = 20(0) - 40\sqrt{0} + 50 = 50$; $f(1) = 20(1) - 40\sqrt{1} + 50 = 30$

　　　　$f(2) = 20(2) - 40\sqrt{2} + 50 \approx 33.43$.

　　The average velocity at 6, 7, and 8 A.M. is 50 mph, 30 mph, and 33.43 mph, respectively.

　　c. $f'(\tfrac{1}{2}) = 20 - 20(\tfrac{1}{2})^{-1/2} \approx -8.28$. $f'(1) = 20 - 20(1)^{-1/2} \approx 0$.

　　　　$f'(2) = 20 - 20(2)^{-1/2} \approx 5.86$.

　　At 6:30 A.M. the average velocity is decreasing at the rate of 8.28 mph/hr; at 7 A.M., it is unchanged, and at 8 A.M., it is increasing at the rate of 5.86 mph.

69.　a. $\dfrac{d}{dx}[0.075t^3 + 0.025t^2 + 2.45t + 2.4] = 0.225t^2 + 0.05t + 2.45$

　　b. $f'(3) = 0.225(3)^2 + 0.05(3) + 2.45 = 4.625$, or $4.625 billion/yr.

　　c. $f(3) = 0.075(3)^3 + 0.025(3)^2 + 2.45(3) + 2.4 = 12$, or $12 billion/yr.

71.　True. $\dfrac{d}{dx}[2f(x) - 5g(x)] = \dfrac{d}{dx}[2f(x)] - \dfrac{d}{dx}[5g(x)] = 2f'(x) - 5g'(x)$.

73.　$\dfrac{d}{dx}\left(x^3\right) = \lim_{h \to 0} \dfrac{(x+h)^3 - x^3}{h} = \lim_{h \to 0} \dfrac{x^3 + 3x^2h + 3xh^2 + h^3 - x^3}{h}$

　　　$= \lim_{h \to 0} \dfrac{h(3x^2 + 3xh + h^2)}{h} = \lim_{h \to 0}(3x^2 + 3xh + h^2) = 3x^2$.

USING TECHNOL0GY EXERCISES 3.1, page 171

1.　1　　　　　　　　　3. 0.4226　　　　　　　　5.　　0.1613

7.　　a.　　　　　　　　　　　b. 3.4295 parts/million;
　　　　　　　　　　　　　　　　　105.4332 parts/million

9.   a.

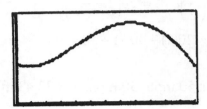

b. decreasing at the rate of 9 days/yr
increasing at the rate of 13 days/yr

11.   a.

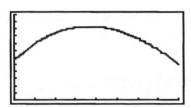

b. Increasing at the rate of 1.1557%/yr
decreasing at the rate of 0.2116%/yr

EXERCISES 3.2, page 205

1. $f(x) = 2x(x^2 + 1)$.

$$f'(x) = 2x\frac{d}{dx}\left(x^2 + 1\right) + (x^2 + 1)\frac{d}{dx}(2x)$$

$$= 2x(2x) + (x^2 + 1)(2) = 6x^2 + 2.$$

3. $f(t) = (t - 1)(2t + 1)$

$$f'(t) = (t - 1)\frac{d}{dt}\left(2t + 1\right) + (2t + 1)\frac{d}{dt}(t - 1)$$

$$= (t - 1)(2) + (2t + 1)(1) \;\; = 4t - 1$$

5. $f(x) = (3x + 1)(x^2 - 2)$

$$f'(x) = (3x + 1)\frac{d}{dx}\left(x^2 - 2\right) + (x^2 - 2)\frac{d}{dx}(3x + 1)$$

$$= (3x + 1)(2x) + (x^2 - 2)(3) = 9x^2 + 2x - 6.$$

7. $f(x) = (x^3 - 1)(x + 1)$.

$$f'(x) = (x^3 - 1)\frac{d}{dx}\left(x + 1\right) + (x + 1)\frac{d}{dx}(x^3 - 1)$$

$$= (x^3 - 1)(1) + (x+1)(3x^2) = 4x^3 + 3x^2 - 1.$$

9. $f(w) = (w^3 - w^2 + w - 1)(w^2 + 2).$

$$f'(w) = (w^3 - w^2 + w - 1)\frac{d}{dw}\left(w^2 + 2\right) + (w^2 + 2)\frac{d}{dw}(w^3 - w^2 + w - 1)$$

$$= (w^3 - w^2 + w - 1)(2w) + (w^2 + 2)(3w^2 - 2w + 1)$$

$$= 2w^4 - 2w^3 + 2w^2 - 2w + 3w^4 - 2w^3 + w^2 + 6w^2 - 4w + 2$$

$$= 5w^4 - 4w^3 + 9w^2 - 6w + 2.$$

11. $f(x) = (5x^2 + 1)(2\sqrt{x} - 1)$

$$f'(x) = (5x^2 + 1)\frac{d}{dx}(2x^{1/2} - 1) + (2x^{1/2} - 1)\frac{d}{dx}(5x^2 + 1)$$

$$= (5x^2 + 1)(x^{-1/2}) + (2x^{1/2} - 1)(10x)$$

$$= 5x^{3/2} + x^{-1/2} + 20x^{3/2} - 10x \ = \frac{25x^2 - 10x\sqrt{x} + 1}{\sqrt{x}}.$$

13. $f(x) = (x^2 - 5x + 2)(x - \frac{2}{x})$

$$f'(x) = (x^2 - 5x + 2)\frac{d}{dx}(x - \frac{2}{x}) + (x - \frac{2}{x})\frac{d}{dx}(x^2 - 5x + 2)$$

$$= \frac{(x^2 - 5x + 2)(x^2 + 2)}{x^2} + \frac{(x^2 - 2)(2x - 5)}{x}$$

$$= \frac{(x^2 - 5x + 2)(x^2 + 2) + x(x^2 - 2)(2x - 5)}{x^2}$$

$$= \frac{x^4 + 2x^2 - 5x^3 - 10x + 2x^2 + 4 + 2x^4 - 5x^3 - 4x^2 + 10x}{x^2}$$

$$= \frac{3x^4 - 10x^3 + 4}{x^2}.$$

15. $f(x) = \dfrac{1}{x-2}.$ $f'(x) = \dfrac{(x-2)\dfrac{d}{dx}(1) - (1)\dfrac{d}{dx}(x-2)}{(x-2)^2} = \dfrac{0 - 1(1)}{(x-2)^2} = -\dfrac{1}{(x-2)^2}.$

17. $f(x) = \dfrac{x-1}{2x+1}.$

$$f'(x) = \frac{(2x+1)\dfrac{d}{dx}(x-1)-(x-1)\dfrac{d}{dx}(2x+1)}{(2x+1)^2}$$

$$= \frac{2x+1-(x-1)(2)}{(2x+1)^2} = \frac{3}{(2x+1)^2}.$$

19. $f(x) = \dfrac{1}{x^2+1}.$

$$f'(x) = \frac{(x^2+1)\dfrac{d}{dx}(1)-(1)\dfrac{d}{dx}(x^2+1)}{(x^2+1)^2}$$

$$= \frac{(x^2+1)(0)-1(2x)}{(x^2+1)^2} = -\frac{2x}{(x^2+1)^2}.$$

21. $f(s) = \dfrac{s^2-4}{s+1}.$

$$f'(s) = \frac{(s+1)\dfrac{d}{ds}(s^2-4)-(s^2-4)\dfrac{d}{ds}(s+1)}{(s+1)^2}$$

$$= \frac{(s+1)(2s)-(s^2-4)(1)}{(s+1)^2} = \frac{s^2+2s+4}{(s+1)^2}.$$

23. $f(x) = \dfrac{\sqrt{x}}{x^2+1}.$

$$f'(x) = \frac{(x^2+1)\dfrac{d}{dx}(x^{1/2})-(x^{1/2})\dfrac{d}{dx}(x^2+1)}{(x^2+1)^2} = \frac{(x^2+1)(\tfrac{1}{2}x^{-1/2})-(x^{1/2})(2x)}{(x^2+1)^2}$$

$$= \frac{(\tfrac{1}{2}x^{-1/2})[(x^2+1)-4x^2]}{(x^2+1)^2} = \frac{1-3x^2}{2\sqrt{x}(x^2+1)^2}.$$

25. $f(x) = \dfrac{x^2+2}{x^2+x+1}.$

$$f'(x) = \frac{(x^2 + x + 1)\dfrac{d}{dx}(x^2 + 2) - (x^2 + 2)\dfrac{d}{dx}(x^2 + x + 1)}{(x^2 + x + 1)^2}$$

$$= \frac{(x^2 + x + 1)(2x) - (x^2 + 2)(2x + 1)}{(x^2 + x + 1)^2}$$

$$= \frac{2x^3 + 2x^2 + 2x - 2x^3 - x^2 - 4x - 2}{(x^2 + x + 1)^2} = \frac{x^2 - 2x - 2}{(x^2 + x + 1)^2}.$$

27. $f(x) = \dfrac{(x + 1)(x^2 + 1)}{x - 2} = \dfrac{(x^3 + x^2 + x + 1)}{x - 2}.$

$$f'(x) = \frac{(x - 2)\dfrac{d}{dx}(x^3 + x^2 + x + 1) - (x^3 + x^2 + x + 1)\dfrac{d}{dx}(x - 2)}{(x - 2)^2}$$

$$= \frac{(x - 2)(3x^2 + 2x + 1) - (x^3 + x^2 + x + 1)}{(x - 2)^2}$$

$$= \frac{3x^3 + 2x^2 + x - 6x^2 - 4x - 2 - x^3 - x^2 - x - 1}{(x - 2)^2} = \frac{2x^3 - 5x^2 - 4x - 3}{(x - 2)^2}.$$

29. $f(x) = \dfrac{x}{x^2 - 4} - \dfrac{x - 1}{x^2 + 4} = \dfrac{x(x^2 + 4) - (x - 1)(x^2 - 4)}{(x^2 - 4)(x^2 + 4)} = \dfrac{x^2 + 8x - 4}{(x^2 - 4)(x^2 + 4)}.$

$$f'(x) = \frac{(x^2 - 4)(x^2 + 4)\dfrac{d}{dx}(x^2 + 8x - 4) - (x^2 + 8x - 4)\dfrac{d}{dx}(x^4 - 16)}{(x^2 - 4)^2(x^2 + 4)^2}$$

$$= \frac{(x^2 - 4)(x^2 + 4)(2x + 8) - (x^2 + 8x - 4)(4x^3)}{(x^2 - 4)^2(x^2 + 4)^2}$$

$$= \frac{2x^5 + 8x^4 - 32x - 128 - 4x^5 - 32x^4 + 16x^3}{(x^2 - 4)^2(x^2 + 4)^2}$$

$$= \frac{-2x^5 - 24x^4 + 16x^3 - 32x - 128}{(x^2 - 4)^2(x^2 + 4)^2}.$$

31. $h'(x) = f(x)g'(x) + f'(x)g(x),$ by the Product Rule. Therefore,

$$h'(1) = f(1)g'(1) + f'(1)g(1) = (2)(3) + (-1)(-2) = 8.$$

33. Using the Quotient Rule followed by the Product Rule, we have

$$h'(x) = \frac{[x+g(x)]\frac{d}{dx}[xf(x)] - xf(x)\frac{d}{dx}[x+g(x)]}{[x+g(x)]^2}$$

$$= \frac{[x+g(x)][xf'(x) + f(x)] - xf(x)[1+g'(x)]}{[x+g(x)]^2}$$

Therefore, $h'(1) = \dfrac{[1+g(1)][f'(1) + f(1)] - f(1)[1+g'(1)]}{[1+g(1)]^2}$

$$= \frac{(1-2)(-1+2) - 2(1+3)}{(1-2)^2} = \frac{-1-8}{1} = -9.$$

35. $f(x) = (2x-1)(x^2+3)$

$$f'(x) = (2x-1)\frac{d}{dx}(x^2+3) + (x^2+3)\frac{d}{dx}(2x-1)$$

$$= (2x-1)(2x) + (x^2+3)(2) = 6x^2 - 2x + 6 = 2(3x^2 - x + 3).$$

At $x = 1, f'(1) = 2[3(1)^2 - (1) + 3] = 2(5) = 10.$

37. $f(x) = \dfrac{x}{x^4 - 2x^2 - 1}.$

$$f'(x) = \frac{(x^4 - 2x^2 - 1)\frac{d}{dx}(x) - x\frac{d}{dx}(x^4 - 2x^2 - 1)}{(x^4 - 2x^2 - 1)^2}$$

$$= \frac{(x^4 - 2x^2 - 1)(1) - x(4x^3 - 4x)}{(x^4 - 2x^2 - 1)^2} = \frac{-3x^4 + 2x^2 - 1}{(x^4 - 2x^2 - 1)^2}.$$

Therefore, $f'(-1) = \dfrac{-3+2-1}{(1-2-1)^2} = -\dfrac{2}{4} = -\dfrac{1}{2}.$

39. $f(x) = (x^3+1)(x^2-2).$

$$f'(x) = (x^3+1)\frac{d}{dx}(x^2-2) + (x^2-2)\frac{d}{dx}(x^3+1)$$

$$= (x^3+1)(2x) + (x^2-2)(3x^2).$$

The slope of the tangent line at $(2,18)$ is $f'(2) = (8 + 1)(4) + (4 - 2)(12) = 60.$

An equation of the tangent line is $y - 18 = 60(x - 2)$, or $y = 60x - 102$.

41. $f(x) = \dfrac{x+1}{x^2+1}$.

$$f'(x) = \frac{(x^2+1)\dfrac{d}{dx}(x+1)-(x+1)\dfrac{d}{dx}(x^2+1)}{(x^2+1)^2}$$

$$= \frac{(x^2+1)(1)-(x+1)(2x)}{(x^2+1)^2} = \frac{-x^2-2x+1}{(x^2+1)^2}.$$

At $x = 1$, $f'(1) = \dfrac{-1-2+1}{4} = -\dfrac{1}{2}$. Therefore, the slope of the tangent line at

$x = 1$ is -1/2. Then an equation of the tangent line is
$$y-1 = -\tfrac{1}{2}(x-1) \quad \text{or} \quad y = -\tfrac{1}{2}x + \tfrac{3}{2}.$$

43. $f(x) = (x^3+1)(3x^2-4x+2)$

$$f'(x) = (x^3+1)\frac{d}{dx}(3x^2-4x+2)+(3x^2-4x+2)\frac{d}{dx}(x^3+1)$$

$$= (x^3+1)(6x-4)+(3x^2-4x+2)(3x^2)$$

$$= 6x^4+6x-4x^3-4+9x^4-12x^3+6x^2$$

$$= 15x^4-16x^3+6x^2+6x-4.$$

At $x = 1$, $f'(1) = 15(1)^4 - 16(1)^3 + 6(1) + 6(1) - 4 = 7$. The slope of the tangent line
at the point $x = 1$ is 7. The equation of the tangent line is
$$y - 2 = 7(x - 1), \quad \text{or} \quad y = 7x - 5.$$

45. $f(x) = (x^2+1)(2-x)$

$$f'(x) = (x^2+1)\frac{d}{dx}(2-x)+(2-x)\frac{d}{dx}(x^2+1)$$

$$= (x^2+1)(-1)+(2-x)(2x) = -3x^2+4x-1.$$

At a point where the tangent line is horizontal, we have
$$f'(x) = -3x^2+4x-1=0$$

or $\quad 3x^2-4x+1 = (3x-1)(x-1) = 0$, giving $x = 1/3$ or $x = 1$.

Since $\quad f(\tfrac{1}{3}) = (\tfrac{1}{9}+1)(2-\tfrac{1}{3}) = \tfrac{50}{27}$, and $f(1) = 2(2 - 1) = 2$, we see that the
required points are $(\tfrac{1}{3}, \tfrac{50}{27})$ and $(1, 2)$.

3 Differentiation

47. $f(x) = (x^2 + 6)(x - 5)$

$$f'(x) = (x^2 + 6)\frac{d}{dx}(x - 5) + (x - 5)\frac{d}{dx}(x^2 + 6)$$

$$= (x^2 + 6)(1) + (x - 5)(2x) = x^2 + 6 + 2x^2 - 10x = 3x^2 - 10x + 6.$$

At a point where the slope of the tangent line is -2, we have

$$f'(x) = 3x^2 - 10x + 6 = -2.$$

This gives $3x^2 - 10x + 8 = (3x - 4)(x - 2) = 0.$ So $x = \frac{4}{3}$ or $x = 2.$

Since $f(\frac{4}{3}) = (\frac{16}{9} + 6)(\frac{4}{3} - 5) = -\frac{770}{27}$ and $f(2) = (4 + 6)(2 - 5) = -30,$

the required points are $(\frac{4}{3}, -\frac{770}{27})$ and $(2, -30).$

49. $y = \dfrac{1}{1 + x^2}. \quad y' = \dfrac{(1 + x^2)\frac{d}{dx}(1) - (1)\frac{d}{dx}(1 + x^2)}{(1 + x^2)^2} = \dfrac{-2x}{(1 + x^2)^2}.$

So, the slope of the tangent line at $(1, \frac{1}{2})$ is

$$y'|_{x=1} = \frac{-2x}{(1 + x^2)^2}\bigg|_{x=1} = \frac{-2}{4} = -\frac{1}{2}$$

and the equation of the tangent line is $y - \frac{1}{2} = -\frac{1}{2}(x - 1),$ or $y = -\frac{1}{2}x + 1.$

Next, the slope of the required normal line is 2 and its equation is

$$y - \frac{1}{2} = 2(x - 1), \quad \text{or} \quad y = 2x - \frac{3}{2}.$$

51. $C(x) = \dfrac{0.5x}{100 - x}. \quad C'(x) = \dfrac{(100 - x)(0.5) - 0.5x(-1)}{(100 - x)^2} = \dfrac{50}{(100 - x)^2}.$

$$C'(80) = \frac{50}{20^2} = 0.125; \qquad C'(90) = \frac{50}{10^2} = 0.5,$$

$$C'(95) = \frac{50}{5^2} = 2; \qquad C'(99) = \frac{50}{1} = 50.$$

The rates of change of the cost in removing 80%, 90%, and 99% of the toxic waste are 0.125, 0.5, 2, and 50 million dollars per 1% more of the waste to be removed, respectively. It is too costly to remove *all* of the pollutant.

53. $N(t) = \dfrac{10,000}{1 + t^2} + 2000$

$$N'(t) = \frac{d}{dt}[10,000(1 + t^2)^{-1} + 2000] = -\frac{10,000}{(1 + t^2)^2}(2t) = -\frac{20,000t}{(1 + t^2)^2}.$$

The rate of change after 1 minute and after 2 minutes is

$$N'(1) = -\frac{20{,}000}{(1+1^2)^2} = -5000; \quad N'(2) = -\frac{20{,}000(2)}{(1+2^2)^2} = -1600.$$

The population of bacteria after one minute is $N(1) = \dfrac{10{,}000}{1+1} + 2000 = 7000$.

The population after two minutes is $N(2) = \dfrac{10{,}000}{1+4} + 2000 = 4000$.

55. a. $N(t) = \dfrac{60t+180}{t+6}$.

$$N'(t) = \frac{(t+6)\dfrac{d}{dt}(60t+180) - (60t+180)\dfrac{d}{dt}(t+6)}{(t+6)^2}$$

$$= \frac{(t+6)(60) - (60t+180)(1)}{(t+6)^2} = \frac{180}{(t+6)^2}.$$

b. $\quad N'(1) = \dfrac{180}{(1+6)^2} = 3.7, \quad N'(3) = \dfrac{180}{(3+6)^2} = 2.2, \quad N'(4) = \dfrac{180}{(4+6)^2} = 1.8,$

$$N'(7) = \frac{180}{(7+6)^2} = 1.1$$

We conclude that the rate at which the average student is increasing his or her speed one week, three weeks, four weeks, and seven weeks into the course is 3.7, 2.2, 1.8, and 1.1 words per minute, respectively.

c. Yes

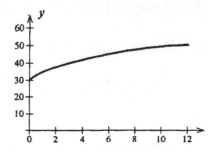

d. $N(12) = \dfrac{60(12)+180}{12+6} = 50$, or 50 words/minute.

57. $f(t) = \dfrac{0.055t + 0.26}{t+2}$; $f'(t) = \dfrac{(t+2)(0.055) - (0.055t + 0.26)(1)}{(t+2)^2} = -\dfrac{0.15}{(t+2)^2}$.

At the beginning, the formaldehyde level is changing at the rate of

$$f'(0) = -\frac{0.15}{4} = -0.0375;$$

that is, it is dropping at the rate of 0.0375 parts per million per year. Next,

$$f'(3) = -\frac{0.15}{5^2} = -0.006,$$

and so the level is dropping at the rate of 0.006 parts per million per year at the beginning of the fourth year ($t = 3$).

59. False. Take $f(x) = x$ and $g(x) = x$. Then $f(x)g(x) = x^2$. So

$$\frac{d}{dx}[f(x)g(x)] = \frac{d}{dx}(x^2) = 2x \neq f'(x)g'(x) = 1.$$

61. False. Let $f(x) = x^3$. Then

$$\frac{d}{dx}\left[\frac{f(x)}{x^2}\right] = \frac{d}{dx}\left(\frac{x^3}{x^2}\right) = \frac{d}{dx}(x) = 1 \neq \frac{f'(x)}{2x} = \frac{3x^2}{2x} = \frac{3}{2}x.$$

63. Let $f(x) = u(x)v(x)$ and $g(x) = w(x)$. Then $h(x) = f(x)g(x)$. Therefore,
$$h'(x) = f'(x)g(x) + f(x)g'(x).$$
But $\quad f'(x) = u(x)v'(x) + u'(x)v(x).$
Therefore, $h'(x) = [u(x)v'(x) + u'(x)v(x)]g(x) + u(x)v(x)w'(x)$
$$= u(x)v(x)w'(x) + u(x)v'(x)w(x) + u'(x)v(x)w(x).$$

USING TECHNOLOGY EXERCISES 3.2, page 185

1. 0.8750   3. 0.0774   5. −0   7. 87,322 per year

EXERCISES 3.3, page 195

1. $\quad f(x) = (2x-1)^4.\ f'(x) = 4(2x-1)^3 \dfrac{d}{dx}(2x-1) = 4(2x-1)^3(2) = 8(2x-1)^3.$

3. $f(x) = (x^2 + 2)^5$. $f'(x) = 5(x^2 + 2)^4 (2x) = 10x(x^2 + 2)^4$.

5. $f(x) = (2x - x^2)^3$.

$$f'(x) = 3(2x - x^2)^2 \frac{d}{dx}(2x - x^2) = 3(2x - x^2)^2(2 - 2x)$$
$$= 6x^2(1 - x)(2 - x)^2.$$

7. $f(x) = (2x + 1)^{-2}$.

$$f'(x) = -2(2x + 1)^{-3} \frac{d}{dx}(2x + 1) = -2(2x + 1)^{-3}(2) = -4(2x + 1)^{-3}.$$

9. $f(x) = (x^2 - 4)^{3/2}$.

$$f'(x) = \tfrac{3}{2}(x^2 - 4)^{1/2} \frac{d}{dx}(x^2 - 4) = \tfrac{3}{2}(x^2 - 4)^{1/2}(2x) = 3x(x^2 - 4)^{1/2}.$$

11. $f(x) = \sqrt{3x - 2} = (3x - 2)^{1/2}$.

$$f'(x) = \frac{1}{2}(3x - 2)^{-1/2}(3) = \frac{3}{2}(3x - 2)^{-1/2} = \frac{3}{2\sqrt{3x - 2}}.$$

13. $f(x) = \sqrt[3]{1 - x^2}$.

$$f'(x) = \frac{d}{dx}(1 - x^2)^{1/3} = \frac{1}{3}(1 - x^2)^{-2/3} \frac{d}{dx}(1 - x^2)$$
$$= \frac{1}{3}(1 - x^2)^{-2/3}(-2x) = -\frac{2}{3}x(1 - x^2)^{-2/3} = \frac{-2x}{3(1 - x^2)^{2/3}}.$$

15. $f(x) = \dfrac{1}{(2x + 3)^3} = (2x + 3)^{-3}$.

$$f'(x) = -3(2x + 3)^{-4}(2) = -6(2x + 3)^{-4} = -\frac{6}{(2x + 3)^4}.$$

17. $f(t) = \dfrac{1}{\sqrt{2t - 3}}$.

$$f'(t) = \frac{d}{dt}(2t - 3)^{-1/2} = -\frac{1}{2}(2t - 3)^{-3/2}(2) = -(2t - 3)^{-3/2} = -\frac{1}{(2t - 3)^{3/2}}.$$

19. $y = \dfrac{1}{(4x^4 + x)^{3/2}}$.

$\dfrac{dy}{dx} = \dfrac{d}{dx}(4x^4 + x)^{-3/2} = -\dfrac{3}{2}(4x^4 + x)^{-5/2}(16x^3 + 1) = -\dfrac{3}{2}(16x^3 + 1)(4x^4 + x)^{-5/2}$.

21. $f(x) = (3x^2 + 2x + 1)^{-2}$.

$f'(x) = -2(3x^2 + 2x + 1)^{-3}\dfrac{d}{dx}(3x^2 + 2x + 1)$

$= -2(3x^2 + 2x + 1)^{-3}(6x + 2) = -4(3x + 1)(3x^2 + 2x + 1)^{-3}$.

23. $f(x) = (x^2 + 1)^3 - (x^3 + 1)^2$.

$f'(x) = 3(x^2 + 1)^2\dfrac{d}{dx}(x^2 + 1) - 2(x^3 + 1)\dfrac{d}{dx}(x^3 + 1)$

$= 3(x^2 + 1)^2(2x) - 2(x^3 + 1)(3x^2)$

$= 6x[(x^2 + 1)^2 - x(x^3 + 1)] = 6x(2x^2 - x + 1)$.

25. $f(t) = (t^{-1} - t^{-2})^3$. $f'(t) = 3(t^{-1} - t^{-2})^2\dfrac{d}{dt}(t^{-1} - t^{-2}) = 3(t^{-1} - t^{-2})^2(-t^{-2} + 2t^{-3})$.

27. $f(x) = \sqrt{x + 1} + \sqrt{x - 1} = (x + 1)^{1/2} + (x - 1)^{1/2}$.

$f'(x) = \frac{1}{2}(x + 1)^{-1/2}(1) + \frac{1}{2}(x - 1)^{-1/2}(1) = \frac{1}{2}[(x + 1)^{-1/2} + (x - 1)^{-1/2}]$.

29. $f(x) = 2x^2(3 - 4x)^4$.

$f'(x) = 2x^2(4)(3 - 4x)^3(-4) + (3 - 4x)^4(4x) = 4x(3 - 4x)^3(-8x + 3 - 4x)$

$= 4x(3 - 4x)^3(-12x + 3) = (-12x)(4x - 1)(3 - 4x)^3$.

31. $f(x) = (x - 1)^2(2x + 1)^4$.

$f'(x) = (x - 1)^2\dfrac{d}{dx}(2x + 1)^4 + (2x + 1)^4\dfrac{d}{dx}(x - 1)^2$ [Product Rule]

$= (x - 1)^2(4)(2x + 1)^3\dfrac{d}{dx}(2x + 1) + (2x + 1)^4(2)(x - 1)\dfrac{d}{dx}(x - 1)$

$= 8(x - 1)^2(2x + 1)^3 + 2(x - 1)(2x + 1)^4$

$= 2(x - 1)(2x + 1)^3(4x - 4 + 2x + 1) = 6(x - 1)(2x - 1)(2x + 1)^3$.

33. $f(x) = \left(\dfrac{x+3}{x-2}\right)^3.$

$$f'(x) = 3\left(\frac{x+3}{x-2}\right)^2 \frac{d}{dx}\left(\frac{x-3}{x-2}\right) = 3\left(\frac{x+3}{x-2}\right)^2 \left[\frac{(x-2)(1)-(x+3)(1)}{(x-2)^2}\right]$$

$$= 3\left(\frac{x+3}{x-2}\right)^2 \left[-\frac{5}{(x-2)^2}\right] = -\frac{15(x+3)^2}{(x-2)^4}.$$

35. $s(t) = \left(\dfrac{t}{2t+1}\right)^{3/2}.$

$$s'(t) = \frac{3}{2}\left(\frac{t}{2t+1}\right)^{1/2} \frac{d}{dt}\left(\frac{t}{2t+1}\right) = \frac{3}{2}\left(\frac{t}{2t+1}\right)^{1/2}\left[\frac{(2t+1)(1)-t(2)}{(2t+1)^2}\right]$$

$$= \frac{3}{2}\left(\frac{t}{2t+1}\right)^{1/2}\left[\frac{1}{(2t+1)^2}\right] = \frac{3t^{1/2}}{2(2t+1)^{5/2}}.$$

37. $g(u) = \left(\dfrac{u+1}{3u+2}\right)^{1/2}.$

$$g'(u) = \frac{1}{2}\left(\frac{u+1}{3u+2}\right)^{-1/2} \frac{d}{du}\left(\frac{u+1}{3u+2}\right)$$

$$= \frac{1}{2}\left(\frac{u+1}{3u+2}\right)^{-1/2}\left[\frac{(3u+2)(1)-(u+1)(3)}{(3u+2)^2}\right] = -\frac{1}{2\sqrt{u+1}(3u+2)^{3/2}}.$$

39. $f(x) = \dfrac{x^2}{(x^2-1)^4}.$

$$f'(x) = \frac{(x^2-1)^4 \dfrac{d}{dx}(x^2) - (x^2)\dfrac{d}{dx}(x^2-1)^4}{\left[(x^2-1)^4\right]^2}$$

$$= \frac{(x^2-1)^4(2x) - x^2(4)(x^2-1)^3(2x)}{(x^2-1)^8}$$

$$= \frac{(x^2-1)^3(2x)(x^2-1-4x^2)}{(x^2-1)^8} = \frac{(-2x)(3x^2+1)}{(x^2-1)^5}.$$

41. $\quad h(x) = \dfrac{(3x^2+1)^3}{(x^2-1)^4}.$

$h'(x) = \dfrac{(x^2-1)^4(3)(3x^2+1)^2(6x) - (3x^2+1)^3(4)(x^2-1)^3(2x)}{(x^2-1)^8}$

$\qquad = \dfrac{2x(x^2-1)^3(3x^2+1)^2[9(x^2-1) - 4(3x^2+1)]}{(x^2-1)^8}$

$\qquad = -\dfrac{2x(3x^2+13)(3x^2+1)^2}{(x^2-1)^5}.$

43. $\quad f(x) = \dfrac{\sqrt{2x+1}}{x^2-1}.$

$f'(x) = \dfrac{(x^2-1)(\frac{1}{2})(2x+1)^{-1/2}(2) - (2x+1)^{1/2}(2x)}{(x^2-1)^2}$

$\qquad = \dfrac{(2x+1)^{-1/2}[(x^2-1) - (2x+1)(2x)]}{(x^2-1)^2} = -\dfrac{3x^2+2x+1}{\sqrt{2x+1}(x^2-1)^2}.$

45. $\quad g(t) = \dfrac{(t+1)^{1/2}}{(t^2+1)^{1/2}}.$

$g'(t) = \dfrac{(t^2+1)^{1/2}\dfrac{d}{dt}(t+1)^{1/2} - (t+1)^{1/2}\dfrac{d}{dt}(t^2+1)^{1/2}}{t^2+1}$

$\qquad = \dfrac{(t^2+1)^{1/2}(\frac{1}{2})(t+1)^{-1/2}(1) - (t+1)^{1/2}(\frac{1}{2})(t^2+1)^{-1/2}(2t)}{t^2+1}$

$\qquad = \dfrac{\frac{1}{2}(t+1)^{-1/2}(t^2+1)^{-1/2}[(t^2+1) - 2t(t+1)]}{t^2+1} = -\dfrac{t^2+2t-1}{2\sqrt{t+1}(t^2+1)^{3/2}}.$

47. $\quad f(x) = (3x+1)^4(x^2-x+1)^3$

$f'(x) = (3x+1)^4 \cdot \dfrac{d}{dx}(x^2-x+1)^3 + (x^2-x+1)^3\dfrac{d}{dx}(3x+1)^4$

$\qquad = (3x+1)^4 \cdot 3(x^2-x+1)^2(2x-1) + (x^2-x+1)^3 \cdot 4(3x+1)^3 \cdot 3$

$\qquad = 3(3x+1)^3(x^2-x+1)^2[(3x+1)(2x-1) + 4(x^2-x+1)]$

$$= 3(3x+1)^3(x^2-x+1)^2(6x^2-3x+2x-1+4x^2-4x+4)$$
$$= 3(3x+1)^3(x^2-x+1)^2(10x^2-5x+3)$$

49. $y = g(u) = u^{4/3}$ and $\dfrac{dy}{du} = \dfrac{4}{3}u^{1/3}$, $u = f(x) = 3x^2 - 1$, and $\dfrac{du}{dx} = 6x$.

So $\dfrac{dy}{dx} = \dfrac{dy}{du}\cdot\dfrac{du}{dx} = \tfrac{4}{3}u^{1/3}(6x) = \tfrac{4}{3}(3x^2-1)^{1/3}6x = 8x(3x^2-1)^{1/3}$.

51. $\dfrac{dy}{du} = -\dfrac{2}{3}u^{-5/3} = -\dfrac{2}{3u^{5/3}}$, $\dfrac{du}{dx} = 6x^2 - 1$.

$\dfrac{dy}{dx} = \dfrac{dy}{du}\cdot\dfrac{du}{dx} = -\dfrac{2(6x^2-1)}{3u^{5/3}} = -\dfrac{2(6x^2-1)}{3(2x^3-x+1)^{5/3}}$.

53. $\dfrac{dy}{du} = \tfrac{1}{2}u^{-1/2} - \tfrac{1}{2}u^{-3/2}$, $\dfrac{du}{dx} = 3x^2 - 1$.

$\dfrac{dy}{dx} = \dfrac{dy}{du}\cdot\dfrac{du}{dx} = \left[\dfrac{1}{2\sqrt{x^3-x}} - \dfrac{1}{2(x^3-x)^{3/2}}\right](3x^2-1)$

$= \dfrac{(3x^2-1)(x^3-x-1)}{2(x^3-x)^{3/2}}$.

55. $F(x) = g(f(x))$; $F'(x) = g'(f(x))f'(x)$ and $F'(2) = g'(3)(-3) = (4)(-3) = -12$

57. Let $g(x) = x^2 + 1$, then $F(x) = f(g(x))$. Next, $F'(x) = f'(g(x))g'(x)$
and $F'(1) = f'(2)(2x) = (3)(2) = 6$.

59. No. Suppose $h = g(f(x))$. Let $f(x) = x$ and $g(x) = x^2$. Then
$h = g(f(x)) = g(x) = x^2$ and $h'(x) = 2x \ne g'(f'(x)) = g'(1) = 2(1) = 2$.

61. $f(x) = (1-x)(x^2-1)^2$.

$f'(x) = (1-x)2(x^2-1)(2x) + (-1)(x^2-1)^2$

$= (x^2-1)(4x-4x^2-x^2+1) = (x^2-1)(-5x^2+4x+1)$.

Therefore, the slope of the tangent line at $(2,-9)$ is

$f'(2) = [(2)^2-1][-5(2)^2+4(2)+1] = -33$.

Then the required equation is $y + 9 = -33(x - 2)$, or $y = -33x + 57$.

63. $f(x) = x\sqrt{2x^2 + 7}$. $f'(x) = \sqrt{2x^2 + 7} + x(\frac{1}{2})(2x^2 + 7)^{-1/2}(4x)$.

The slope of the tangent line is $f'(3) = \sqrt{25} + (\frac{3}{2})(25)^{-1/2}(12) = \frac{43}{5}$.

An equation of the tangent line is $y - 15 = \frac{43}{5}(x - 3)$ or $y = \frac{43}{5}x - \frac{54}{5}$.

65. $N(x) = (60 + 2x)^{2/3}$. $N'(x) = \frac{2}{3}(60 + 2x)^{-1/3}\frac{d}{dx}(60 + 2x) = \frac{4}{3}(60 + 2x)^{-1/3}$.

The rate of increase at the end of the second week is

$N'(2) = \frac{4}{3}(64)^{-1/3} = \frac{1}{3}$, or $\frac{1}{3}$ million/week

At the end of the 12th week, $N'(12) = \frac{4}{3}(84)^{-1/3} \approx 0.3$ million/wk. The number of viewers in the 2nd and 24th week are $N(2) = (60 + 4)^{2/3} = 16$ million and $N(24) = (60 + 48)^{2/3} = 22.7$ million, respectively.

67. $C(t) = 0.01(0.2t^2 + 4t + 64)^{2/3}$.

a. $C'(t) = 0.01(\frac{2}{3})(0.2t^2 + 4t + 64)^{-1/3}\frac{d}{dt}(0.2t^2 + 4t + 64)$

$= (0.01)(0.667)(0.4t + 4)(0.2t^2 + 4t + 4)^{-1/3}$

$= 0.027(0.1t + 1)(0.2t^2 + 4t + 64)^{-1/3}$.

b. $C'(5) = 0.007[0.4(5) + 4][0.2(25) + 4(5) + 64]^{-1/3} \approx 0.009$,

or 0.009 parts per million per year.

69. $P(t) = 33.55(t + 5)^{0.205}$. $P'(t) = 33.55(0.205)(t + 5)^{-0.795}(1) = 6.87775(t + 5)^{-0.795}$

The rate of change at the beginning of 2000 is

$$P'(20) = 6.87775(25)^{-0.795} \approx 0.5322 \text{ or } 0.53\%/\text{yr.}$$

The percent of these mothers was $P(20) = 33.55(25)^{0.205} \approx 64.90$, or 64.9%.

71. a. $A(t) = 0.03t^3(t - 7)^4 + 60.2$

$A'(t) = 0.03[3t^2(t - 7)^4 + t^3(4)(t - 7)^3] = 0.03t^2(t - 7)^3[3(t - 7) + 4t]$

$= 0.21t^2(t - 3)(t - 7)^3$.

b. $A'(1) = 0.21(-2)(-6)^3 = 90.72$; $A'(3) = 0$. $A'(4) = 0.21(16)(1)(-3)^3 = -90.72$.

The amount of pollutant is increasing at the rate of 90.72 units/hr at 8 A.M. Its rate of change is 0 units/hr at 10 A.M.; its rate of change is -90.72 units/hr at 11 A.M.

73.

$$P(t) = \frac{300\sqrt{\frac{1}{2}t^2 + 2t + 25}}{t + 25} = \frac{300(\frac{1}{2}t^2 + 2t + 25)^{1/2}}{t + 25}.$$

$$P'(t) = 300\left[\frac{(t+25)\frac{1}{2}(\frac{1}{2}t^2 + 2t + 25)^{-1/2}(t+2) - (\frac{1}{2}t^2 + 2t + 25)^{1/2}(1)}{(t+25)^2}\right]$$

$$= 300\left[\frac{(\frac{1}{2}t^2 + 2t + 25)^{-1/2}[(t+25)(t+2) - 2(\frac{1}{2}t^2 + 2t + 25)]}{(t+25)^2}\right]$$

$$= \frac{3450t}{(t+25)^2\sqrt{\frac{1}{2}t^2 + 2t + 25}}.$$

Ten seconds into the run, the athlete's pulse rate is increasing at

$$P'(10) = \frac{3450(10)}{(35)^2\sqrt{50 + 20 + 25}} \approx 2.9, \text{ or approximately 2.9 beats per minute per}$$

minute. Sixty seconds into the run, it is increasing at

$$P'(60) = \frac{3450(60)}{(85)^2\sqrt{1800 + 120 + 25}} \approx 0.65, \text{ or approximately 0.7 beats per minute}$$

per minute. Two minutes into the run, it is increasing at

$$P'(120) = \frac{3450(120)}{(145)^2\sqrt{7200 + 240 + 25}} \approx 0.23, \text{ or approximately 0.2 beats per}$$

minute per minute. The pulse rate two minutes into the run is given by

$$P(120) = \frac{300\sqrt{7200 + 240 + 25}}{120 + 25} \approx 178.8, \text{ or approximately 179 beats per minute.}$$

75.   The area is given by $A = \pi r^2$. The rate at which the area is increasing is given by

dA/dt, that is, $\dfrac{dA}{dt} = \dfrac{d}{dt}(\pi r^2) = \dfrac{d}{dt}(\pi r^2)\dfrac{dr}{dt} = 2\pi r \dfrac{dr}{dt}.$

If $r = 40$ and $dr/dt = 2$, then $\dfrac{dA}{dt} = 2\pi(40)(2) = 160\pi$, that is, it is increasing at

the rate of 160π, or approximately 503, sq ft/sec.

77. $f(t) = 6.25t^2 + 19.75t + 74.75.\;\; g(x) = -0.00075x^2 + 67.5.$

$\dfrac{dS}{dt} = g'(x)f'(t) = (-0.0015x)(12.5t + 19.75).$

When $t = 4$, we have $x = f(4) = 6.25(16) + 19.75(4) + 74.75 = 253.75$

and $\dfrac{dS}{dt}\Big|_{t=4} = (-0.0015)(253.75)[12.5(4)+19.75] \approx -26.55;$

that is, the average speed will be dropping at the rate of approximately 27 mph per decade. The average speed of traffic flow at that time will be

$S = g(f(4)) = -0.00075(253.75^2) + 67.5 = 19.2$, or approximately 19 mph.

79. $N(x) = 1.42x$ and $x(t) = \dfrac{7t^2 + 140t + 700}{3t^2 + 80t + 550}$. The number of construction jobs as a function of time is $n(t) = N[x(t)]$. Using the Chain Rule,

$$n'(t) = \dfrac{dN}{dx} \cdot \dfrac{dx}{dt} = 1.42 \dfrac{dx}{dt}$$

$$= (1.42)\left[\dfrac{(3t^2 + 80t + 550)(14t + 140) - (7t^2 + 140t + 700)(6t + 80)}{(3t^2 + 80t + 550)^2} \right]$$

$$= \dfrac{1.42(140t^2 + 3500t + 21000)}{(3t^2 + 80t + 550)^2}.$$

$$n'(1) = \dfrac{1.42(140 + 3500 + 21000)}{(3 + 80 + 550)^2} \approx 0.0873216, \text{ or approximately } 87{,}322$$

jobs/year.

81. $x = f(p) = 10\sqrt{\dfrac{50-p}{p}};$

$$\dfrac{dx}{dp} = \dfrac{d}{dp}\left[10\left(\dfrac{50-p}{p} \right)^{1/2} \right] = (10)(\tfrac{1}{2})\left(\dfrac{50-p}{p} \right)^{-1/2} \dfrac{d}{dp}\left(\dfrac{50-p}{p} \right)$$

$$= 5\left(\dfrac{50-p}{p} \right)^{-1/2} \cdot \dfrac{d}{dp}\left(\dfrac{50}{p} - 1 \right)$$

$$= 5\left(\dfrac{50-p}{p} \right)^{-1/2}\left(-\dfrac{50}{p^2} \right) = -\dfrac{250}{p^2 \left(\dfrac{50-p}{p} \right)^{1/2}}$$

$$\dfrac{dx}{dp}\Big|_{p=25} = -\dfrac{250}{p^2 \left(\dfrac{50-p}{p} \right)^{1/2}} = -\dfrac{250}{(625)\left(\dfrac{25}{25} \right)^{1/2}} = -0.4$$

So the quantity demanded is falling at the rate of 0.4(1000) or 400 wristwatches per dollar increase in price.

83.   True. This is just the statement of the Chain Rule.

85.   True. $\dfrac{d}{dx}\sqrt{f(x)} = \dfrac{d}{dx}[f(x)]^{1/2} = \dfrac{1}{2}[f(x)]^{-1/2}f'(x) = \dfrac{f'(x)}{2\sqrt{f(x)}}$.

87.   Let $f(x) = x^{1/n}$ so that $[f(x)]^n = x$.
      Differentiating both sides with respect to x, we get
      $$n[f(x)]^{n-1}f'(x) = 1$$
      $$f'(x) = \dfrac{1}{n[f(x)]^{n-1}} = \dfrac{1}{n[x^{1/n}]^{n-1}} = \dfrac{1}{nx^{1-(1/n)}} = \dfrac{1}{n}x^{(1/n)-1}.$$
      as was to be shown.

USING TECHNOLOGY EXERCISES 3.3 page 196

1.   0.5774   3.   0.9390   5.   –4.9498

7.   a. 10,146,200/decade   b. 7,810,520/decade

EXERCISES 3.4, page 212

1.   a. $C(x)$ is always increasing because as x, the number of units produced, increases, the greater the amount of money that must be spent on production.
     b. This occurs at $x = 4$, or a production level of 4000. You can see this by looking at the slopes of the tangent lines for x less than, equal to, and a little larger then $x = 4$.

3.   a. The actual cost incurred in the production of the 1001st record is given by
     $$C(1001) - C(1000) = [2000 + 2(1001) - 0.0001(1001)^2]$$
     $$-[2000 + 2(1000) - 0.0001(1000)^2]$$
     $$= 3901.7999 - 3900 = 1.7999,$$
     or $1.80. The actual cost incurred in the production of the 2001st record is given by $C(2001) - C(2000) = [2000 + 2(2001) - 0.0001(2001)^2]$
     $$-[2000 + 2(2000) - 0.0001(2000)^2]$$
     $$= 5601.5999 - 5600 = 1.5999, \text{ or } \$1.60.$$

b. The marginal cost is $C'(x) = 2 - 0.0002x$. In particular
$$C'(1000) = 2 - 0.0002(1000) = 1.80$$
and $\qquad C'(2000) = 2 - 0.0002(2000) = 1.60.$

5.   a. $\bar{C}(x) = \dfrac{C(x)}{x} = \dfrac{100x + 200,000}{x} = 100 + \dfrac{200,000}{x}.$

      b. $\bar{C}'(x) = \dfrac{d}{dx}(100) + \dfrac{d}{dx}(200,000x^{-1}) = -200,000x^{-2} = -\dfrac{200,000}{x^2}.$

      c. $\displaystyle\lim_{x\to\infty} \bar{C}(x) = \lim_{x\to\infty}\left[100 + \dfrac{200,000}{x}\right] = 100$

and this says that the average cost approaches \$100 per unit if the production level is very high.

7. $\bar{C}(x) = \dfrac{C(x)}{x} = \dfrac{2000 + 2x - 0.0001x^2}{x} = \dfrac{2000}{x} + 2 - 0.0001x.$

      $\bar{C}'(x) = -\dfrac{2000}{x^2} + 0 - 0.0001 = -\dfrac{2000}{x^2} - 0.0001.$

9.   a. $R'(x) = \dfrac{d}{dx}(8000x - 100x^2) = 8000 - 200x.$

      b. $R'(39) = 8000 - 200(39) = 200.$ $R'(40) = 8000 - 200(40) = 0$
         $R'(41) = 8000 - 200(41) = -200$
This suggests the total revenue is maximized if the price charged/ passenger is \$40.

11.   a. $P(x) = R(x) - C(x) = (-0.04x^2 + 800x) - (200x + 300,000)$
         $= -0.04x^2 + 600x - 300,000.$
      b. $P'(x) = -0.08x + 600$
      c. $P'(5000) = -0.08(5000) + 600 = 200$ $P'(8000) = -0.08(8000) + 600 = -40.$
      d.

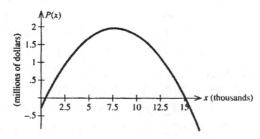

The profit realized by the company increases as production increases, peaking at a level of production of 7500 units. Beyond this level of production, the profit begins to fall.

13. a. The revenue function is $R(x) = px = (600 - 0.05x)x = 600x - 0.05x^2$
    and the profit function is
    $$P(x) = R(x) - C(x)$$
    $$= (600x - 0.05x^2) - (0.000002x^3 - 0.03x^2 + 400x + 80{,}000)$$
    $$= -0.000002x^3 - 0.02x^2 + 200x - 80{,}000.$$

    b. $C'(x) = \dfrac{d}{dx}(0.000002x^3 - 0.03x^2 + 400x + 80{,}000) = 0.000006x^2 - 0.06x + 400.$

    $$R'(x) = \frac{d}{dx}(600x - 0.05x^2) = 600 - 0.1x.$$

    $$P'(x) = \frac{d}{dx}(-0.000002x^3 - 0.02x^2 + 200x - 80{,}000) = -0.000006x^2 - 0.04x + 200.$$

    c. $C'(2000) = 0.000006(2000)^2 - 0.06(2000) + 400 = 304$, and this says that at a level of production of 2000 units, the cost for producing the 2001st unit is $304. $R'(2000) = 600 - 0.1(2000) = 400$ and this says that the revenue realized in selling the 2001st unit is $400. $P'(2000) = R'(2000) - C'(2000) = 400 - 304 = 96$, and this says that the revenue realized in selling the 2001st unit is $96.

    d.

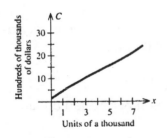
Units of a thousand

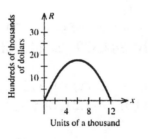
Units of a thousand

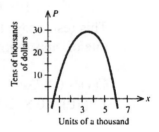
Units of a thousand

15. $\overline{C}(x) = \dfrac{C(x)}{x} = \dfrac{0.000002x^3 - 0.03x^2 + 400x + 80{,}000}{x}$

    $$= 0.000002x^2 - 0.03x + 400 + \frac{80{,}000}{x}.$$

    a. $\overline{C}'(x) = 0.000004x - 0.03 - \dfrac{80{,}000}{x^2}.$

    b. $\overline{C}'(5000) = 0.000004(5000) - 0.03 - \dfrac{80{,}000}{5000^2} \approx -0.0132,$

3 Differentiation

and this says that, at a level of production of 5000 units, the average cost of production is dropping at the rate of approximately a penny per unit.

$$\overline{C}'(10,000) = 0.000004(10000) - 0.03 - \frac{80,000}{10,000^2} \approx 0.0092,$$

and this says that, at a level of production of 10,000 units, the average cost of production is increasing at the rate of approximately a penny per unit.

c.

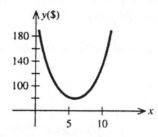

17. a. $R(x) = px = \dfrac{50x}{0.01x^2 + 1}$.

  b. $R'(x) = \dfrac{(0.01x^2 + 1)50 - 50x(0.02x)}{(0.01x^2 + 1)^2} = \dfrac{50 - 0.5x^2}{(0.01x^2 + 1)^2}$

  c. $R'(2) = \dfrac{50 - 0.5(4)}{[0.01(4) + 1]^2} \approx 44.379$.

This result says that at a level of sale of 2000 units, the revenue increases at the rate of approximately $44,379 per sales of 1000 units.

19. $C(x) = 0.873x^{1.1} + 20.34;\ \ C'(x) = 0.873(1.1)x^{0.1}$
$C'(10) = 0.873(1.1)(10)^{0.1} = 1.21$, or $1.21 billion per billion dollars.

21. The consumption function is given by $C(x) = 0.712x + 95.05$. The marginal propensity to consume is given by $\dfrac{dC}{dx} = 0.712$. The marginal propensity to save is given by $\dfrac{dS}{dx} = 1 - \dfrac{dC}{dx} = 1 - 0.712 = 0.288$, or $0.288 billion per billion dollars.

23. Here $x = f(p) = -\frac{5}{4}p + 20$ and so $f'(p) = -\frac{5}{4}$. Therefore,

$$E(p) = -\frac{pf'(p)}{f(p)} = -\frac{p(-\frac{5}{4})}{-\frac{5}{4}p + 20} = \frac{5p}{80 - 5p}.$$

$$E(10) = \frac{5(10)}{80 - 5(10)} = \frac{50}{30} = \frac{5}{3} > 1, \quad \text{and so the demand is elastic.}$$

25. $f(p) = -\frac{1}{3}p + 20$; $f'(p) = -\frac{1}{3}$.

Then the elasticity of demand is given by

$$E(p) = -\frac{p\left(-\frac{1}{3}\right)}{-\frac{1}{3}p + 20}, \quad \text{and} \quad E(30) = -\frac{30\left(-\frac{1}{3}\right)}{-\frac{1}{3}(30) + 20} = 1,$$

and we conclude that the demand is unitary at this price.

27. $x^2 = 169 - p$ and $f(p) = (169 - p)^{1/2}$.

Next, $f'(p) = \frac{1}{2}(169 - p)^{-1/2}(-1) = -\frac{1}{2}(169 - p)^{-1/2}$.

Then the elasticity of demand is given by

$$E(p) = -\frac{pf'(p)}{f(p)} = -\frac{p\left(-\frac{1}{2}\right)(169 - p)^{-1/2}}{(169 - p)^{1/2}} = \frac{\frac{1}{2}p}{169 - p}.$$

Therefore, when $p = 29$,

$$E(p) = \frac{\frac{1}{2}(29)}{169 - 29} = \frac{14.5}{140} = 0.104.$$

Since $E(p) < 1$, we conclude that demand is inelastic at this price.

29. $f(p) = \frac{1}{5}(225 - p^2)$; $f'(p) = \frac{1}{5}(-2p) = -\frac{2}{5}p$.

Then the elasticity of demand is given by

$$E(p) = -\frac{pf'(p)}{f(p)} = -\frac{p\left(-\frac{2}{5}p\right)}{\frac{1}{5}(225 - p^2)} = \frac{2p^2}{225 - p^2}.$$

a. When $p = 8$, $E(8) = \frac{2(64)}{225 - 64} = 0.8 < 1$ and the demand is inelastic. When $p = 10$,

$$E(10) = \frac{2(100)}{225 - 100} = 1.6 > 1$$

and the demand is elastic.

b. The demand is unitary when $E = 1$. Solving $\frac{2p^2}{225 - p^2} = 1$ we find $2p^2 = 225 - p^2$,

$3p^2 = 225$, and $p = 8.66$. So the demand is unitary when $p = 8.66$.

c. Since demand is elastic when $p = 10$, lowering the unit price will cause the revenue to increase.

3 Differentiation

d. Since the demand is inelastic at $p = 8$, a slight increase in the unit price will cause the revenue to increase.

31. $f(p) = \frac{2}{3}(36 - p^2)^{1/2}$

$f'(p) = \frac{2}{3}(\frac{1}{2})(36 - p^2)^{-1/2}(-2p) = -\frac{2}{3}p(36 - p^2)^{-1/2}$.

Then the elasticity of demand is given by

$$E(p) = -\frac{pf'(p)}{f(p)} = -\frac{-\frac{2}{3}p(36 - p^2)^{-1/2}p}{\frac{2}{3}(36 - p^2)^{1/2}} = \frac{p^2}{36 - p^2}.$$

When $p = 2$, $E(2) = \frac{4}{36 - 4} = \frac{1}{8} < 1$, and we conclude that the demand is inelastic.

b. Since the demand is inelastic, the revenue will increase when the rental price is increased.

33. We first solve the demand equation for x in terms of p. Thus,

$p = \sqrt{9 - 0.02x}$

$p^2 = 9 - 0.02x$

or $x = -50p^2 + 450$. With $f(p) = -50p^2 + 450$, we find

$$E(p) = -\frac{pf'(p)}{f(p)} = -\frac{p(-100p)}{-50p^2 + 450} = \frac{2p^2}{9 - p^2}.$$

Setting $E(p) = 1$ gives $2p^2 = 9 - p^2$, so $p = \sqrt{3}$. So the demand is inelastic in $[0, \sqrt{3}]$, unitary when $p = \sqrt{3}$, and elastic in $(\sqrt{3}, 3)$.

35. True. $\overline{C}'(x) = \frac{d}{dx}\left[\frac{C(x)}{x}\right] = \frac{xC'(x) - C(x)\frac{d}{dx}(x)}{x^2} = \frac{xC'(x) - C(x)}{x^2}$.

EXERCISES 3.5, page 219

1. $f(x) = 4x^2 - 2x + 1$; $f'(x) = 8x - 2$; $f''(x) = 8$.

3. $f(x) = 2x^3 - 3x^2 + 1$; $f'(x) = 6x^2 - 6x$; $f''(x) = 12x - 6 = 6(2x - 1)$.

5. $h(t) = t^4 - 2t^3 + 6t^2 - 3t + 10$; $h'(t) = 4t^3 - 6t^2 + 12t - 3$

$$h''(t) = 12t^2 - 12t + 12 = 12(t^2 - t + 1).$$

7. $f(x) = (x^2 + 2)^5$; $f'(x) = 5(x^2 + 2)^4(2x) = 10x(x^2 + 2)^4$ and
   $f''(x) = 10(x^2 + 2)^4 + 10x(x^2 + 2)^3(2x)$
   $$= 10(x^2 + 2)^3[(x^2 + 2) + 8x^2] = 10(9x^2 + 2)(x^2 + 2)^3.$$

9. $g(t) = (2t^2 - 1)^2(3t^2)$;
   $g'(t) = 2(2t^2 - 1)(4t)(3t^2) - (2t^2 - 1)^2(6t)$
   $$= 6t(2t^2 - 1)[4t^2 + (2t^2 - 1)] = 6t(2t^2 - 1)(6t^2 - 1)$$
   $$= 6t(12t^4 - 8t^2 + 1) = 72t^5 - 48t^3 + 6t.$$
   $g''(t) = 360t^4 - 144t^2 + 6 = 6(60t^4 - 24t^2 + 1)$

11. $f(x) = (2x^2 + 2)^{7/2}$; $f'(x) = \frac{7}{2}(2x^2 + 2)^{5/2}(4x) = 14x(2x^2 + 2)^{5/2}$;
    $f''(x) = 14(2x^2 + 2)^{5/2} + 14x(\frac{5}{2})(2x^2 + 2)^{3/2}(4x)$
    $$= 14(2x^2 + 2)^{3/2}[(2x^2 + 2) + 10x^2] = 28(6x^2 + 1)(2x^2 + 2)^{3/2}.$$

13. $f(x) = x(x^2 + 1)^2$;
    $f'(x) = (x^2 + 1)^2 + x(2)(x^2 + 1)(2x)$
    $$= (x^2 + 1)[(x^2 + 1) + 4x^2] = (x^2 + 1)(5x^2 + 1);$$
    $f''(x) = 2x(5x^2 + 1) + (x^2 + 1)(10x) = 2x(5x^2 + 1 + 5x^2 + 5)$
    $$= 4x(5x^2 + 3).$$

15. $f(x) = \dfrac{x}{2x + 1}$; $f'(x) = \dfrac{(2x + 1)(1) - x(2)}{(2x + 1)^2} = \dfrac{1}{(2x + 1)^2}$;
    $$f''(x) = \frac{d}{dx}(2x + 1)^{-2} = -2(2x + 1)^{-3}(2) = -\frac{4}{(2x + 1)^3}.$$

17. $f(s) = \dfrac{s - 1}{s + 1}$; $f'(s) = \dfrac{(s + 1)(1) - (s - 1)(1)}{(s + 1)^2} = \dfrac{2}{(s + 1)^2}$.
    $$f''(s) = 2\frac{d}{ds}(s + 1)^{-2} = -4(s + 1)^{-3} = -\frac{4}{(s + 1)^3}.$$

19. $f(u) = \sqrt{4-3u} = (4-3u)^{1/2}$. $f'(u) = \frac{1}{2}(4-3u)^{-1/2}(-3) = -\frac{3}{2\sqrt{4-3u}}$.

$f''(u) = -\frac{3}{2} \cdot \frac{d}{du}(4-3u)^{-1/2} = -\frac{3}{2}\left(-\frac{1}{2}\right)(4-3u)^{-3/2}(-3) = -\frac{9}{4(4-3u)^{3/2}}$.

21. $f(x) = 3x^4 - 4x^3$; $f'(x) = 12x^3 - 12x^2$; $f''(x) = 36x^2 - 24x$; $f'''(x) = 72x - 24$.

23. $f(x) = \frac{1}{x}$; $f'(x) = \frac{d}{dx}(x^{-1}) = -x^{-2}$; $f''(x) = 2x^{-3}$; $f'''(x) = -6x^{-4} = -\frac{6}{x^4}$.

25. $g(s) = (3s-2)^{1/2}$; $g'(s) = \frac{1}{2}(3s-2)^{-1/2}(3) = \frac{3}{2(3s-2)^{1/2}}$;

$g''(s) = \frac{3}{2}\left(-\frac{1}{2}\right)(3s-2)^{-3/2}(3) = -\frac{9}{4}(3s-2)^{-3/2} = -\frac{9}{4(3s-2)^{3/2}}$;

$g'''(s) = \frac{27}{8}(3s-2)^{-5/2}(3) = \frac{81}{8}(3s-2)^{-5/2} = \frac{81}{8(3s-2)^{5/2}}$.

27. $f(x) = (2x-3)^4$; $f'(x) = 4(2x-3)^3(2) = 8(2x-3)^3$
$f''(x) = 24(2x-3)^2(2) = 48(2x-3)^2$; $f'''(x) = 96(2x-3)(2) = 192(2x-3)$.

29. Its velocity at any time t is $v(t) = \frac{d}{dt}(16t^2) = 32t$. The hammer strikes the ground when $16t^2 = 256$ or $t = 4$ (we reject the negative root). Therefore, its velocity at the instant it strikes the ground is $v(4) = 32(4) = 128$ ft/sec. Its acceleration at time t is $a(t) = \frac{d}{dt}(32t) = 32$. In particular, its acceleration at $t = 4$ is 32 ft/sec^2.

31. $N(t) = -0.1t^3 + 1.5t^2 + 100$.
a. $N'(t) = -0.3t^2 + 3t = 0.3t(10 - t)$. Since $N'(t) > 0$ for $t = 0, 1, 2, ..., 7$, it is evident that $N(t)$ (and therefore the crime rate) was increasing from 1988 through 1995.
b. $N''(t) = -0.6t + 3 = 0.6(5 - t)$. Now $N''(4) = 0.6 > 0$, $N''(5) = 0$, $N''(6) = -0.6 < 0$ and $N''(7) = -1.2 < 0$. This shows that the rate of the rate of change was decreasing beyond $t = 5$ (1990). This shows that the program was

working.

33. $N(t) = 0.00037t^3 - 0.0242t^2 + 0.52t + 5.3 \qquad (0 \le t \le 10)$

   $N'(t) = 0.00111t^2 - 0.0484t + 0.52$

   $N''(t) = 0.00222t - 0.0484$

   So $N(8) = 0.00037(8)^3 - 0.0242(8)^2 + 5.3 = 8.1$

   $N'(8) = 0.00111(8)^2 - 0.0484(8) + 0.52 \approx 0.204.$

   $N''(8) = 0.00222(8) - 0.0484 = -0.031.$

   We conclude that at the beginning of 1998, there were 8.1 million persons receiving disability benefits, the number is increasing at the rate of 0.2 million/yr, and the rate of the rate of change of the number of persons is decreasing at the rate of 0.03 million persons/yr^2.

35. $N(t) = -0.00233t^4 + 0.00633t^3 - 0.05417t^2 + 1.3467t + 25$

   $N'(t) = -0.00932t^3 + 0.01899t^2 - 0.10834t + 1.3467$

   $N''(t) = -0.02796t^2 + 0.03798t - 0.10834$

   So $N'(10) = -7.158$ and $N''(10) = -2.5245.$

   Our computations show that at the beginning of the year 2000, the number of Americans aged 45 to 54 was decreasing at the rate of 7 million people per year, and the number decreases at a rate of approximately 2.5 million people per year per year.

37. $f(t) = 10.72(0.9t + 10)^{0.3}$.

   $f'(t) = 10.72(0.3)(0.9t + 10)^{-0.7}(0.9) = 2.8944(0.9t + 10)^{-0.7}$

   $f''(t) = 2.8944(-0.7)(0.9t + 10)^{-1.7}(0.9) = -1.823472(0.9t + 10)^{-1.7}$

   So $f''(10) = -1.823472(19)^{-1.7} \approx -0.01222$. And this says that the rate of the rate of change of the population is decreasing at the rate of 0.01%/ yr^2.

39. False. If f has derivatives of order two at $x = a$, then $f''(a) = [f'(a)]^2$.

41. True. If $f(x)$ is a polynomial function of degree n, then $f^{(n+1)}(x) = 0$.

43. True. Using the chain rule, $h'(x) = f'(2x) \cdot \dfrac{d}{dx}(2x) = f'(x) \cdot 2 = 2f'(2x)$

Using the chain rule again, $h''(x) = 2f''(2x) \cdot 2 = 4f''(2x)$.

45. Consider the function $f(x) = x^{(2n+1)/2} = x^{n+(1/2)}$.

   Then $\quad f'(x) = (n+\frac{1}{2})x^{n-(1/2)}$

   $\qquad f''(x) = (n+\frac{1}{2})(n-\frac{1}{2})x^{n-(3/2)}$

   ....

   $\qquad f^{(n)}(x) = (n+\frac{1}{2})(n-\frac{1}{2}) \cdots \frac{3}{2}x^{1/2}$

   $\qquad f^{(n+1)}(x) = (n+\frac{1}{2})(n-\frac{1}{2}) \cdots \frac{1}{2}x^{-1/2}$.

   The first n derivatives exist at $x = 0$, but the $(n+1)$st derivative fails to be defined there.

USING TECHNOLOGY EXERCISES 3.5, page 223

1. -18 3. 15.2762

5. -0.6255 7. 0.1973

9. $f''(6) = -68.46214$ and it tells us that at the beginning of 1988, the rate of the rate of the rate at which banks were failing was 68 banks per year per year per year.

EXERCISES 3.6, page 231

1. a. Solving for y in terms of x, we have $y = -\frac{1}{2}x + \frac{5}{2}$. Therefore, $y' = -\frac{1}{2}$.

    b. Next, differentiating $x + 2y = 5$ implicitly, we have $1 + 2y' = 0$, or $y' = -\frac{1}{2}$.

3. a. $xy = 1$, $y = \dfrac{1}{x}$, and $\dfrac{dy}{dx} = -\dfrac{1}{x^2}$.

    b. $\qquad x\dfrac{dy}{dx} + y = 0$

    $\qquad\qquad x\dfrac{dy}{dx} = -y$

    $\qquad\qquad \dfrac{dy}{dx} = -\dfrac{y}{x} = \dfrac{-\frac{1}{x}}{x} = -\dfrac{1}{x^2}$.

5. $x^3 - x^2 - xy = 4$.

   a. $-xy = 4 - x^3 + x^2$

$$y = -\frac{4}{x} + x^2 - x \quad \text{and} \quad y' = \frac{4}{x^2} + 2x - 1.$$

   b. $x^3 - x^2 - xy = 4$

$$-x\frac{dy}{dx} = -3x^2 + 2x + y$$

$$\frac{dy}{dx} = 3x - 2 - \frac{y}{x}$$

$$= 3x - 2 - \frac{1}{x}\left(-\frac{4}{x} + x^2 - x\right) = 3x - 2 + \frac{4}{x^2} - x + 1$$

$$= \frac{4}{x^2} + 2x - 1.$$

7.   a. $\dfrac{x}{y} - x^2 = 1$ is equivalent to $\dfrac{x}{y} = x^2 + 1$, or $y = \dfrac{x}{x^2 + 1}$. Therefore,

$$y' = \frac{(x^2 + 1) - x(2x)}{(x^2 + 1)^2} = \frac{1 - x^2}{(x^2 + 1)^2}.$$

   b. Next, differentiating the equation $x - x^2 y = y$ implicitly, we obtain

$$1 - 2xy - x^2 y' = y', \; y'(1 + x^2) = 1 - 2xy, \text{ or } \; y' = \frac{1 - 2xy}{(1 + x^2)}.$$

(This may also be written in the form $-2y^2 + \dfrac{y}{x}$.) To show that this is equivalent to

the results obtained earlier, use the value of y obtained before, to get

$$y' = \frac{1 - 2x\left(\dfrac{x}{x^2 + 1}\right)}{1 + x^2} = \frac{x^2 + 1 - 2x^2}{(1 + x^2)^2} = \frac{1 - x^2}{(1 + x^2)^2}.$$

9. $x^2 + y^2 = 16$. Differentiating both sides of the equation implicitly, we obtain

$$2x + 2yy' = 0 \text{ and so } y' = -\frac{x}{y}.$$

11. $x^2 - 2y^2 = 16$. Differentiating implicitly with respect to x, we have

$$2x - 4y\frac{dy}{dx} = 0 \ \text{ and } \ \frac{dy}{dx} = \frac{x}{2y}.$$

13. $x^2 - 2xy = 6$. Differentiating both sides of the equation implicitly, we obtain

$2x - 2y - 2xy' = 0$ and so $y' = \dfrac{x - y}{x} = 1 - \dfrac{y}{x}$.

15. $x^2y^2 - xy = 8$. Differentiating both sides of the equation implicitly, we obtain

$$2xy^2 + 2x^2yy' - y - xy' = 0, \ 2xy^2 - y + y'(2x^2y - x) = 0$$

and so $\qquad\qquad y' = \dfrac{y(1 - 2xy)}{x(2xy - 1)} = -\dfrac{y}{x}.$

17. $x^{1/2} + y^{1/2} = 1$. Differentiating implicitly with respect to x, we have

$$\tfrac{1}{2}x^{-1/2} + \tfrac{1}{2}y^{-1/2}\frac{dy}{dx} = 0. \ \text{ Therefore, } \ \frac{dy}{dx} = -\frac{x^{-1/2}}{y^{-1/2}} = -\frac{\sqrt{y}}{\sqrt{x}}.$$

19. $\sqrt{x + y} = x$. Differentiating both sides of the equation implicitly, we obtain

$$\tfrac{1}{2}(x + y)^{-1/2}(1 + y') = 1, \ \ 1 + y' = 2(x + y)^{1/2},$$

or $\qquad\qquad\qquad y' = 2\sqrt{x + y} - 1.$

21. $\dfrac{1}{x^2} + \dfrac{1}{y^2} = 1$. Differentiating both sides of the equation implicitly, we obtain

$$-\frac{2}{x^3} - \frac{2}{y^3}y' = 0, \ \text{ or } \ y' = -\frac{y^3}{x^3}.$$

23. $\sqrt{xy} = x + y$. Differentiating both sides of the equation implicitly, we obtain

$$\tfrac{1}{2}(xy)^{-1/2}(xy' + y) = 1 + y'$$
$$xy' + y = 2\sqrt{xy}(1 + y')$$
$$y'(x - 2\sqrt{xy}) = 2\sqrt{xy} - y$$

or $\qquad\qquad\qquad y' = -\dfrac{(2\sqrt{xy} - y)}{(2\sqrt{xy} - x)} = \dfrac{2\sqrt{xy} - y}{x - 2\sqrt{xy}}.$

25. $\dfrac{x+y}{x-y} = 3x$, or $x+y = 3x^2 - 3xy$. Differentiating both sides of the equation

implicitly, we obtain $\quad 1 + y' = 6x - 3xy' - 3y$ or $y' = \dfrac{6x - 3y - 1}{3x + 1}$.

27. $xy^{3/2} = x^2 + y^2$. Differentiating implicitly with respect to x, we obtain

$$y^{3/2} + x\left(\tfrac{3}{2}\right)y^{1/2}\frac{dy}{dx} = 2x + 2y\frac{dy}{dx}$$

$$2y^{3/2} + 3xy^{1/2}\frac{dy}{dx} = 4x + 4y\frac{dy}{dx} \qquad \text{(Multiplying by 2.)}$$

$$(3xy^{1/2} - 4y)\frac{dy}{dx} = 4x - 2y^{3/2}$$

$$\frac{dy}{dx} = \frac{2(2x - y^{3/2})}{3xy^{1/2} - 4y}.$$

29. $(x+y)^3 + x^3 + y^3 = 0$. Differentiating implicitly with respect to x, we obtain

$$3(x+y)^2\left(1 + \frac{dy}{dx}\right) + 3x^2 + 3y^2\frac{dy}{dx} = 0$$

$$(x+y)^2 + (x+y)^2\frac{dy}{dx} + x^2 + y^2\frac{dy}{dx} = 0$$

$$[(x+y)^2 + y^2]\frac{dy}{dx} = -[(x+y)^2 + x^2]$$

$$\frac{dy}{dx} = -\frac{2x^2 + 2xy + y^2}{x^2 + 2xy + 2y^2}.$$

31. $4x^2 + 9y^2 = 36$. Differentiating the equation implicitly, we obtain
$$8x + 18yy' = 0.$$
At the point $(0,2)$, we have $0 + 36y' = 0$ and the slope of the tangent line is 0.
Therefore, an equation of the tangent line is $y = 2$.

33. $x^2y^3 - y^2 + xy - 1 = 0$. Differentiating implicitly with respect to x, we have

$$2xy^3 + 3x^2y^2\frac{dy}{dx} - 2y\frac{dy}{dx} + y + x\frac{dy}{dx} = 0.$$

3 Differentiation

At $(1,1)$, $2+3\dfrac{dy}{dx}-2\dfrac{dy}{dx}+1+\dfrac{dy}{dx}=0$, and

$$2\dfrac{dy}{dx}=-3 \quad\text{and}\quad \dfrac{dy}{dx}=-\dfrac{3}{2}.$$

Using the point-slope form of an equation of a line, we have

$$y-1=-\tfrac{3}{2}(x-1)$$

and the equation of the tangent line to the graph of the function f at $(1,1)$ is

$$y=-\tfrac{3}{2}x+\tfrac{5}{2}.$$

35. $xy=1$. Differentiating implicitly, we have $xy'+y=0$, or $y'=-\dfrac{y}{x}$.

Differentiating implicitly once again, we have $xy''+y'+y'=0$.

Therefore, $\quad y''=-\dfrac{2y'}{x}=\dfrac{2\left(\dfrac{y}{x}\right)}{x}=\dfrac{2y}{x^2}$.

37. $y^2-xy=8$. Differentiating implicitly we have $2yy'-y-xy'=0$

and $y'=\dfrac{y}{2y-x}$. Differentiating implicitly again, we have

$$2(y')^2+2yy''-y'-y'-xy''=0, \quad\text{or}\quad y''=\dfrac{2y'-2(y')^2}{2y-x}.$$

Then $\quad y''=\dfrac{2\left(\dfrac{y}{2y-x}\right)\left(1-\dfrac{y}{2y-x}\right)}{2y-x}=\dfrac{2y(2y-x-y)}{(2y-x)^3}=\dfrac{2y(y-x)}{(2y-x)^3}$.

39. a. Differentiating the given equation with respect to t, we obtain

$$\dfrac{dV}{dt}=\pi r^2\dfrac{dh}{dt}+2\pi rh\dfrac{dr}{dt}=\pi r\left(r\dfrac{dh}{dt}+2h\dfrac{dr}{dt}\right).$$

b. Substituting $r=2$, $h=6$, $\dfrac{dr}{dt}=0.1$ and $\dfrac{dh}{dt}=0.3$ into the expression for $\dfrac{dV}{dt}$

we obtain $\dfrac{dV}{dt}=\pi(2)[2(0.3)+2(6)(0.1)]=3.6\pi$

and so the volume is increasing at the rate of 3.6π cu in/sec.

41. We are given $\dfrac{dp}{dt} = 2$ and are required to find $\dfrac{dx}{dt}$ when $x = 9$ and $p = 63$.

Differentiating the equation $p + x^2 = 144$ with respect to t, we obtain

$$\frac{dp}{dt} + 2x\frac{dx}{dt} = 0.$$

When $x = 9$, $p = 63$, and $\dfrac{dp}{dt} = 2$,

$$2 + 2(9)\frac{dx}{dt} = 0$$

and

$$\frac{dx}{dt} = -\frac{1}{9} \approx -0.111,$$

or the quantity demanded is decreasing at the rate of 111 tires per week.

43. $100x^2 + 9p^2 = 3600$. Differentiating the given equation implicitly with respect to t,

we have $200x\dfrac{dx}{dt} + 18p\dfrac{dp}{dt} = 0$. Next, when $p = 14$, the given equation yields

$$100x^2 + 9(14)^2 = 3600$$
$$100x^2 = 1836,$$

or $x = 4.2849$. When $p = 14$, $\dfrac{dp}{dt} = -0.15$, and $x = 4.2849$, we have

$$200(4.2849)\frac{dx}{dt} + 18(14)(-0.15) = 0$$

$$\frac{dx}{dt} = 0.0441.$$

So the quantity demanded is increasing at the rate of 44 ten–packs per week.

45. From the results of Problem 44, we have

$$1250p\frac{dp}{dt} - 2x\frac{dx}{dt} = 0.$$

When $p = 1.0770$, $x = 25$, and $\dfrac{dx}{dt} = -1$, we find that

$$1250(1.077)\frac{dp}{dt} - 2(25)(-1) = 0,$$

and $$\frac{dp}{dt} = -\frac{50}{1250(1.077)} = -0.037.$$

We conclude that the price is decreasing at the rate of 3.7 cents per carton.

47. $p = -0.01x^2 - 0.2x + 8$. Differentiating the given equation implicitly with respect to p, we have

$$1 = -0.02x\frac{dx}{dp} - 0.2\frac{dx}{dp} = [0.02x + 0.2]\frac{dx}{dp}$$

or $$\frac{dx}{dp} = -\frac{1}{0.02x + 0.2}.$$

When $x = 15$, $p = -0.01(15)^2 - 0.2(15) + 8 = 2.75$

and $$\frac{dx}{dp} = -\frac{1}{0.02(15) + 0.2} = -2.$$

Therefore, $E(p) = -\frac{pf'(p)}{f(p)} = -\frac{(2.75)(-2)}{15} = 0.37 < 1,$

and the demand is inelastic.

49. $A = \pi r^2$. Differentiating with respect to t, we obtain
$$\frac{dA}{dt} = 2\pi r \frac{dr}{dt}.$$
When the radius of the circle is 40 ft and increasing at the rate of 2 ft/sec,
$$\frac{dA}{dt} = 2\pi(40)(2) = 160\pi \ \text{ft}^2 / \text{sec}.$$

51. Let D denote the distance between the two cars, x the distance traveled by the car heading east, and y the distance traveled by the car heading north as shown in the diagram at the right. Then
$D^2 = x^2 + y^2$. Differentiating with respect to t, we have

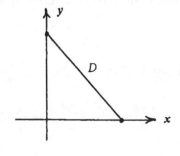

$$2D\frac{dD}{dt} = 2x\frac{dx}{dt} + 2y\frac{dy}{dt},$$

or $$\frac{dD}{dt} = \frac{x\frac{dx}{dt} + y\frac{dy}{dt}}{D}.$$

When $t = 5$, $x = 30$, $y = 40$, $\dfrac{dx}{dt} = 2(5) + 1 = 11$, and $\dfrac{dy}{dt} = 2(5) + 3 = 13$.

Therefore, $\dfrac{dD}{dt} = \dfrac{(30)(11) + (40)(13)}{\sqrt{900 + 1600}} = 17$ ft/sec.

53. Referring to the diagram at the right, we see that

$$D^2 = 120^2 + x^2.$$

Differentiating this last equation with respect to t, we have

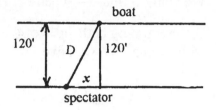

$$2D\frac{dD}{dt} = 2x\frac{dx}{dt} \quad \text{and} \quad \frac{dD}{dt} = \frac{x\dfrac{dx}{dt}}{D}.$$

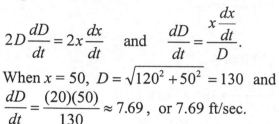

When $x = 50$, $D = \sqrt{120^2 + 50^2} = 130$ and

$$\frac{dD}{dt} = \frac{(20)(50)}{130} \approx 7.69, \text{ or } 7.69 \text{ ft/sec.}$$

55. Let V and S denote its volume and surface area. Then we are given that $\dfrac{dV}{dt} = -kS$, where k is the constant of proportionality. But from $V = \left(\dfrac{4}{3}\right)\pi r^3$,

we find, upon differentiating both sides with respect to t, that

$$\frac{dV}{dt} = \left(\frac{4}{3}\right)\pi(3\pi r^2)\frac{dr}{dt} = 4\pi^2 r^2 \frac{dr}{dt}$$

and using the fact stated earlier,

$$\frac{dV}{dt} = 4\pi^2 r^2 \frac{dr}{dt} = -kS = -k(4\pi r^2).$$

Therefore, $\dfrac{dr}{dt} = -\dfrac{k(4\pi r^2)}{4\pi^2 r^2} = -\dfrac{k}{\pi}$ and this proves that the radius is decreasing at the constant rate of (k/π) units/unit time.

57. Refer to the figure at the right.

We are given that $\dfrac{dx}{dt} = 264$. Using the

Pythagorean Theorem,
$$s^2 = x^2 + 1000^2 = x^2 + 1000000.$$

We want to find $\dfrac{ds}{dt}$ when $s = 1500$.

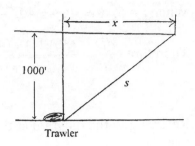

1000'

Trawler

Differentiating both sides of the equation
with respect to t, we have

$$2s\frac{ds}{dt} = 2x\frac{dx}{dt} \quad \text{and so} \quad \frac{ds}{dt} = \frac{x\dfrac{dx}{dt}}{s}.$$

Now, when $s = 1500$, we have
$$1500^2 = x^2 + 10000 \quad \text{or} \quad x = \sqrt{1250000}.$$

Therefore, $\quad \dfrac{ds}{dt} = \dfrac{\sqrt{1250000}\cdot(264)}{1500} \approx 196.8$, that is, the aircraft is receding

from the trawler at the speed of approximately 196.8 ft/sec.

59. Refer to the diagram at the right.
$$\frac{y}{6} = \frac{y+x}{18}, \quad 18y = 6(y+x)$$
$$3y = y+x, \quad 2y = x, \quad y = \tfrac{1}{2}x.$$
Then $D = y + x = \tfrac{3}{2}x$. Differentiating

implicitly, we have $\quad \dfrac{dD}{dt} = \dfrac{3}{2}\bullet\dfrac{dx}{dt}$

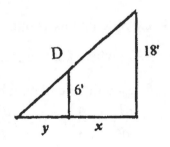

D

18'

6'

y   x

and when $\dfrac{dx}{dt} = 6$, $\dfrac{dD}{dt} = \dfrac{3}{2}(6) = 9$, or 9 ft / sec.

61. Differentiating $x^2 + y^2 = 13^2 = 169$ with
respect to t gives
$$2x\frac{dx}{dt} + 2y\frac{dy}{dt} = 0.$$
When $x = 12$, we have
$$144 + y^2 = 169 \quad \text{or} \quad y = 5.$$

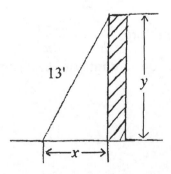

13'

y

x

Therefore, with $x = 12$, $y = 5$, and $\dfrac{dx}{dt} = 8$, we find $\quad 2(12)(8) + 2(5)\dfrac{dy}{dt} = 0$

or $\dfrac{dy}{dt} = -19.2$, that is, the top of the ladder is sliding down the wall at 19.2 ft/sec.

63. True. Differentiating both sides of the equation with respect to x, we have

$$\frac{d}{dx}[f(x)g(y)] = \frac{d}{dx}(0)$$

$$f(x)g'(y)\frac{dy}{dx} + f'(x)g(y) = 0$$

$$\frac{dy}{dx} = -\frac{f'(x)g(y)}{f(x)g'(y)}$$

provided $f(x) \neq 0$ and $g'(y) \neq 0$.

EXERCISES 3.7, page 241

1. $f(x) = 2x^2$ and $dy = 4x\,dx$.

3. $f(x) = x^3 - x$ and $dy = (3x^2 - 1)\,dx$.

5. $f(x) = \sqrt{x+1} = (x+1)^{1/2}$ and $dy = \dfrac{1}{2}(x+1)^{-1/2}\,dx = \dfrac{dx}{2\sqrt{x+1}}$.

7. $f(x) = 2x^{3/2} + x^{1/2}$ and $dy = (3x^{1/2} + \tfrac{1}{2}x^{-1/2})\,dx = \tfrac{1}{2}x^{-1/2}(6x+1)dx = \dfrac{6x+1}{2\sqrt{x}}\,dx$.

9. $f(x) = x + \dfrac{2}{x}$ and $dy = \left(1 - \dfrac{2}{x^2}\right)dx = \dfrac{x^2 - 2}{x^2}\,dx$.

11. $f(x) = \dfrac{x-1}{x^2+1}$ and $dy = \dfrac{x^2 + 1 - (x-1)2x}{(x^2+1)^2}\,dx = \dfrac{-x^2 + 2x + 1}{(x^2+1)^2}\,dx$.

13. $f(x) = \sqrt{3x^2 - x} = (3x^2 - x)^{1/2}$ and

$dy = \dfrac{1}{2}(3x^2 - x)^{-1/2}(6x-1)dx = \dfrac{6x-1}{2\sqrt{3x^2-x}}\,dx$.

15. $f(x) = x^2 - 1$.
    a. $dy = 2x\,dx$.   b. $dy \approx 2(1)(0.02) = 0.04$.
    c. $\Delta y = [(1.02)^2 - 1] - [1 - 1] = 0.0404$.

17. $f(x) = \dfrac{1}{x}$.

    a. $dy = -\dfrac{dx}{x^2}$.   b. $dy \approx -0.05$ c. $\Delta y = \dfrac{1}{-0.95} - \dfrac{1}{-1} = -0.05263$.

19. $y = \sqrt{x}$ and $dy = \dfrac{dx}{2\sqrt{x}}$. Therefore, $\sqrt{10} = 3 + \dfrac{1}{2 \cdot \sqrt{9}} = 3.167$.

21. $y = \sqrt{x}$ and $dy = \dfrac{dx}{2\sqrt{x}}$. Therefore, $\sqrt{49.5} = 7 + \dfrac{0.5}{2 \cdot 7} = 7.0358$.

23. $y = x^{1/3}$ and $dy = \tfrac{1}{3}x^{-2/3}\,dx$. Therefore, $\sqrt[3]{7.8} = 2 - \dfrac{0.2}{3 \cdot 4} = 1.983$.

25. $y = \sqrt{x}$ and $dy = \dfrac{dx}{2\sqrt{x}}$. Therefore, $\sqrt{0.089} = \tfrac{1}{10}\sqrt{8.9} = \tfrac{1}{10}\left[3 - \dfrac{0.1}{2 \cdot 3}\right] \approx 0.298$.

27. $y = f(x) = \sqrt{x} + \dfrac{1}{\sqrt{x}} = x^{1/2} + x^{-1/2}$. Therefore,

$$\frac{dy}{dx} = \frac{1}{2}x^{-1/2} - \frac{1}{2}x^{-3/2}$$

$$dy = \left(\frac{1}{2x^{1/2}} - \frac{1}{2x^{3/2}}\right)dx.$$

Letting $x = 4$ and $dx = 0.02$, we find

$$\sqrt{4.02} + \frac{1}{\sqrt{4.02}} - f(4) = f(4.02) - f(4) = \Delta y \approx dy$$

$$\sqrt{4.02} + \frac{1}{\sqrt{4.02}} = f(4) + dy\Big|_{\substack{x=4 \\ dx=0.02}}$$

$$\approx 2 + \frac{1}{2} + \left(\frac{1}{2 \cdot 2} - \frac{1}{2 \cdot 2\sqrt{2}} \right)(0.02) \approx 2.50146.$$

29. The volume of the cube is given by $V = x^3$. Then $dV = 3x^2 \, dx$ and
    when $x = 12$ and $dx = 0.02$, $dV = 3(144)(\pm 0.02) = \pm 8.64$,
    and the possible error that might occur in calculating the volume is ± 8.64 cm^3.

31. The volume of the hemisphere is given by $V = \frac{2}{3}\pi r^3$. The amount of rust-proofer
    needed is
    $$\Delta V = \frac{2}{3}\pi(r + \Delta r)^3 - \frac{2}{3}\pi r^3$$
    $$\approx dV = \left(\frac{2}{3} \right)(3\pi r^2)dr.$$
    So, with $r = 60$, and $dr = \frac{1}{12}(0.01)$, we have
    $$\Delta V \approx 2\pi(60^2)\left(\frac{1}{12} \right)(0.01) \approx 18.85.$$
    So we need approximately 18.85 ft^3 of rust-proofer.

33. $dR = \dfrac{d}{dr}(k\ell r^{-4})dr = -4k\ell r^{-5}\, dr$. With $\dfrac{dr}{r} = 0.1$, we find
    $$\frac{dR}{R} = -\frac{4k\ell r^{-5}}{k\ell r^{-4}}\, dr = -4\frac{dr}{r} = -4(0.1) = -0.4.$$
    In other words, the resistance will drop by 40%.

35. $f(n) = 4n\sqrt{n-4} = 4n(n-4)^{1/2}$.
    Then $df = 4[(n-4)^{1/2} + \frac{1}{2}n(n-4)^{-1/2}]dn$
    When $n = 85$ and $dn = 5$, $df = 4[9 + \frac{85}{2 \cdot 9}]5 \approx 274$ seconds.

37. $N(r) = \dfrac{7}{1 + 0.02r^2}$ and $dN = -\dfrac{0.28r}{(1 + 0.02r^2)^2}\, dr$. To estimate the decrease in the
    number of housing starts when the mortgage rate is increased from 12 to 12.5
    percent, we compute

$$dN = -\frac{(0.28)(12)(0.5)}{(3.88)^2} \approx -0.111595 \quad (r = 12, \, dr = 0.5)$$

or 111,595 fewer housing starts.

39. $p = \dfrac{30}{0.02x^2 + 1}$ and $dp = -\dfrac{(1.2x)}{(0.02x^2 + 1)^2} dx$. To estimate the change in the price p when the quantity demanded changed from 5000 to 5500 units ($x = 5$ to $x = 5.5$) per week, we compute $dp = \dfrac{(-1.2)(5)(0.5)}{[0.02(25) + 1]^2} \approx -1.33$, or a decrease of $1.33.

41. $P(x) = -0.000032x^3 + 6x - 100$ and $dP = (-0.000096x^2 + 6) \, dx$. To determine the error in the estimate of Trappee's profits corresponding to a maximum error in the forecast of 15 percent [$dx = \pm 0.15(200)$], we compute
$$dP = [(-0.000096)(200)^2 + 6] \, (\pm 30) = (2.16)(30) = \pm 64.80$$
or $\pm$ 64,800.

43. $N(x) = \dfrac{500(400 + 20x)^{1/2}}{(5 + 0.2x)^2}$ and

$$N'(x) = \frac{(5 + 0.2x)^2 \, 250(400 + 20x)^{-1/2}(20) - 500(400 + 20x)^{1/2}(2)(5 + 0.2x)(0.2)}{(5 + 0.2x)^4} \, dx.$$

To estimate the change in the number of crimes if the level of reinvestment changes from 20 cents per dollars deposited to 22 cents per dollar deposited, we compute
$$dN = \frac{(5 + 4)^2 (250)(800)^{-1/2}(20) - 500(400 + 400)^{1/2}(2)(9)(0.2)}{(5 + 4)^4}(2)$$

$$= \frac{(14318.91 - 50911.69)}{9^4}(2) \approx -11$$

or a decrease of approximately 11 crimes per year.

45. $A = 10{,}000\left(1 + \dfrac{r}{12}\right)^{120}$.

    a. $dA = 10{,}000(120)\left(1 + \dfrac{r}{12}\right)^{119}\left(\dfrac{1}{12}\right) dr = 100{,}000\left(1 + \dfrac{r}{12}\right)^{119} dr.$

b. At 8.1%, it will be worth $100,000\left(1+\dfrac{0.08}{12}\right)^{119}(0.001)$, or $220.49 more.

At 8.2%, it will be worth $100,000\left(1+\dfrac{0.08}{12}\right)^{119}(0.002)$, or $440.99 more.

At 8.3%, it will be worth $100,000\left(1+\dfrac{0.08}{12}\right)^{119}(0.003)$, or $661.48 more.

47. True. $dy = f'(x)\,dx = \dfrac{d}{dx}(ax+b)\,dx = a\,dx$. On the other hand,

$$\Delta y = f(x+\Delta x) - f(x) = [a(x+\Delta x)+b] - (ax+b) = a\Delta x = a\,dx.$$

USING TECHNOLOGY EXERCISES 3.7, page 243

1. $dy = f'(3)\,dx = 757.87(0.01) \approx 7.5787$.

3. $dy = f'(1)\,dx = 1.04067285926(0.03) \approx 0.031220185778$.

5. $dy = f'(4)(0.1) = -0.198761598(0.1) = -0.0198761598$.

7. If the interest rate changes from 10% to 10.3% per year, the monthly payment will increase by
$$dP = f'(0.1)(0.003) \approx 26.60279,$$
or approximately $26.60 per month. If the interest rate changes from 10% to 10.4% per year, it will be $35.47 per month. If the interest rate changes from 10% to 10.5% per year, it will be $44.34 per month.

9. $dx = f'(40)(2) \approx -0.625$. That is, the quantity demanded will decrease by 625 watches per week.

CHAPTER 3 REVIEW, page 247

1. $f'(x) = \dfrac{d}{dx}(3x^5 - 2x^4 + 3x^2 - 2x + 1) = 15x^4 - 8x^3 + 6x - 2$.

3. $g'(x) = \dfrac{d}{dx}(-2x^{-3} + 3x^{-1} + 2) = 6x^{-4} - 3x^{-2}$.

5. $g'(t) = \dfrac{d}{dt}(2t^{-1/2} + 4t^{-3/2} + 2) = -t^{-3/2} - 6t^{-5/2}$.

7. $f'(t) = \dfrac{d}{dt}(t + 2t^{-1} + 3t^{-2}) = 1 - 2t^{-2} - 6t^{-3} = 1 - \dfrac{2}{t^2} - \dfrac{6}{t^3}$.

9. $h'(x) = \dfrac{d}{dx}(x^2 - 2x^{-3/2}) = 2x + 3x^{-5/2} = 2x + \dfrac{3}{x^{5/2}}$.

11. $g(t) = \dfrac{t^2}{2t^2 + 1}$.

$$g'(t) = \dfrac{(2t^2 + 1)\dfrac{d}{dt}(t^2) - t^2\dfrac{d}{dt}(2t^2 + 1)}{(2t^2 + 1)^2}$$

$$= \dfrac{(2t^2 + 1)(2t) - t^2(4t)}{(2t^2 + 1)^2} = \dfrac{2t}{(2t^2 + 1)^2}.$$

13. $f(x) = \dfrac{\sqrt{x} - 1}{\sqrt{x} + 1} = \dfrac{x^{1/2} - 1}{x^{1/2} + 1}$.

$$f'(x) = \dfrac{(x^{1/2} + 1)(\frac{1}{2}x^{-1/2}) - (x^{1/2} - 1)(\frac{1}{2}x^{-1/2})}{(x^{1/2} + 1)^2}$$

$$= \dfrac{\frac{1}{2} + \frac{1}{2}x^{-1/2} - \frac{1}{2} + \frac{1}{2}x^{-1/2}}{(x^{1/2} + 1)^2} = \dfrac{x^{-1/2}}{(x^{1/2} + 1)^2} = \dfrac{1}{\sqrt{x}(\sqrt{x} + 1)^2}.$$

15. $f(x) = \dfrac{x^2(x^2 + 1)}{x^2 - 1}$.

$$f'(x) = \dfrac{(x^2 - 1)\dfrac{d}{dx}(x^4 + x^2) - (x^4 + x^2)\dfrac{d}{dx}(x^2 - 1)}{(x^2 - 1)^2}$$

$$= \frac{(x^2-1)(4x^3+2x)-(x^4+x^2)(2x)}{(x^2-1)^2}$$

$$= \frac{4x^5+2x^3-4x^3-2x-2x^5-2x^3}{(x^2-1)^2}$$

$$= \frac{2x^5-4x^3-2x}{(x^2-1)^2} = \frac{2x(x^4-2x^2-1)}{(x^2-1)^2}.$$

17. $f(x)=(3x^3-2)^8; f'(x)=8(3x^3-2)^7(9x^2)=72x^2(3x^3-2)^7.$

19. $f'(t)=\dfrac{d}{dt}(2t^2+1)^{1/2}=\dfrac{1}{2}(2t^2+1)^{-1/2}\dfrac{d}{dt}(2t^2+1)$

$$=\frac{1}{2}(2t^2+1)^{-1/2}(4t)=\frac{2t}{\sqrt{2t^2+1}}.$$

21. $s(t)=(3t^2-2t+5)^{-2}$

$$s'(t)=-2(3t^2-2t+5)^{-3}(6t-2)=-4(3t^2-2t+5)^{-3}(3t-1)$$

$$=-\frac{4(3t-1)}{(3t^2-2t+5)^3}.$$

23. $h(x)=\left(x+\dfrac{1}{x}\right)^2=(x+x^{-1})^2.$

$$h'(x)=2(x+x^{-1})(1-x^{-2})=2\left(x+\frac{1}{x}\right)\left(1-\frac{1}{x^2}\right)$$

$$=2\left(\frac{x^2+1}{x}\right)\left(\frac{x^2-1}{x^2}\right)=\frac{2(x^2+1)(x^2-1)}{x^3}.$$

25. $h'(t)=(t^2+t)^4\dfrac{d}{dt}(2t^2)+2t^2\dfrac{d}{dt}(t^2+t)^4$

$$=(t^2+t)^4(4t)+2t^2\cdot 4(t^2+t)^3(2t+1)$$

$$=4t(t^2+t)^3[(t^2+t)+4t^2+2t]=4t^2(5t+3)(t^2+t)^3.$$

27. $g(x) = x^{1/2}(x^2 - 1)^3$.

$$g'(x) = \frac{d}{dx}[x^{1/2}(x^2 - 1)^3] = x^{1/2} \cdot 3(x^2 - 1)^2(2x) + (x^2 - 1)^3 \cdot \tfrac{1}{2}x^{-1/2}$$

$$= \tfrac{1}{2}x^{-1/2}(x^2 - 1)^2[12x^2 + (x^2 - 1)]$$

$$= \frac{(13x^2 - 1)(x^2 - 1)^2}{2\sqrt{x}}.$$

29. $h(x) = \dfrac{(3x + 2)^{1/2}}{4x - 3}$.

$$h'(x) = \frac{(4x - 3)\tfrac{1}{2}(3x + 2)^{-1/2}(3) - (3x + 2)^{1/2}(4)}{(4x - 3)^2}$$

$$= \frac{\tfrac{1}{2}(3x + 2)^{-1/2}[3(4x - 3) - 8(3x + 2)]}{(4x - 3)^2} = -\frac{12x + 25}{2\sqrt{3x + 2}(4x - 3)^2}.$$

31. $f(x) = 2x^4 - 3x^3 + 2x^2 + x + 4$.

$$f'(x) = \frac{d}{dx}(2x^4 - 3x^3 + 2x^2 + x + 4) = 8x^3 - 9x^2 + 4x + 1.$$

$$f''(x) = \frac{d}{dx}(8x^3 - 9x^2 + 4x + 1) = 24x^2 - 18x + 4 = 2(12x^2 - 9x + 2).$$

33. $h(t) = \dfrac{t}{t^2 + 4}$. $\quad h'(t) = \dfrac{(t^2 + 4)(1) - t(2t)}{(t^2 + 4)^2} = \dfrac{4 - t^2}{(t^2 + 4)^2}.$

$$h''(t) = \frac{(t^2 + 4)^2(-2t) - (4 - t^2)2(t^2 + 4)(2t)}{(t^2 + 4)^4}$$

$$= \frac{-2t(t^2 + 4)[(t^2 + 4) + 2(4 - t^2)]}{(t^2 + 4)^4} = \frac{2t(t^2 - 12)}{(t^2 + 4)^3}.$$

35. $f'(x) = \dfrac{d}{dx}(2x^2 + 1)^{1/2} = \dfrac{1}{2}(2x^2 + 1)^{-1/2}(4x) = 2x(2x^2 + 1)^{-1/2}.$

$$f''(x) = 2(2x^2 + 1)^{-1/2} + 2x \cdot (-\tfrac{1}{2})(2x^2 + 1)^{-3/2}(4x)$$

$$= 2(2x^2 + 1)^{-3/2}[(2x^2 + 1) - 2x^2] = \frac{2}{(2x^2 + 1)^{3/2}}.$$

37. $6x^2 - 3y^2 = 9$ so $12x - 6y\dfrac{dy}{dx} = 0$ and $-6y\dfrac{dy}{dx} = -12x$.

Therefore, $\dfrac{dy}{dx} = \dfrac{-12x}{-6y} = \dfrac{2x}{y}$.

39. $y^3 + 3x^2 = 3y$, so $3y^2 y' + 6x = 3y'$, $3y^2 y' - 3y' = -6x$,

and $y'(3y^2 - 3) = -6x$. Therefore, $y' = -\dfrac{6x}{3(y^2 - 1)} = -\dfrac{2x}{y^2 - 1}$.

41. $x^2 - 4xy - y^2 = 12$ so $2x - 4xy' - 4y - 2yy' = 0$ and $y'(-4x - 2y) = -2x + 4y$.

So $y' = \dfrac{-2(x - 2y)}{-2(2x + y)} = \dfrac{x - 2y}{2x + y}$.

43. $df = f'(x)dx = (2x - 2x^{-3})dx = \left(2x - \dfrac{2}{x^3}\right)dx = \dfrac{2(x^4 - 1)}{x^3}dx$

45. a. $df = f'(x)dx = \dfrac{d}{dx}(2x^2 + 4)^{1/2}\,dx = \dfrac{1}{2}(2x^2 + 4)^{-1/2}(4x) = \dfrac{2x}{\sqrt{2x^2 + 4}}dx$

b. $\Delta f \approx df\big|_{\substack{x=4 \\ dx=0.1}} = \dfrac{2(4)(0.1)}{\sqrt{2(16) + 4}} = \dfrac{0.8}{6} = \dfrac{8}{60} = \dfrac{2}{15}$.

c. $\Delta f = f(4.1) - f(4) = \sqrt{2(4.1)^2 + 4} - \sqrt{2(16) + 4} = 0.1335$

From (b), $\Delta f \approx \dfrac{2}{15} \approx 0.1333$.

47. $f(x) = 2x^3 - 3x^2 - 16x + 3$ and $f'(x) = 6x^2 - 6x - 16$.

a. To find the point(s) on the graph of f where the slope of the tangent line is equal to -4, we solve
$$6x^2 - 6x - 16 = -4,\ 6x^2 - 6x - 12 = 0,\ 6(x^2 - x - 2) = 0$$
$$6(x - 2)(x + 1) = 0$$
and $x = 2$ or $x = -1$. Then $f(2) = 2(2)^3 - 3(2)^2 - 16(2) + 3 = -25$ and

$f(-1) = 2(-1)^3 - 3(-1)^2 - 16(-1) + 3 = 14$ and the points are $(2,-25)$ and $(-1,14)$.

b. Using the point-slope form of the equation of a line, we find

that $\quad y - (-25) = -4(x - 2),\ y + 25 = -4x + 8,$ or $y = -4x - 17$

and $\quad y - 14 = -4(x + 1),$ or $y = -4x + 10$

are the equations of the tangent lines at $(2,-25)$ and $(-1,14)$.

49. $y = (4 - x^2)^{1/2}.\ y' = \frac{1}{2}(4 - x^2)^{-1/2}(-2x) = -\dfrac{x}{\sqrt{4 - x^2}}.$

The slope of the tangent line is obtained by letting $x = 1$, giving

$$m = -\frac{1}{\sqrt{3}} = -\frac{\sqrt{3}}{3}.$$

Therefore, an equation of the tangent line is

$$y - \sqrt{3} = -\frac{\sqrt{3}}{3}(x - 1),\quad \text{or}\quad y = -\frac{\sqrt{3}}{3}x + \frac{4\sqrt{3}}{3}.$$

51. $f(x) = (2x - 1)^{-1};\ f'(x) = -2(2x - 1)^{-2},\ f''(x) = 8(2x - 1)^{-3} = \dfrac{8}{(2x - 1)^3}.$

$$f'''(x) = -48(2x - 1)^4 = -\frac{48}{(2x - 1)^4}.$$

Since $(2x - 1)^4 = 0$ when $x = 1/2$, we see that the domain of f''' is $(-\infty, \frac{1}{2}) \cup (\frac{1}{2}, \infty).$

53. $x = \dfrac{25}{\sqrt{p}} - 1;\quad f'(p) = -\dfrac{25}{2p^{3/2}};\quad E(p) = -\dfrac{p\left(-\frac{25}{2p^{3/2}}\right)}{\frac{25}{p^{1/2}} - 1} = \dfrac{\frac{25}{2p^{1/2}}}{\frac{25 - p^{1/2}}{p^{1/2}}} = \dfrac{25}{2(25 - p^{1/2})}$

Since $\quad E(p) = 1,$

$$2(25 - p^{1/2}) = 25,$$

$$25 - p^{1/2} = \tfrac{25}{2},\quad p^{1/2} = \tfrac{25}{2},\ \text{and}\ p = \tfrac{625}{4}.$$

$E(p) > 1$ and demand is elastic if $p > 156.25$; $E(p) = 1$ and demand is unitary if $p = 156.25$; and $E(p) < 1$ and demand is inelastic, if $p < 156.25.$

55. $p = 9\sqrt[3]{1000 - x}\ ;\quad \sqrt[3]{1000 - x} = \dfrac{p}{9};\quad 1000 - x = \dfrac{p^3}{729};\ x = 1000 - \dfrac{p^3}{729}$

Therefore, $\quad x = f(p) = \dfrac{729,000 - p^3}{729}$ and $\quad f'(x) = -\dfrac{3p^2}{729} = -\dfrac{p^2}{243}$.

Then $\quad E(p) = -\dfrac{p(-\frac{p^2}{243})}{\frac{729,000-p^3}{729}} = \dfrac{3p^3}{729,000 - p^3}$.

So $\quad E(60) = \dfrac{3(60)^3}{729,000 - 60^3} = \dfrac{648,000}{513,000} = \dfrac{648}{513} > 1$, and so demand is elastic.

Therefore, raising the price slightly will cause the revenue to decrease.

57. $N(x) = 1000(1 + 2x)^{1/2}$. $\quad N'(x) = 1000(\frac{1}{2})(1+2x)^{-1/2}(2) = \dfrac{1000}{\sqrt{1+2x}}$.

The rate of increase at the end of the twelfth week is $N'(12) = \dfrac{1000}{\sqrt{25}} = 200$,

or 200 subscribers/week.

59. He can expect to live $f(100) = 46.9[1 + 1.09(100)]^{0.1} \approx 75.0433$, or approximately
75.04 years. $\quad f'(t) = 46.9(0.1)(1 + 1.09t)^{-0.9}(1.09) = 5.1121(1 + 1.09t)^{-0.9}$
So the required rate of change is $f'(100) = 5.1121(1 + 1.09)^{-0.9} = 0.074$, or
approximately 0.07 yr/yr.

61. a. $R(x) = px = (-0.02x + 600)x = -0.02x^2 + 600x$
    b. $R'(x) = -0.04x + 600$
    c. $R'(10,000) = -0.04(10,000) + 600 = 200$ and this says that the sale of the
       10,001st phone will bring a revenue of $200.

CHAPTER 4

EXERCISES 4.1, page 262

1. f is decreasing on $(-\infty,0)$ and increasing on $(0, \infty)$.

3. f is increasing on $(-\infty,-1) \cup (1,\infty)$, and decreasing on $(-1,1)$.

5. f is increasing on $(0,2)$ and decreasing on $(-\infty,0) \cup (2,\infty)$.

7. f is decreasing on $(-\infty,-1) \cup (1,\infty)$ and increasing on $(-1,1)$.

9. Increasing on $(20.2, 20.6) \cup (21.7, 21.8)$, constant on $(19.6, 20.2) \cup (20.6, 21.1)$, and decreasing on $(21.1, 21.7) \cup (21.8, 22.7)$,

11. $f(x) = 3x + 5; f'(x) = 3 > 0$ for all x and so f is increasing on $(-\infty,\infty)$.

13. $f(x) = x^2 - 3x.$ $f'(x) = 2x - 3$ is continuous everywhere and is equal to zero when $x = 3/2$. From the following sign diagram

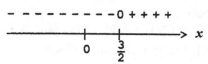

we see that f is decreasing on $(-\infty, \frac{3}{2})$ and increasing on $(\frac{3}{2}, \infty)$.

15. $g(x) = x - x^3.$ $g'(x) = 1 - 3x^2$ is continuous everywhere and is equal to zero when $1 - 3x^2 = 0$, or $x = \pm\frac{\sqrt{3}}{3}$. From the following sign diagram

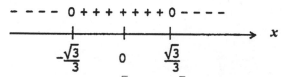

we see that f is decreasing on $(-\infty, -\frac{\sqrt{3}}{3}) \cup (\frac{\sqrt{3}}{3}, \infty)$ and increasing on $(-\frac{\sqrt{3}}{3}, \frac{\sqrt{3}}{3})$.

17. $g(x) = x^3 + 3x^2 + 1$; $g'(x) = 3x^2 + 6x = 3x(x + 2)$.
    From the following sign diagram

$$+ + + + + + \; 0 \; - - \; -0 \; + + + + +$$

we see that g is increasing on $(-\infty,-2) \cup (0,\infty)$ and decreasing on $(-2,0)$.

19. $f(x) = \frac{1}{3}x^3 - 3x^2 + 9x + 20$; $f'(x) = x^2 - 6x + 9 = (x - 3)^2 > 0$ for all x except
    $x = 3$, at which point $f'(3) = 0$. Therefore, f is increasing on $(-\infty,3) \cup (3,\infty)$.

21. $h(x) = x^4 - 4x^3 + 10$; $h'(x) = 4x^3 - 12x^2 = 4x^2(x - 3)$ if $x = 0$ or 3. From the sign
    diagram of h',

$$- \; - \; - \; - \; - \; - \; 0 \; - \; - \; - \; - \; - \; -0 \; + + + + +$$

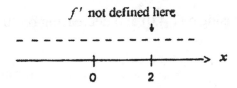

we see that h is increasing on $(3,\infty)$ and decreasing on $(-\infty,0) \cup (0,3)$.

23. $f(x) = \dfrac{1}{x-2} = (x-2)^{-1}$. $f'(x) = -1(x-2)^{-2}(1) = -\dfrac{1}{(x-2)^2}$ is discontinuous at
    $x = 2$ and is continuous everywhere else. From the sign diagram

$$f' \text{ not defined here}$$
$$\downarrow$$
$$- \; - \; - \; - \; - \; - \; - \; - \; - \; - \; - \; -$$

we see that f is decreasing on $(-\infty,2) \cup (2,\infty)$.

25. $h(t) = \dfrac{t}{t-1}$. $h'(t) = \dfrac{(t-1)(1) - t(1)}{(t-1)^2} = -\dfrac{1}{(t-1)^2}$.
    From the following sign diagram,

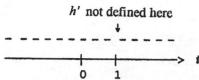

we see that $h'(t) < 0$ whenever it is defined. We conclude that h is decreasing on $(-\infty,1) \cup (1,\infty)$.

27. $f(x) = x^{3/5}$. $f'(x) = \dfrac{3}{5}x^{-2/5} = \dfrac{3}{5x^{2/5}}$. Observe that $f'(x)$ is not defined at $x = 0$, but is positive everywhere else and therefore increasing on $(-\infty,0) \cup (0,\infty)$.

29. $f(x) = \sqrt{x+1}$. $f'(x) = \dfrac{d}{dx}(x+1)^{1/2} = \dfrac{1}{2}(x+1)^{-1/2} = \dfrac{1}{2\sqrt{x+1}}$ and we see that $f'(x) > 0$ if $x > -1$. Therefore, f is increasing on $(-1, \infty)$.

31. $f(x) = \sqrt{16-x^2} = (16-x^2)^{1/2}$. $f'(x) = \dfrac{1}{2}(16-x^2)^{-1/2}(-2x) = -\dfrac{x}{\sqrt{16-x^2}}$.

Since the domain of f is $[-4,4]$, we consider the sign diagram for f' on this interval. Thus,

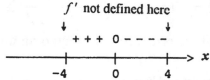

and we see that f is increasing on $(-4,0)$ and decreasing on $(0,4)$.

33. $f'(x) = \dfrac{d}{dx}(x - x^{-1}) = 1 + \dfrac{1}{x^2} = \dfrac{x^2+1}{x^2}$ and so $f'(x) > 0$ for all $x \ne 0$.

Therefore, f is increasing on $(-\infty,0) \cup (0,\infty)$.

35. $f'(x) = \dfrac{d}{dx}(x-1)^{-2} = -2(x-1)^{-3} = -\dfrac{2}{(x-1)^3}$. From the sign diagram of f'

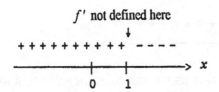

f' not defined here

we see that f is increasing on $(-\infty,1)$ and decreasing on $(1,\infty)$.

37. f has a relative maximum of $f(0) = 1$ and relative minima of $f(-1) = 0$ and $f(1) = 0$.

39. f has a relative maximum of $f(-1) = 2$ and a relative minimum of $f(1) = -2$.

41. f has a relative maximum of $f(1) = 3$ and a relative minimum of $f(2) = 2$.

43. f has a relative minimum at (0,2).

45. a   47. d

49. $f(x) = x^2 - 4x$. $f'(x) = 2x - 4 = 2(x - 2)$ has a critical point at $x = 2$. From the following sign diagram

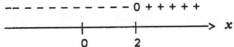

we see that $f(2) = -4$ is a relative minimum by the First Derivative Test.

51. $h(t) = -t^2 + 6t + 6$; $h'(t) = -2t + 6 = -2(t - 3) = 0$ if $t = 3$, a critical point. The sign diagram

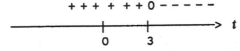

and the First Derivative Test imply that h has a relative maximum at 3 with value $f(3) = -9 + 18 + 6 = 15$.

53. $f(x) = x^{5/3}$. $f'(x) = \frac{5}{3}x^{2/3}$ giving $x = 0$ as the critical point of f.

From the sign diagram

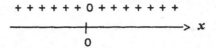

we see that f' does not change sign as we move across $x = 0$, and conclude that f has no relative extremum.

55. $g(x) = x^3 - 3x^2 + 4.$ $g'(x) = 3x^2 - 6x = 3x(x - 2) = 0$ if $x = 0$ or 2. From the sign

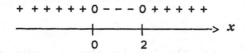

diagram, we see that the critical point $x = 0$ gives a relative maximum, whereas, $x = 2$ gives a relative minimum. The values are $g(0) = 4$ and $g(2) = 8 - 12 + 4 = 0.$

57. $f(x) = \frac{1}{2}x^4 - x^2.$ $f'(x) = 2x^3 - 2x = 2x(x^2 - 1) = 2x(x + 1)(x - 1)$ is continuous everywhere and has zeros as $x = -1$, $x = 0$, and $x = 1$, the critical points of f. Using the First Derivative Test and the following sign diagram of f'

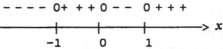

we see that $f(-1) = -1/2$ and $f(1) = -1/2$ are relative minima of f and $f(0) = 0$ is a relative maximum of f.

59. $F(x) = \frac{1}{3}x^3 - x^2 - 3x + 4.$ Setting $F'(x) = x^2 - 2x - 3 = (x - 3)(x + 1) = 0$ gives $x = -1$ and $x = 3$ as critical points. From the sign diagram

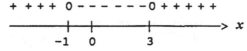

we see that $x = -1$ gives a relative maximum and $x = 3$ gives a relative minimum. The values are
$$F(-1) = -\tfrac{1}{3} - 1 + 3 + 4 = \tfrac{17}{3} \quad \text{and} \quad F(3) = 9 - 9 - 9 + 4 = -5,$$
respectively.

61. $g(x) = x^4 - 4x^3 + 8$. Setting $g'(x) = 4x^3 - 12x^2 = 4x^2(x - 3) = 0$ gives $x = 0$ and $x = 3$ as critical points. From the sign diagram

```
- - - - - 0 - - -  0 + + + + +
_____> x
        +         +
        0         3
```

we see that $x = 3$ gives a relative minimum. Its value is $g(3) = 3^4 - 4(3)^3 + 8 = -19$.

63. $g'(x) = \dfrac{d}{dx}\left(1 + \dfrac{1}{x}\right) = -\dfrac{1}{x^2}$. Observe that g' is never zero for all values of x.

Furthermore, g' is undefined at $x = 0$, but $x = 0$ is not in the domain of g. Therefore g has no critical points and so g has no relative extrema.

65. $f(x) = x + \dfrac{9}{x} + 2$. Setting $f'(x) = 1 - \dfrac{9}{x^2} = \dfrac{x^2 - 9}{x^2} = \dfrac{(x+3)(x-3)}{x^2} = 0$

gives $x = -3$ and $x = 3$ as critical points. From the sign diagram

```
            f' not defined here
                    ↓
+ + + + 0 - - - -  0  - - -0+ + + + +
_____> x
        +          +    +
       -3          0    3
```

we see that $(-3,-4)$ is a relative maximum and $(3,8)$ is a relative minimum.

67. $f(x) = \dfrac{x}{1+x^2}$. $f'(x) = \dfrac{(1+x^2)(1) - x(2x)}{(1+x^2)^2} = \dfrac{1-x^2}{(1+x^2)^2} = \dfrac{(1-x)(1+x)}{(1+x^2)^2} = 0$ if $x = \pm 1$,

and these are critical points of f. From the sign diagram of f'

```
- - - - 0+ + +  + + + 0- - - -
_____> x
       +       +     +
      -1       0     1
```

we see that f has a relative minimum at $(-1, -\tfrac{1}{2})$ and a relative maximum at $(1, \tfrac{1}{2})$.

69. $f(x) = \dfrac{x^2}{x^2 - 4}$. $f'(x) = \dfrac{(x^2 - 4)(2x) - x^2(2x)}{(x^2 - 4)^2} = -\dfrac{8x}{(x^2 - 4)^2}$ is continuous

everywhere except at $x \pm 2$ and has a zero at $x = 0$. Therefore, $x = 0$ is the only critical point of f (the points $x = \pm 2$ do not lie in the domain of f).Using the following sign diagram of f'

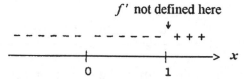

f' not defined here

and the First Derivative Test, we conclude that $f(0) = 0$ is a relative maximum of f.

71. $f(x) = (x - 1)^{2/3}$. $f'(x) = \dfrac{2}{3}(x-1)^{-1/3} = \dfrac{2}{3(x-1)^{1/3}}$.

$f'(x)$ is discontinuous at $x = 1$. The sign diagram for f' is

f' not defined here

We conclude that $f(1) = 0$ is a relative minimum.

73. $h(t) = -16t^2 + 64t + 80$. $h'(t) = -32t + 64 = -32(t - 2)$ and has sign diagram

This tells us that the stone is rising on the time interval $(0,2)$ and falling when $t > 2$. It hits the ground when $h(t) = -16t^2 + 64t + 80 = 0$
or $t^2 - 4t - 5 = (t - 5)(t + 1) = 0$ or $t = 5$ (we reject the root $t = -1$.)

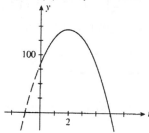

75. $P'(x) = \dfrac{d}{dx}(0.0726x^2 + 0.7902x + 4.9623) = 0.1452x + 0.7902$.

Since $P'(x) > 0$ on $(0, 25)$, we see that P is increasing on the interval in question. Our result tells us that the percent of the population afflicted with Alzheimer's disease increases with age for those that are 65 and over.

77. $I(t) = \frac{1}{3}t^3 - \frac{5}{2}t^2 + 80$; $I'(t) = t^2 - 5t = t(t - 5) = 0$ if $t = 0$ or 5. From the sign

```
        - - - - - -0 + + + + + +
  ┼────────────┼───────────┼──────────> t
  0            5          10
```

diagram, we see that I is decreasing on $(0,5)$ and increasing on $(5,10)$. After declining from 1984 through 1989, the index begins to increase after 1989.

79. $\overline{C}(x) = -0.0001x + 2 + \dfrac{2000}{x}$. $\overline{C}'(x) = -0.0001 - \dfrac{2000}{x^2} < 0$ for all values of x and

so $\overline{C}$ is always decreasing.

81. $A(t) = -96.6t^4 + 403.6t^3 + 660.9t^2 + 250$
$A'(t) = -386.4t^3 + 1210.8t^2 + 1321.8t = t(386.4t^2 + 1210.8t + 1321.8)$.
Solving $A'(t) = 0$, we find $t = 0$ and

$$t = \frac{-1210.8 \pm \sqrt{(1210.8)^2 - 4(-386.4)(1321.8)}}{-2(386.4)} = \frac{-1210.8 \pm 1873.2}{-2(386.4)} \approx 4.$$

Since t lies in the interval $[0,5]$, we see that the continuous function A' has zeros at $t = 0$ and $t = 4$. From the sign diagram

```
        + + + + 0 - -
  ┼────────────┼─────┼───┼────> t
  0            4     5
```

we see that f is increasing on $(0,4)$ and decreasing on $(4,5)$. We conclude that the cash in the Central Provident Trust Funds will be increasing from 1995 to 2035 and decreasing from 2035 to 2045.

83. a. $f'(t) = \dfrac{d}{dt}(-0.05t^3 + 0.56t^2 + 5.47t + 7.5) = -0.15t^2 + 1.12t + 5.47.$

Setting $f'(t) = 0$ gives $-0.15t^2 + 1.12t + 5.47 = 0$. Using the quadratic formula, we find

$$t = \frac{-1.12 \pm \sqrt{(1.12)^2 - 4(-0.15)(5.47)}}{-0.3}$$

that is, $t = -3.37$, or 10.83. Since f' is continuous, the only critical points of f are $t = -3.4$ and $t = 10.8$, both of which lie outside the interval of interest. Nevertheless this result can be used to tell us that f' does not change sign in the interval $(-3.4, 10.8)$. Using $t = 0$ as the test point, we see that $f'(0) = 5.47 > 0$ and so we see that f is increasing on $(-3.4, 10.8)$, and , in particular, in the interval $(0, 6)$. Thus, we conclude that f is increasing on $(0, 6)$.

b. The result of part (a) tells us that sales in the Web-hosting industry will be increasing throughout the years from 1999 through 2005.

85. a. $N'(t) = \dfrac{d}{dt}(0.09444t^3 - 1.44167t^2 + 10.65695t + 52)$

   $= (0.28332t^2 - 2.88334t + 10.65695)$

Observe that N' is continuous everywhere. Setting $N'(t) = 0$ gives

$$t = \frac{2.88334 \pm \sqrt{2.88334^2 - 4(0.28332)(10.65695)}}{2(0.28332)}$$

Since the expression under the radical is $-3.76 < 0$, we see that there is no solution. Now, $N'(0) = 10.65695 > 0$, and we conclude that N is always increasing on $(0, 6)$.

b. The result of part (a) shows that the number of subscribers is always increasing over the period in question.

87. $C(t) = \dfrac{t^2}{2t^3 + 1}$; $C'(t) = \dfrac{(2t^3 + 1)(2t) - t^2(6t^2)}{(2t^3 + 1)^2} = \dfrac{2t - 2t^4}{(2t^3 + 1)^2} = \dfrac{2t(1 - t^3)}{(2t^3 + 1)^2}$.

From the sign diagram of C' on $(0, \infty)$,

```
        + + 0 - - - - - - -
   ────────┼──┼──────────┼──→ t
           0  1          4
```

We see that the drug concentration is increasing on $(0,1)$ and decreasing on $(1,4)$.

89. $A(t) = \dfrac{136}{1 + 0.25(t - 4.5)^2} + 28.$

$$A'(t) = 136\frac{d}{dt}[1+0.25(t-4.5)^2]^{-1} = -136[1+0.25(t-4.5)^2]^{-2}2(0.25)(t-4.5)$$

$$= -\frac{68(t-4.5)}{[1+0.25(t-4.5)^2]^2}.$$

Observe that $A'(t) > 0$ if $t < 4.5$ and $A'(t) < 0$ if $t > 4.5$, so the pollution is increasing from 7 A.M. to 11:30 A.M. and decreasing from 11:30 A.M. to 6 P.M.

91. We compute $f'(x) = m$. If $m > 0$, then $f'(x) > 0$ for all x and f is increasing; if $m < 0$, then $f'(x) < 0$ for all x and f is decreasing; if $m = 0$, then $f'(x) = 0$ for all x and f is a constant function.

93. False. The function $f(x) = \begin{cases} -x+1 & x < 0 \\ -\dfrac{1}{2}x+1 & x \geq 0 \end{cases}$

is decreasing on $(-1,1)$, but $f'(0)$ does not exist.

95. False. Let $f(x) = -x$ and $g(x) = -2x$. then both f and g are decreasing on $(-\infty, \infty)$, but $f(x) - g(x) = x - (-2x) = x$ is increasing on $(-\infty, \infty)$.

97. False. Let $f(x) = x^3$. then $f'(0) = 3x^2\big|_{x=0} = 0$. But f does not have a relative extremum at $x = 0$.

99. $f'(x) = 3x^2 + 1$ is continuous on $(-\infty, \infty)$ and is always greater than or equal to 1. So f has no critical points in $(-\infty, \infty)$. Therefore f has no relative extrema on $(-\infty, \infty)$.

101. a. $f'(x) = -2x$ if $x \neq 0$. $f'(-1) = 2$ and $f'(1) = -2$ so $f'(x)$ changes sign from positive to negative as we move across $x = 0$.
b. f does not have a relative maximum at $x = 0$ because $f(0) = 2$ but a neighborhood of $x = 0$, for example $(-\frac{1}{2}, \frac{1}{2})$, contains points with values larger than 2. This does not contradict the First Derivative Test because f is not

continuous at $x = 0$.

103. $f(x) = ax^2 + bx + c$. Setting $f'(x) = 2ax + b = 2a(x + \frac{b}{2a}) = 0$ gives $x = -\frac{b}{2a}$ as the only critical point of f. If $a < 0$, we have the sign diagram

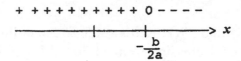

from which we see that $x = -b/2a$ gives a relative maximum. Similarly, you can show that if $a > 0$, then $x = -b/2a$ gives a relative minimum.

105. a. $f'(x) = 3x^2 + 1$ and so $f'(x) > 1$ on the interval $(0,1)$. Therefore, f is increasing on $(0,1)$.
b. $f(0) = -1$ and $f(1) = 1 + 1 - 1 = 1$. So the Intermediate Value Theorem guarantees that there is at least one root of $f(x) = 0$ in $(0,1)$. Since f is increasing on $(0,1)$, the graph of f can cross the x-axis at only one point in $(0,1)$. So $f(x) = 0$ has exactly one root.

USING TECHNOLOGY EXERCISES 4.1, page 267

1. a. f is decreasing on $(-\infty,-0.2934)$ and increasing on $(-0.2934,\infty)$.
   b. Relative minimum: $f(-0.2934) = -2.5435$

3. a. f is increasing on $(-\infty,-1.6144) \cup (0.2390,\infty)$ and decreasing on $(-1.6144, 0.2390)$
   b. Relative maximum: $f(-1.6144) = 26.7991$; relative minimum: $f(0.2390) = 1.6733$

5. a. f is decreasing on $(-\infty,-1) \cup (0.33,\infty)$ and increasing on $(-1,0.33)$
   b. Relative maximum: $f(0.33) = 1.11$; relative minimum: $f(-1) = -0.63$

7. a. f is decreasing on $(-1,-0.71)$ and increasing on $(-0.71,1)$.
   b. f has a relative minimum at $(-0.71,-1.41)$.

9. a.

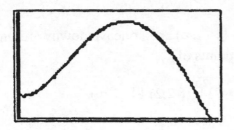

  b. *f* is decreasing on (0,0.2398) ∪ (6.8758,12) and increasing on (0.2398,6.8758)
  c. (6.8758, 200.14); The rate at which the number of banks were failing reached a
     peak of 200/yr during the latter part of 1988 (*t* = 6.8758).

11. a.

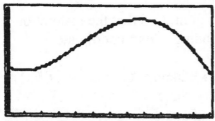

  b. *f* is decreasing on (0, 0.8343) ∪ (7.6726, 12) and increasing on
  (0.8343, 7.6726).   The rate at which single-family homes in the greater Boston
  area were selling was decreasing during most of 1984, but started increasing in
  late 1984 and continued increasing until mid 1991 when it started decreasing again
  until 1996.

13. *f* is decreasing on the interval (0,1) and increasing on (1,4). The relative minimum
    occurs at the point (1,32). These results indicate that the speed of traffic flow drops
    between 6 A.M. and 7 A.M. reaching a low of 32 mph. Thereafter, it increases till
    10 A.M.

EXERCISES 4.2, page 280

1. *f* is concave downward on (-∞,0) and concave upward on (0,∞). *f* has an inflection

point at (0,0).

3. f is concave downward on $(-\infty,0) \cup (0,\infty)$.

5. f is concave upward on $(-\infty,0) \cup (1,\infty)$ and concave downward on $(0,1)$. $(0,0)$ and $(1,-1)$ are inflection points of f.

7. f is concave downward on $(-\infty,-2) \cup (-2,2) \cup (2,\infty)$.

9. a   11. b

13. a. $D_1'(t) > 0$, $D_2'(t) > 0$, $D_1''(t) > 0$, and $D_2''(t) < 0$ on $(0,12)$.
   b. With or without the proposed promotional campaign, the deposits will increase, but with the promotion, the deposits will increase at an increasing rate whereas without the promotion, the deposits will increase at a decreasing rate.

15. The significance of the inflection point Q is that the restoration process is working at its peak at the time t_0 corresponding to its t-coordinate.

17. $f(x) = 4x^2 - 12x + 7$. $f'(x) = 8x - 12$ and $f''(x) = 8$. So, $f''(x) > 0$ everywhere and therefore f is concave upward everywhere.

19. $f(x) = \dfrac{1}{x^4} = x^{-4}$; $f'(x) = -\dfrac{4}{x^5}$ and $f''(x) = \dfrac{20}{x^6} > 0$ for all values of x in $(-\infty,0) \cup (0,\infty)$ and so f is concave upward everywhere.

21. $f(x) = 2x^2 - 3x + 4$; $f'(x) = 4x - 3$ and $f''(x) = 4 > 0$ for all values of x. So f is concave upward on $(-\infty,\infty)$.

23. $f(x) = x^3 - 1$. $f'(x) = 3x^2$ and $f''(x) = 6x$. The sign diagram of f'' follows.

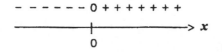

We see that f is concave downward on $(-\infty,0)$ and concave upward on $(0,\infty)$.

25. $f(x) = x^4 - 6x^3 + 2x + 8$; $f'(x) = 4x^3 - 18x^2 + 2$ and $f''(x) = 12x^2 - 36x = 12x(x - 3)$.

The sign diagram of f''

$$+ + + + + + 0 - - - - - 0 + + +$$

shows that f is concave upward on $(-\infty,0) \cup (3,\infty)$ and concave downward on $(0,3)$.

27. $f(x) = x^{4/7}$. $f'(x) = \dfrac{4}{7}x^{-3/7}$ and $f''(x) = -\dfrac{12}{49}x^{-10/7} = -\dfrac{12}{49x^{10/7}}$.

Observe that $f''(x) < 0$ for all x different from zero. So f is concave downward on $(-\infty,0) \cup (0,\infty)$.

29. $f(x) = (4-x)^{1/2}$. $f'(x) = \dfrac{1}{2}(4-x)^{-1/2}(-1) = -\dfrac{1}{2}(4-x)^{-1/2}$;

$f''(x) = \dfrac{1}{4}(4-x)^{-3/2}(-1) = -\dfrac{1}{4(4-x)^{3/2}} < 0$.

whenever it is defined. So f is concave downward on $(-\infty,4)$.

31. $f'(x) = \dfrac{d}{dx}(x-2)^{-1} = -(x-2)^{-2}$ and $f''(x) = 2(x-2)^{-3} = \dfrac{2}{(x-2)^3}$.

The sign diagram of f'' shows that f is concave downward on $(-\infty,2)$ and concave upward on $(2,\infty)$.

$$f'' \text{ is not defined here}$$
$$\downarrow$$
$$- - - - - - - - - + + + + +$$

33. $f'(x) = \dfrac{d}{dx}(2+x^2)^{-1} = -(2+x^2)^{-2}(2x) = -2x(2+x^2)^{-2}$ and

$f''(x) = -2(2+x^2)^{-2} - 2x(-2)(2+x^2)^{-3}(2x)$

$= 2(2+x^2)^{-3}[-(2+x^2)+4x^2] = \dfrac{2(3x^2-2)}{(2+x^2)^3} = 0$ if $x = \pm\sqrt{2/3}$.

145

From the sign diagram of f''

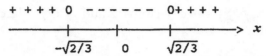

we see that f is concave upward on $(-\infty, -\sqrt{2/3}\,) \cup (\sqrt{2/3}\,, \infty)$ and concave downward on $(-\sqrt{2/3}\,, \sqrt{2/3}\,)$.

35. $h(t) = \dfrac{t^2}{t-1}$; $h'(t) = \dfrac{(t-1)(2t) - t^2(1)}{(t-1)^2} = \dfrac{t^2 - 2t}{(t-1)^2}$;

$h''(t) = \dfrac{(t-1)^2(2t-2) - (t^2 - 2t)2(t-1)}{(t-1)^4}$

$= \dfrac{(t-1)(2t^2 - 4t + 2 - 2t^2 + 4t)}{(t-1)^4} = \dfrac{2}{(t-1)^3}$.

The sign diagram of h'' is

h'' is not defined here
$\downarrow$

$-\ -\ -\ -\ -\ -\qquad +\ +\ +\ +\ +\ +$

$\xrightarrow{} t$
$\qquad\qquad 0 \qquad 1$

and tells us that h is concave downward on $(-\infty, 1)$ and concave upward on $(1, \infty)$.

37. $g(x) = x + \dfrac{1}{x^2}$. $g'(x) = 1 - 2x^{-3}$ and $g''(x) = 6x^{-4} = \dfrac{6}{x^4} > 0$ whenever $x \neq 0$.

Therefore, g is concave upward on $(-\infty, 0) \cup (0, \infty)$.

39. $g(t) = (2t - 4)^{1/3}$. $g'(t) = \ = \dfrac{1}{3}(2t - 4)^{-2/3}(2) = \dfrac{2}{3}(2t - 4)^{-2/3}$.

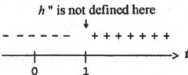

$g''(t) = -\dfrac{4}{9}(2t - 4)^{-5/3} = -\dfrac{4}{9(2t-4)^{5/3}}$. The sign diagram of g'' tells us that g is concave upward on $(-\infty, 2)$ and concave downward on $(2, \infty)$.

41. $f(x) = x^3 - 2$. $f'(x) = 3x^2$ and $f''(x) = 6x$. $f''(x)$ is continuous everywhere and has a zero at $x = 0$. From the sign diagram of f''

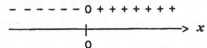

we conclude that $(0,-2)$ is an inflection point of f.

43. $f(x) = 6x^3 - 18x^2 + 12x - 15$; $f'(x) = 18x^2 - 36x + 12$ and $f''(x) = 36x - 36 = 36(x - 1) = 0$ if $x = 1$. The sign diagram of f''

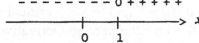

tells us that f has an inflection point at $(1,-15)$.

45. $f(x) = 3x^4 - 4x^3 + 1$. $f'(x) = 12x^3 - 12x^2$ and $f''(x) = 36x^2 - 24x = 12x(3x - 2) = 0$ if $x = 0$ or $2/3$. These are candidates for inflection points. The sign diagram of f''

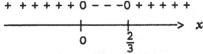

shows that $(0,1)$ and $(\frac{2}{3},\frac{11}{27})$ are inflection points of f.

47. $g(t) = t^{1/3}$, $g'(t) = \frac{1}{3}t^{-2/3}$ and $g''(t) = -\frac{2}{9}t^{-5/3} = -\dfrac{2}{9t^{5/3}}$. Observe that $t = 0$ is in the domain of g. Next, since $g''(t) > 0$ if $t < 0$ and $g''(t) < 0$, if $t > 0$, we see that $(0,0)$ is an inflection point of g.

49. $f(x) = (x - 1)^3 + 2$. $f'(x) = 3(x - 1)^2$ and $f''(x) = 6(x - 1)$. Observe that $f''(x) < 0$ if $x < 1$ and $f''(x) > 0$ if $x > 1$ and so $(1,2)$ is an inflection point of f.

51. $f(x) = \dfrac{2}{1+x^2} = 2(1+x^2)^{-1}$. $f'(x) = -2(1+x^2)^{-2}(2x) = -4x(1+x^2)^{-2}$.

$$f''(x) = -4(1+x^2)^{-2} - 4x(-2)(1+x^2)^{-3}(2x)$$

$$= 4(1+x^2)^{-3}[-(1+x^2)+4x^2] = \frac{4(3x^2-1)}{(1+x^2)^3},$$

is continuous everywhere and has zeros at $x = \pm\frac{\sqrt{3}}{3}$. From the sign diagram of f'' we conclude that $(-\frac{\sqrt{3}}{3}, \frac{3}{2})$ and $(\frac{\sqrt{3}}{3}, \frac{3}{2})$ are inflection points of f.

53. $f(x) = -x^2 + 2x + 4$ and $f'(x) = -2x + 2$. The critical point of f is $x = 1$. Since $f''(x) = -2$ and $f''(1) = -2 < 0$, we conclude that $f(1) = 5$ is a relative maximum of f.

55. $f(x) = 2x^3 + 1$; $f'(x) = 6x^2 = 0$ if $x = 0$ and this is a critical point of f. Next, $f''(x) = 12x$ and so $f''(0) = 0$. Thus, the Second Derivative Test fails. But the First Derivative Test shows that $(0,0)$ is not a relative extremum.

57. $f(x) = \frac{1}{3}x^3 - 2x^2 - 5x - 10$. $f'(x) = x^2 - 4x - 5 = (x - 5)(x + 1)$ and this gives $x = -1$ and $x = 5$ as critical points of f. Next, $f''(x) = 2x - 4$. Since $f''(-1) = -6 < 0$, we see that $(-1, -\frac{22}{3})$ is a relative maximum. Next, $f''(5) = 6 > 0$ and this shows that $(5, -\frac{130}{3})$ is a relative minimum.

59. $g(t) = t + \dfrac{9}{t}$. $g'(t) = 1 - \dfrac{9}{t^2} = \dfrac{t^2-9}{t^2} = \dfrac{(t+3)(t-3)}{t^2}$ and this shows that $t = \pm 3$ are critical points of g. Now, $g''(t) = 18t^{-3} = \dfrac{18}{t^3}$. Since $g''(-3) = -\dfrac{18}{27} < 0$ the Second Derivative Test implies that g has a relative maximum at $(-3,-6)$. Also, $g''(3) = \dfrac{18}{27} > 0$ and so g has a relative minimum at $(3,6)$.

61. $f(x) = \dfrac{x}{1-x}$. $f'(x) = \dfrac{(1-x)(1) - x(-1)}{(1-x)^2} = \dfrac{1}{(1-x)^2}$ is never zero.

So there are no critical points and f has no relative extrema.

63. $f(t) = t^2 - \dfrac{16}{t}$. $f'(t) = 2t + \dfrac{16}{t^2} = \dfrac{2t^3 + 16}{t^2} = \dfrac{2(t^3+8)}{t^2}$. Setting

$f'(t) = 0$ gives $t = -2$ as a critical point. Next, we compute

$$f''(t) = \frac{d}{dt}(2t + 16t^{-2}) = 2 - 32t^{-3} = 2 - \frac{32}{t^3}. \text{ Since } f''(-2) = 2 - \frac{32}{(-8)} = 6 > 0, \text{ we}$$

see that $(-2, 12)$ is a relative minimum.

65. $g(s) = \dfrac{s}{1+s^2}$; $g'(s) = \dfrac{(1+s^2)(1) - s(2s)}{(1+s^2)^2} = \dfrac{1-s^2}{(1+s^2)^2} = 0$ gives $s = -1$ and $s = 1$

as critical points of g. Next, we compute

$$g''(s) = \frac{(1+s^2)^2(-2s) - (1-s^2)2(1+s^2)(2s)}{(1+s^2)^4}$$

$$= \frac{2s(1+s^2)(-1-s^2-2+2s^2)}{(1+s^2)^4} = \frac{2s(s^2-3)}{(1+s^2)^3}.$$

Now, $g''(-1) = \frac{1}{2} > 0$ and so $g(-1) = -\frac{1}{2}$ is a relative minimum of g. Next, $g''(1) = -\frac{1}{2} < 0$ and so $g(1) = \frac{1}{2}$ is a relative maximum of g.

67. $f(x) = \dfrac{x^4}{x-1}$.

$$f'(x) = \frac{(x-1)(4x^3) - x^4(1)}{(x-1)^2} = \frac{4x^4 - 4x^3 - x^4}{(x-1)^2} = \frac{3x^4 - 4x^3}{(x-1)^2} = \frac{x^3(3x-4)}{(x-1)^2}$$

and so $x = 0$ and $x = 4/3$ are critical points of f. Next,

$$f''(x) = \frac{(x-1)^2(12x^3 - 12x^2) - (3x^4 - 4x^3)(2)(x-1)}{(x-1)^4}$$

$$= \frac{(x-1)(12x^4 - 12x^3 - 12x^3 + 12x^2 - 6x^4 + 8x^3)}{(x-1)^4}$$

$$= \frac{6x^4 - 16x^3 + 12x^2}{(x-1)^3} = \frac{2x^2(3x^2 - 8x + 6)}{(x-1)^3}.$$

Since $f''(\frac{4}{3}) > 0$, we see that $f(\frac{4}{3}) = \frac{256}{27}$ is a relative minimum. Since $f''(0) = 0$, the Second Derivative Test fails. Using the sign diagram for f',

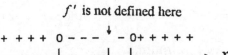

$$+ + + +\ 0\ -\ -\ -\quad-\ 0+ + + + +$$

and the First Derivative Test, we see that $f(0) = 0$ is a relative maximum.

69.

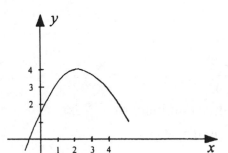

71.

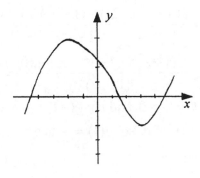

73.

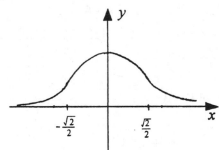

75. a. $N'(t)$ is positive because N is increasing on $(0,12)$.

b. $N''(t) < 0$ on $(0,6)$ and $N''(t) > 0$ on $(6,12)$.

c. The rate of growth of the number of help-wanted advertisements was decreasing over the first six months of the year and increasing over the last six months.

77. $f(t)$ increases at an increasing rate until the water level reaches the middle of the vase at which time (and this corresponds to the inflection point of f), $f(t)$ is

increasing at the fastest rate. Though $f(t)$ still increases until the vase is filled, it does so at a decreasing rate.

79. a. $f'(t) = \dfrac{d}{dt}(0.0117t^3 + 0.0037t^2 + 0.7563t + 4.1)$

$= 0.0351t^2 + 0.0074t + 0.7563 \geq 0.7563$

for all t in the interval $[0, 9]$. This shows that f is increasing on $(0, 9)$. It tells us that the projected amount of AMT will keep on increasing over the years in question.

b. $f''(t) = \dfrac{d}{dt}(0.0351t^2 + 0.0074t + 0.7563) = 0.0702t + 0.0074 \geq 0.0074.$

This shows that f' is increasing on $(0, 9)$. Out result tells us that not only is the amount of AMT paid increasing over the period in question, but it is actually accelerating!

81. $S(x) = -0.002x^3 + 0.6x^2 + x + 500;\; S'(x) = -0.006x^2 + 1.2x + 1;$
$S''(x) = -0.012x + 1.2.\; x = 100$ is a candidate for an inflection point of S.
The sign diagram for S'' is

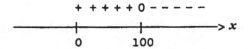

$$+ + + + + \; 0 \; - - - - -$$

We see that $(100, 4600)$ is an inflection point of S.

83. We wish to find the inflection point of the function $N(t) = -t^3 + 6t^2 + 15t$. Now,
$N'(t) = -3t^2 + 12t + 15$ and $N''(t) = -6t + 12 = -6(t - 2)$ giving $t = 2$ as the only candidate for an inflection point of N. From the sign diagram

$$+ + + \; 0 \; - - - - \;-$$

for N'', we conclude that $t = 2$ gives an inflection point of N. Therefore, the average worker is performing at peak efficiency at 10 A.M.

85. $s = f(t) = -t^3 + 54t^2 + 480t + 6.$ The velocity of the rocket is
$v = f'(t) = -3t^2 + 108t + 480$

and its acceleration is $a = f''(t) = -6t + 108 = -6(t - 18)$. From the sign diagram

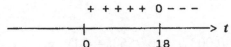

we see that $(18, 20{,}310)$ is an inflection point of f. Our computations reveal that the maximum velocity of the rocket is attained when $t = 18$. The maximum velocity is
$$f'(18) = -3(18)^2 + 108(18) + 480 = 1452, \text{ or } 1452 \text{ ft/sec}$$

87. $A(t) = 1.0974t^3 - 0.0915t^4$. $A'(t) = 3.2922t^2 - 0.366t^3$ and $A''(t) = 6.5844t - 1.098t^2$.
     Setting $A'(t) = 0$, we obtain $t^2(3.2922 - 0.366t) = 0$, and this gives $t = 0$ or
     $t \approx 8.995 \approx 9$. Using the Second Derivative Test, we find
     $A''(9) = 6.5844(9) - 1.098(81) = -29.6784 < 0$, and this tells us that $t \approx 9$ gives rise
     to a relative maximum of A. Our analysis tells us that on that May day, the level of
     ozone peaked at approximately 4 P.M. in the afternoon.

89. True. If f' is increasing on (a,b), then $-f'$ is decreasing on (a,b), and so if the
     graph of f is concave upward on (a,b), the graph of $-f$ must be concave
     downward on (a,b).

91. True. The given conditions imply that $f''(0) < 0$ and the Second Derivative Test
     gives the desired conclusion.

93. $f(x) = ax^2 + bx + c$. $f'(x) = 2ax + b$ and $f''(x) = 2a$. So $f''(x) > 0$ if $a > 0$, and the
     parabola opens upward. If $a < 0$, then $f''(x) < 0$ and the parabola opens downward.

USING TECHNOLOGY EXERCISES 4.2, page 285

1. a. f is concave upward on $(-\infty, 0) \cup (1.1667, \infty)$ and concave downward on
    $(0, 1.1667)$.
    b. $(1.1667, 1.1153)$; $(0,2)$

3. a. f is concave downward on $(-\infty, 0)$ and concave upward on $(0, \infty)$.
    b. $(0,2)$

5. a. f is concave downward on $(-\infty, 0)$ and concave upward on $(0, \infty)$. b. $(0,0)$

7. a. f is concave downward on $(-\infty, -2.4495) \cup (0, 2.4495)$; f is concave upward on $(-2.4495, 0) \cup (2.4495, \infty)$.   b. $(-2.4495, -0.3402)$; $(2.4495, 0.3402)$

9. a.

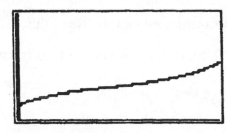

    b. $(5.5318, 35.9483)$
    c. $t = 5.5318$

11. a.

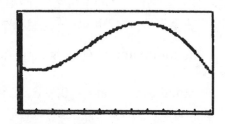

    b. $(3.9024, 77.0919)$;
       sales of houses were increasing
       at the fastest rate in late 1988.

13. a.

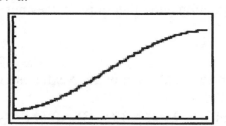

    b. April 1993 $(t = 7.36)$

EXERCISES 4.3, page 296

1. $y = 0$ is a horizontal asymptote.

3. $y = 0$ is a horizontal asymptote and $x = 0$ is a vertical asymptote.

5. $y = 0$ is a horizontal asymptote and $x = -1$ and $x = 1$ are vertical asymptotes.

7. $y = 3$ is a horizontal asymptote and $x = 0$ is a vertical asymptote.

9. $y = 1$ and $y = -1$ are horizontal asymptotes.

11. $\lim\limits_{x\to\infty}\dfrac{1}{x}=0$ and so $y=0$ is a horizontal asymptote. Next, since the numerator of the rational expression is not equal to zero and the denominator is zero at $x=0$, we see that $x=0$ is a vertical asymptote.

13. $f(x)=-\dfrac{2}{x^2}$. $\lim\limits_{x\to\infty}-\dfrac{2}{x^2}=0$, so $y=0$ is a horizontal asymptote. Next, the denominator of $f(x)$ is equal to zero at $x=0$. Since the numerator of $f(x)$ is not

   equal to zero at $x=0$, we see that $x=0$ is a vertical asymptote.

15. $\lim\limits_{x\to\infty}\dfrac{x-1}{x+1}=\lim\limits_{x\to\infty}\dfrac{1-\frac{1}{x}}{1+\frac{1}{x}}=1$, and so $y=1$ is a horizontal asymptote. Next, the denominator is equal to zero at $x=-1$ and the numerator is not equal to zero at this point, so $x=-1$ is a vertical asymptote.

17. $h(x)=x^3-3x^2+x+1$. $h(x)$ is a polynomial function and, therefore, it does not have any horizontal or vertical asymptotes.

19. $\lim\limits_{t\to\infty}\dfrac{t^2}{t^2-9}=\lim\limits_{t\to\infty}\dfrac{1}{1-\frac{9}{t^2}}=1$, and so $y=1$ is a horizontal asymptote. Next, observe

   that the denominator of the rational expression $t^2-9=(t+3)(t-3)=0$ if $t=-3$ and $t=3$. But the numerator is not equal to zero at these points. Therefore, $t=-3$ and $t=3$ are vertical asymptotes.

21. $\lim\limits_{x\to\infty}\dfrac{3x}{x^2-x-6}=\lim\limits_{x\to\infty}\dfrac{\frac{3}{x}}{1-\frac{1}{x}-\frac{6}{x^2}}=0$ and so $y=0$ is a horizontal asymptote. Next,

   observe that the denominator $x^2-x-6=(x-3)(x+2)=0$ if $x=-2$ or $x=3$. But the numerator $3x$ is not equal to zero at these points. Therefore, $x=-2$ and $x=3$ are vertical asymptotes.

23. $\lim\limits_{t\to\infty}\left[2+\dfrac{5}{(t-2)^2}\right]=2$, and so $y=2$ is a horizontal asymptote. Next observe that

$$\lim_{t \to 2^+} g(t) = \lim_{t \to 2^-} \left[2 + \frac{5}{(t-2)^2} \right] = \infty \,, \text{ and so } t = 2 \text{ is a vertical asymptote.}$$

25. $\lim_{x \to \infty} \dfrac{x^2 - 2}{x^2 - 4} = \lim_{x \to \infty} \dfrac{1 - \frac{2}{x^2}}{1 - \frac{4}{x^2}} = 1$ and so $y = 1$ is a horizontal asymptote. Next, observe

that the denominator $x^2 - 4 = (x + 2)(x - 2) = 0$ if $x = -2$ or 2. Since the numerator $x^2 - 2$ is not equal to zero at these points, the lines $x = -2$ and $x = 2$ are vertical asymptotes.

27. $g(x) = \dfrac{x^3 - x}{x(x+1)}$; Rewrite $g(x)$ as $g(x) = \dfrac{x^2 - 1}{x + 1}$ $(x \neq 0)$ and note that

$\lim_{x \to -\infty} g(x) = \lim_{x \to -\infty} \dfrac{x - \frac{1}{x}}{1 + \frac{1}{x}} = -\infty$ and $\lim_{x \to \infty} g(x) = \infty$. Therefore, there are no horizontal

asymptotes. Next, note that the denominator of $g(x)$ is equal to zero at $x = 0$ and $x = -1$. However, since the numerator of $g(x)$ is also equal to zero when $x = 0$, we see that $x = 0$ is not a vertical asymptote. Also, the numerator of $g(x)$ is equal to zero when $x = -1$, so $x = -1$ is not a vertical asymptote.

29. f is the derivative function of the function g. Observe that at a relative maximum (relative minimum) of g, $f(x) = 0$.

31.

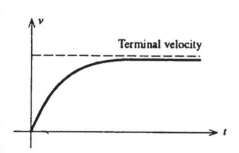

33.

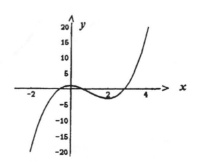

35.

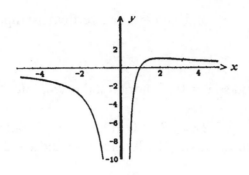

37. $g(x) = 4 - 3x - 2x^3$.

We first gather the following information on the graph of f.

1. The domain of f is $(-\infty, \infty)$.

2. Setting $x = 0$ gives $y = 4$ as the y-intercept. Setting $y = g(x) = 0$ gives a cubic equation which is not easily solved and we will not attempt to find the x-intercepts.

3. $\lim_{x \to -\infty} g(x) = \infty$ and $\lim_{x \to \infty} g(x) = -\infty$. 4. There are no asymptotes of g.

5. $g'(x) = -3 - 6x^2 = -3(2x^2 + 1) < 0$ for all values of x and so g is decreasing on $(-\infty, \infty)$.

6. The results of 5 show that g has no critical points and hence has no relative extrema.

7. $g''(x) = -12x$. Since $g''(x) > 0$ for $x < 0$ and $g''(x) < 0$ for $x > 0$, we see that g is concave upward on $(-\infty, 0)$ and concave downward on $(0, \infty)$.

8. From the results of (7), we see that $(0,4)$ is an inflection point of g.

The graph of g follows.

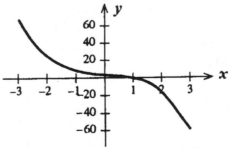

39. $h(x) = x^3 - 3x + 1$

We first gather the following information on the graph of h.

1. The domain of h is $(-\infty, \infty)$.

2. Setting $x = 0$ gives 1 as the y-intercept. We will not find the x-intercept.

3. $\lim\limits_{x \to -\infty} (x^3 - 3x + 1) = -\infty$ and $\lim\limits_{x \to \infty} (x^3 - 3x + 1) = \infty$

4. There are no asymptotes since $h(x)$ is a polynomial.

5. $h'(x) = 3x^2 - 3 = 3(x + 1)(x - 1)$, and we see that $x = -1$ and $x = 1$ are critical points. From the sign diagram

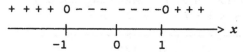

we see that h is increasing on $(-\infty, -1) \cup (1, \infty)$ and decreasing on $(-1, 1)$.

6. The results of (5) shows that $(-1, 3)$ is a relative maximum and $(1, -1)$ is a relative minimum.

7. $h''(x) = 6x$ and $h''(x) < 0$ if $x < 0$ and $h''(x) > 0$ if $x > 0$. So the graph of h is concave downward on $(-\infty, 0)$ and concave upward on $(0, \infty)$.

8. The results of (7) show that $(0, 1)$ is an inflection point of h.
The graph of h follows.

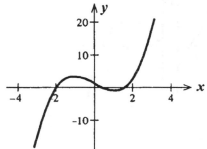

41. $f(x) = -2x^3 + 3x^2 + 12x + 2$

We first gather the following information on the graph of f.

1. The domain of f is $(-\infty, \infty)$.

2. Setting $x = 0$ gives 2 as the y-intercept.

3. $\lim\limits_{x \to -\infty} (-2x^3 + 3x^2 + 12x + 2) = \infty$ and $\lim\limits_{x \to \infty} (-2x^3 + 3x^2 + 12x + 2) = -\infty$

4. There are no asymptotes because $f(x)$ is a polynomial function.

5. $f'(x) = -6x^2 + 6x + 12 = -6(x^2 - x - 2) = -6(x - 2)(x + 1) = 0$ if $x = -1$ or $x = 2$, the critical points of f. From the sign diagram

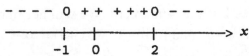

we see that f is decreasing on $(-\infty,-1) \cup (2, \infty)$ and increasing on $(-1,2)$.

6. The results of (5) show that $(-1,-5)$ is a relative minimum and $(2,22)$ is a relative maximum.

7. $f''(x) = -12x + 6 = 0$ if $x = 1/2$. The sign diagram of f''

$$+ + + + + + \ \ + + \ 0 \ - - - -$$

$$\xrightarrow{\hspace{4cm}} x$$
$$\quad\quad 0 \quad\ 1/2$$

shows that the graph of f is concave upward on $(-\infty,1/2)$ and concave downward on $(1/2, \infty)$.

8. The results of (7) show that $(\frac{1}{2}, \frac{17}{2})$ is an inflection point.

The graph of f follows.

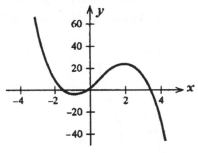

43. $h(x) = \frac{3}{2}x^4 - 2x^3 - 6x^2 + 8$

We first gather the following information on the graph of h.

1. The domain of h is $(-\infty, \infty)$.

2. Setting $x = 0$ gives 8 as the y-intercept.

3. $\lim_{x \to -\infty} h(x) = \lim_{x \to \infty} h(x) = \infty$

4. There are no asymptotes.

5. $h'(x) = 6x^3 - 6x^2 - 12x = 6x(x^2 - x - 2) = 6x(x-2)(x+1) = 0$ if $x = -1, 0,$ or 2,

$$\cdot - - -0 + + 0 - - - \ 0 + \ + +$$

$$\xrightarrow{\hspace{4cm}} x$$
$$\quad\ -1 \quad 0 \quad\quad 2$$

and these are the critical points of h. The sign diagram of h' is
and this tells us that h is increasing on $(-1, 0) \cup (2, \infty)$ and decreasing on

$(-\infty,-1) \cup (0,2)$.

6. The results of (5) show that $(-1, \frac{11}{2})$ and $(2,-8)$ are relative minima of h and $(0,8)$ is a relative maximum of h.

7. $h''(x) = 18x^2 - 12x - 12 = 6(3x^2 - 2x - 2)$. The zeros of h'' are

$$x = \frac{2 \pm \sqrt{4 + 24}}{6} \approx -0.5 \text{ or } 1.2.$$

The sign diagram of h'' is

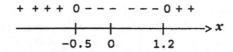

and tells us that the graph of h is concave upward on $(-\infty,-0.5) \cup (1.2, \infty)$ and is concave downward on $(0.5,1.2)$.

8. The results of (7) also show that $(-0.5,6.8)$ and $(1.2,-1)$ are inflection points. The graph of h follows.

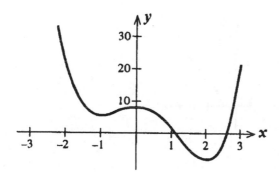

45. $f(t) = \sqrt{t^2 - 4}$.

We first gather the following information on f.

1. The domain of f is found by solving $t^2 - 4 \geq 0$ giving it as $(-\infty,-2] \cup [2,\infty)$.

2. Since $t \neq 0$, there is no y-intercept. Next, setting $y = f(t) = 0$ gives the t-intercepts as -2 and 2.

3. $\displaystyle\lim_{t\to-\infty} f(t) = \lim_{t\to\infty} f(t) = \infty$ 4. There are no asymptotes.

5. $f'(t) = \dfrac{1}{2}(t^2-4)^{-1/2}(2t) = t(t^2-4)^{-1/2} = \dfrac{t}{\sqrt{t^2-4}}.$

Setting $f'(t) = 0$ gives $t = 0$. But $t = 0$ is not in the domain of f and so there are no critical points. The sign diagram for f' is

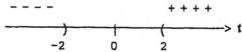

We see that f is increasing on $(2,\infty)$ and decreasing on $(-\infty,-2)$.

6. From the results of (5) we see that there are no relative extrema.

7. $f''(t) = (t^2-4)^{-1/2} + t(-\frac{1}{2})(t^2-4)^{-3/2}(2t) = (t^2-4)^{-3/2}(t^2-4-t^2)$

$$= -\dfrac{4}{(t^2-4)^{3/2}}.$$

8. Since $f''(t) < 0$ for all t in the domain of f, we see that f is concave downward

everywhere. From the results of (7), we see that there are no inflection points. The graph of f follows.

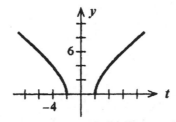

47. $g(x) = \frac{1}{2}x - \sqrt{x}.$

We first gather the following information on g.

1. The domain of g is $[0,\infty)$.

2. The y-intercept is 0. To find the x-intercept, set $y = 0$, giving

$$\tfrac{1}{2}x - \sqrt{x} = 0$$

$$x = 2\sqrt{x}$$

$$x^2 = 4x$$

$$x(x-4) = 0, \text{ and } x = 0 \text{ or } x = 4$$

3. $\lim_{x \to \infty} (\tfrac{1}{2}x - \sqrt{x}) = \lim_{x \to \infty} \tfrac{1}{2}x(1 - \tfrac{2}{\sqrt{x}}) = \infty.$

4. There are no asymptotes.

5. $g'(x) = \tfrac{1}{2} - \tfrac{1}{2}x^{-1/2} = \tfrac{1}{2}x^{-1/2}(x^{1/2} - 1) = \dfrac{\sqrt{x} - 1}{2\sqrt{x}}$

    which is zero when $x = 1$. From the sign diagram for g'

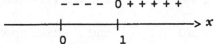

    we see that g is decreasing on $(0,1)$ and increasing on $(1,\infty)$.

6. From the sign diagram of g', we see that $g(1) = -1/2$ is a relative minimum.

7. $g''(x) = (-\tfrac{1}{2})(-\tfrac{1}{2})x^{-3/2} = \dfrac{1}{4x^{3/2}} > 0$ for $x > 0$, and so g is concave upward on

$(0,\infty)$.

8. There are no inflection points.
    The graph of g follows.

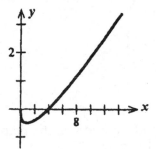

49. $g(x) = \dfrac{2}{x-1}.$ We first gather the following information on g.

1. The domain of g is $(-\infty,1) \cup (1,\infty)$.

2. Setting $x = 0$ gives -2 as the y-intercept. There are no x-intercepts since

    $\dfrac{2}{x-1} \neq 0$ for all values of x.

3. $\lim_{x \to -\infty} \dfrac{2}{x-1} = 0$ and $\lim_{x \to \infty} \dfrac{2}{x-1} = 0$.

4. The results of (3) show that $y = 0$ is a horizontal asymptote. Furthermore, the denominator of $g(x)$ is equal to zero at $x = 1$ but the numerator is not equal to zero

there. Therefore, $x = 1$ is a vertical asymptote.

5. $g'(x) = -2(x-1)^{-2} = -\dfrac{2}{(x-1)^2} < 0$ for all $x \neq 1$ and so g is decreasing on

$(-\infty,1)$ and $(1,\infty)$.

6. Since g has no critical points, there are no relative extrema.

7. $g''(x) = \dfrac{4}{(x-1)^3}$ and so $g''(x) < 0$ if $x < 1$ and $g''(x) > 0$ if $x > 1$. Therefore, the

graph of g is concave downward on $(-\infty,1)$ and concave upward on $(1,\infty)$.

8. Since $g''(x) \neq 0$, there are no inflection points.

The graph of g follows.

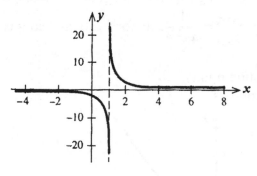

51. $h(x) = \dfrac{x+2}{x-2}$.

We first gather the following information on the graph of h.

1. The domain of h is $(-\infty,2) \cup (2,\infty)$.

2. Setting $x = 0$ gives $y = -1$ as the y-intercept. Next, setting $y = 0$ gives $x = -2$ as the x-intercept.

3. $\displaystyle\lim_{x\to\infty} h(x) = \lim_{x\to-\infty} \dfrac{1+\dfrac{2}{x}}{1-\dfrac{2}{x}} = \lim_{x\to-\infty} h(x) = 1.$

4. Setting $x - 2 = 0$ gives $x = 2$. Furthermore,

$$\lim_{x\to 2^+} \dfrac{x+2}{x-2} = \infty \quad \text{and} \quad \lim_{x\to 2^+} \dfrac{x+2}{x-2} = -\infty$$

So $x = 2$ is a vertical asymptote of h. Also, from the resultsof (3), we see that $y = 1$ is a horizontal asymptote of h.

5. $h'(x) = \dfrac{(x-2)(1)-(x+2)(1)}{(x-2)^2} = -\dfrac{4}{(x-2)^2}$.

We see that there are no critical points of h. (Note $x = 2$ does not belong to the domain of h.) The sign diagram of h' follows.

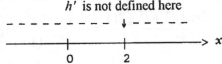

We see that h is decreasing on $(-\infty,2) \cup (2,\infty)$.

6. From the results of (5), we see that there is no relative extremum.

7. $h''(x) = \dfrac{8}{(x-2)^3}$. Note that $x = 2$ is not a candidate for an inflection point

because $h(2)$ is not defined. Since $h''(x) < 0$ for $x < 2$ and $h''(x) > 0$ for $x > 2$, we see that h is concave downward on $(-\infty,2)$ and concave upward on $(2,\infty)$.

8. From the results of (7), we see that there are no inflection points.
The graph of h follows.

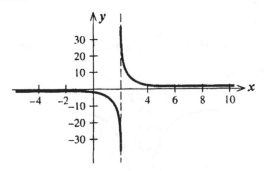

53. $f(t) = \dfrac{t^2}{1+t^2}$.

We first gather the following information on the graph of f.

1. The domain of f is $(-\infty, \infty)$.

2. Setting $t = 0$ gives the y-intercept as 0. Similarly, setting $y = 0$ gives the t-intercept as 0.

3. $\displaystyle\lim_{t \to -\infty} \dfrac{t^2}{1+t^2} = \lim_{t \to \infty} \dfrac{t^2}{1+t^2} = 1$.

4. The results of (3) show that $y = 1$ is a horizontal asymptote. There are no vertical asymptotes since the denominator is not equal to zero.

5. $f'(t) = \dfrac{(1+t^2)(2t) - t^2(2t)}{(1+t^2)^2} = \dfrac{2t}{(1+t^2)^2} = 0$, if $t = 0$, the only critical point of f.

Since $f'(t) < 0$ if $t < 0$ and $f'(t) > 0$ if $t > 0$, we see that f is decreasing on $(-\infty, 0)$ and increasing on $(0, \infty)$.

6. The results of (5) show that $(0,0)$ is a relative minimum.

7. $f''(t) = \dfrac{(1+t^2)^2(2) - 2t(2)(1+t^2)(2t)}{(1+t^2)^4} = \dfrac{2(1+t^2)[(1+t^2) - 4t^2]}{(1+t^2)^4}$

$= \dfrac{2(1-3t^2)}{(1+t^2)^3} = 0$ if $t = \pm\dfrac{\sqrt{3}}{3}$.

The sign diagram of f'' is

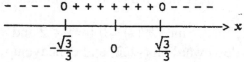

and shows that f is concave downward on $(-\infty, -\tfrac{\sqrt{3}}{3}) \cup (\tfrac{\sqrt{3}}{3}, \infty)$ and concave upward on $(-\tfrac{\sqrt{3}}{3}, \tfrac{\sqrt{3}}{3})$.

8. The results of (7) show that $(-\tfrac{\sqrt{3}}{3}, \tfrac{1}{4})$ and $(\tfrac{\sqrt{3}}{3}, \tfrac{1}{4})$ are inflection points. The graph of f follows.

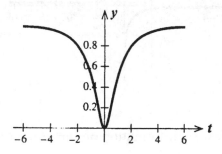

55. $g(t) = -\dfrac{t^2 - 2}{t - 1}$.

First we obtain the following information on g.
1. The domain of g is $(-\infty, 1) \cup (1, \infty)$.
2. Setting $t = 0$ gives -2 as the y-intercept.

3. $\lim\limits_{t \to -\infty} -\dfrac{t^2 - 2}{t - 1} = \infty$ and $\lim\limits_{t \to \infty} -\dfrac{t^2 - 2}{t - 1} = -\infty.$

4. There are no horizontal asymptotes. The denominator is equal to zero at $t = 1$ at which point the numerator is not equal to zero. Therefore $t = 1$ is a vertical asymptote.

5. $g'(t) = -\dfrac{(t-1)(2t) - (t^2 - 2)(1)}{(t-1)^2} = -\dfrac{t^2 - 2t + 2}{(t-1)^2} \neq 0$

for all values of t. The sign diagram of g'

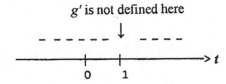

shows that g is decreasing on $(-\infty, 1) \cup (1, \infty)$.

6. Since there are no critical points, g has no relative extrema.

7. $g''(t) = -\dfrac{(t-1)^2(2t-2) - (t^2 - 2t + 2)(2)(t-1)}{(t-1)^4}$

$= \dfrac{-2(t-1)(t^2 - 2t + 1 - t^2 + 2t - 2)}{(t-1)^4} = \dfrac{2}{(t-1)^3}.$

The sign diagram of g''

shows that the graph of g is concave upward on $(1, \infty)$ and concave downward on $(-\infty, 1)$.

8. There are no inflection points since $g''(x) \neq 0$ for all x.

The graph of g follows.

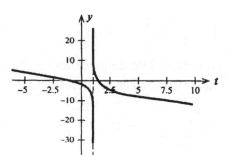

57. $g(t) = \dfrac{t+1}{t^2 - 2t - 1}$.

1. Since $t^2 - 2t - 1 = 0$ if $t = \dfrac{2 \pm \sqrt{4+4}}{2} = 1 \pm \sqrt{2}$, we see that

the domain of g is $(-\infty, 1 - \sqrt{2}) \cup (1 - \sqrt{2}, 1 + \sqrt{2}) \cup (1 + \sqrt{2}, \infty)$.

2. Setting $t = 0$ gives -1 as the y-intercept. Setting $y = 0$ gives -1 as the t-intercept.

3. $\lim\limits_{t \to -\infty} g(t) = \lim\limits_{t \to \infty} g(t) = 0$.

4. The results of (3) show that $y = 0$ is a horizontal asymptote. Since the denominator (but not the numerator) is zero at $t = 1 \pm \sqrt{2}$, we see that $t = 1 - \sqrt{2}$ and $t = 1 + \sqrt{2}$ are vertical asymptotes.

5. $g'(t) = \dfrac{(t^2 - 2t - 1)(1) - (t+1)(2t-2)}{(t^2 - 2t - 1)^2} = -\dfrac{(t^2 + 2t - 1)}{(t^2 - 2t - 1)^2} = 0$

if $\quad t = \dfrac{-2 \pm \sqrt{4+4}}{2} = -1 \pm \sqrt{2}.$

The sign diagram of g' is

g' is not defined here

$\quad \quad \quad \quad \quad \downarrow \quad \quad \quad \quad \quad \quad \downarrow$

$- - - - \ 0 + + + + + + \quad + \ + \ + \ 0 - - - - - \cdot - - - -$

$\xrightarrow{\hspace{6cm}} t$

$\quad \quad -1-\sqrt{2} \quad 1 - \sqrt{2} \ \ 0 \ \ -1 + \sqrt{2} \quad \quad 1 + \sqrt{2}$

We see that g is decreasing on $\ (-\infty, -1 - \sqrt{2}) \cup (-1 + \sqrt{2}, 1 + \sqrt{2}) \cup$

$(1 + \sqrt{2}, \infty)$ and increasing on $(-1 - \sqrt{2}, -1 + \sqrt{2})$.

6. From the results of (5), we see that g has a relative maximum at $t = -1 + \sqrt{2}$ and a relative minimum at $t = -1 - \sqrt{2}$.

7.

$$g''(t) = \frac{(t^2 - 2t - 1)^2(-2t - 2) + (t^2 + 2t - 1)2(t^2 - 2t - 1)(2t - 2)}{(t^2 - 2t - 1)^4}$$

$$= \frac{-2(t^3 - 2t^2 - t + t^2 - 2t - 1 - t^3 - 2t^2 + t + t^2 + 2t - 1)}{(t^2 - 2t - 1)^3}$$

$$= \frac{-2(-2t^2 - 2)}{(t^2 - 2t - 1)^3} = \frac{4(t^2 + 1)}{(t^2 - 2t - 1)^3}$$

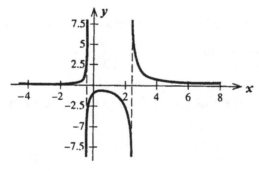

From the sign diagram we see that the graph of g is concave up on $(-\infty, 1 - \sqrt{2}) \cup (1 + \sqrt{2}, \infty)$ and concave down on $(1 - \sqrt{2}, 1 + \sqrt{2})$.

8. From (7), we see that the graph of g has no inflection points. The graph of g follows.

59. $h(x) = (x - 1)^{2/3} + 1$.

We begin by obtaining the following information on h.
1. The domain of h is $(-\infty, \infty)$.
2. Setting $x = 0$ gives 2 as the y-intercept; since $h(x) \neq 0$ there is no x-intercept.
3. $\lim_{x \to \infty} [(x - 1)^{2/3} + 1] = \infty$. Similarly, $\lim_{x \to -\infty} [(x - 1)^{2/3} + 1] = \infty$.
4. There are no asymptotes.
5. $h'(x) = \frac{2}{3}(x - 1)^{-1/3}$ and is positive if $x > 1$ and negative if $x < 1$. So h is

increasing on $(1,\infty)$, and decreasing on $(-\infty,1)$.

6. From (5), we see that h has a relative minimum at $(1,1)$.

7. $h''(x) = \dfrac{2}{3}(-\dfrac{1}{3})(x-1)^{-4/3} = -\dfrac{2}{9}(x-1)^{-4/3} = -\dfrac{2}{(x-1)^{4/3}}$. Since $h''(x) < 0$ on

$(-\infty,1) \cup (1,\infty)$, we see that h is concave downward on $(-\infty,1) \cup (1,\infty)$. Note that $h''(x)$ is not defined at $x = 1$.

8. From the results of (7), we see h has no inflection points.

The graph of h follows.

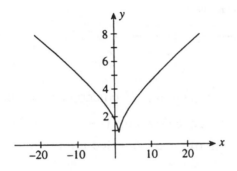

61. a. The denominator of $C(x)$ is equal to zero if $x = 100$. Also,

$$\lim_{x \to 100^-} \frac{0.5x}{100-x} = \infty \quad \text{and} \quad \lim_{x \to 100^+} \frac{0.5x}{100-x} = -\infty$$

Therefore, $x = 100$ is a vertical asymptote of C.

b. No, because the denominator will be equal to zero in that case.

63. a. Since $\lim_{t \to \infty} C(t) = \lim_{t \to \infty} \dfrac{0.2t}{t^2+1} = \lim_{t \to \infty} \left[\dfrac{0.2}{t+\frac{1}{t^2}} \right] = 0$, $y = 0$ is a horizontal asymptote.

b. Our results reveal that as time passes, the concentration of the drug decreases and approaches zero.

65. $G(t) = -0.2t^3 + 2.4t^2 + 60$.

We first gather the following information on the graph of G.

1. The domain of G is $(0,\infty)$.

2. Setting $t = 0$ gives 60 as the y-intercept.

Note that Step 3 is not necessary in this case because of the restricted domain.

4. There are no asymptotes since G is a polynomial function.

5. $G'(t) = -0.6t^2 + 4.8t = -0.6t(t - 8) = 0$, if $t = 0$ or $t = 8$. But these points do not lie in the interval $(0,8)$, so they are not critical points. The sign diagram of G'

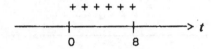

shows that G is increasing on $(0,8)$.
6. The results of (5) tell us that there are no relative extrema.
7. $G''(t) = -1.2t + 4.8 = -1.2(t - 4)$. The sign diagram of G'' is

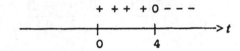

and shows that G is concave upward on $(0,4)$ and concave downward on $(4,8)$.
8.   The results of (7) shows that $(4,85.6)$ is an inflection point.
The graph of G follows.

67. $C(t) = \dfrac{0.2t}{t^2 + 1}$.

We first gather the following information on the function C.
1. The domain of C is $[0,\infty)$.
2. If $t = 0$, then $y = 0$. Also, if $y = 0$, then $t = 0$.
3. $\displaystyle\lim_{t \to \infty} \dfrac{0.2t}{t^2 + 1} = 0$.

4. The results of (3) imply that $y = 0$ is a horizontal asymptote.

5. $\quad C'(t) = \dfrac{(t^2+1)(0.2) - 0.2t(2t)}{(t^2+1)^2} = \dfrac{0.2(t^2+1-2t^2)}{(t^2+1)^2} = \dfrac{0.2(1-t^2)}{(t^2+1)^2}$

and this is equal to zero at $t = \pm 1$, so $t = 1$ is a critical point of C. The sign diagram of C' is

$$+ + + 0 - - - -$$

$$\xrightarrow{\hspace{1cm}\underset{0}{|}\hspace{1cm}\underset{1}{|}\hspace{2cm}} t$$

and tells us that C is decreasing on $(1,\infty)$ and increasing on $(0,1)$.

6. The results of (5) tell us that $(1, 0.1)$ is a relative maximum.

7. $C''(t) = 0.2\left[\dfrac{(t^2+1)^2(-2t) - (1-t^2)2(t^2+1)(2t)}{(t^2+1)^4}\right]$

$= \dfrac{0.2(t^2+1)(2t)(-t^2-1-2+2t^2)}{(t^2+1)^4} = \dfrac{0.4t(t^2-3)}{(t^2+1)^3}.$

The sign diagram of C'' is

$$0 - - 0 + + +$$

$$\xrightarrow{\hspace{1cm}\underset{-\sqrt{3}}{|}\hspace{0.6cm}\underset{0}{}\hspace{0.6cm}\underset{\sqrt{3}}{|}\hspace{1.5cm}} t$$

and so the graph of C is concave downward on $(0, \sqrt{3})$ and concave upward on $(\sqrt{3}, \infty)$.

8. The results of (7) show that $(\sqrt{3}, 0.05\sqrt{3})$ is an inflection point.

The graph of C follows.

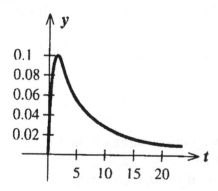

69. $T(x) = \dfrac{120x^2}{x^2 + 4}$.

We first gather the following information on the function T.
1. The domain of T is $[0,\infty)$.
2. Setting $x = 0$ gives 0 as the y-intercept.
3. $\displaystyle\lim_{x\to\infty} \dfrac{120x^2}{x^2 + 4} = 120$.
4. The results of (3) show that $y = 120$ is a horizontal asymptote.
5. $T'(x) = 120\left[\dfrac{(x^2 + 4)2x - x^2(2x)}{(x^2 + 4)^2}\right] = \dfrac{960x}{(x^2 + 4)^2}$. Since $T'(x) > 0$

if $x > 0$, we see that T is increasing on $(0,\infty)$.
6. There are no relative extrema in $(0,\infty)$.
7. $T''(x) = 960\left[\dfrac{(x^2 + 4)^2 - x(2)(x^2 + 4)(2x)}{(x^2 + 4)^4}\right]$

$= \dfrac{960(x^2 + 4)[(x^2 + 4) - 4x^2]}{(x^2 + 4)^4} = \dfrac{960(4 - 3x^2)}{(x^2 + 4)^3}$.

The sign diagram for T'' is

$$+ + + 0 - - -$$
$$\begin{array}{ccc} & & \\ 0 & & \frac{2\sqrt{3}}{3} \end{array} \longrightarrow x$$

We see that T is concave downward on $\left(\frac{2\sqrt{3}}{3},\infty\right)$ and concave upward on $\left(0, \frac{2\sqrt{3}}{3}\right)$.

8. We see from the results of (7) that $\left(\frac{2\sqrt{3}}{3}, 30\right)$ is an inflection point.

The graph of T follows.

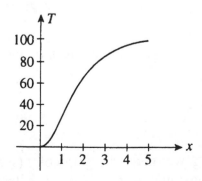

4 Applications of the Derivative

1.

3.

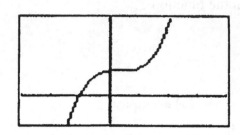

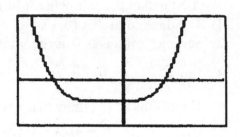

5.   -0.9733; 2.3165, 4.6569   7.   -1.1301; 2.9267   9.   1.5142

EXERCISES 4.4, page 311

1. f has no absolute extrema.

3. f has an absolute minimum at $(0,0)$.

5. f has an absolute minimum at $(0,-2)$ and an absolute maximum at $(1,3)$.

7. f has an absolute minimum at $(\frac{3}{2}, -\frac{27}{16})$ and an absolute maximum at $(-1,3)$.

9.   The graph of $f(x) = 2x^2 + 3x - 4$ is a parabola that opens upward. Therefore, the vertex of the parabola is the absolute minimum of f. To find the vertex, we solve the equation $f'(x) = 4x + 3 = 0$ giving $x = -3/4$. We conclude that the absolute minimum value is $f(-\frac{3}{4}) = -\frac{41}{8}$.

11. Since $\lim_{x \to -\infty} x^{1/3} = -\infty$ and $\lim_{x \to \infty} x^{1/3} = \infty$, we see that h is unbounded. Therefore it has no absolute extrema.

13. $f(x) = \dfrac{1}{1+x^2}$.

Using the techniques of graphing, we sketch the graph of f (see Fig. 4.40, page 297, in the text). The absolute maximum of f is $f(0) = 1$. Alternatively, observe that $1 + x^2 \geq 1$ for all real values of x. Therefore, $f(x) \leq 1$ for all x, and we see that the absolute maximum is attained when $x = 0$.

15. $f(x) = x^2 - 2x - 3$ and $f'(x) = 2x - 2 = 0$, so $x = 1$ is a critical point. From the table,

x	-2	1	3
$f(x)$	5	-4	0

we conclude that the absolute maximum value is $f(-2) = 5$ and the absolute minimum value is $f(1) = -4$.

17. $f(x) = -x^2 + 4x + 6$; The function f is continuous and defined on the closed interval $[0,5]$. $f'(x) = -2x + 4$ and $x = 2$ is a critical point. From the table

x	0	2	5
$f(x)$	6	10	1

we conclude that $f(2) = 10$ is the absolute maximum value and $f(5) = 1$ is the absolute minimum value.

19. The function $f(x) = x^3 + 3x^2 - 1$ is continuous and defined on the closed interval $[-3,2]$ and differentiable in $(-3,2)$. The critical points of f are found by solving
$$f'(x) = 3x^2 + 6x = 3x(x + 2)$$
giving $x = -2$ and $x = 0$. Next, we compute the values of f given in the following table.

x	-3	-2	0	2
$f(x)$	-1	3	-1	19

From the table, we see that the absolute maximum value of f is $f(2) = 19$ and the absolute minimum value is $f(-3) = -1$ and $f(0) = -1$.

21. The function $g(x) = 3x^4 + 4x^3$ is continuous and differentiable on the closed interval $[-2,1]$ and differentiable in $(-2,1)$. The critical points of g are found by solving

$$g'(x) = 12x^3 + 12x^2 = 12x^2(x+1)$$

giving $x = 0$ and $x = -1$. We next compute the values of g shown in the following table.

x	-2	-1	0	1
$g(x)$	16	-1	0	7

From the table we see that $g(-2) = 16$ is the absolute maximum value of g and $g(-1) = -1$ is the absolute minimum value of g.

23. $f(x) = \dfrac{x+1}{x-1}$ on [2,4]. Next, we compute,

$$f'(x) = \frac{(x-1)(1)-(x+1)(1)}{(x-1)^2} = -\frac{2}{(x-1)^2}.$$

Since there are no critical points, ($x = 1$ is not in the domain of f), we need only test the endpoints. From the table

x	2	4
$g(x)$	3	5/3

we conclude that $f(4) = 5/3$ is the absolute minimum value and $f(2) = 3$ is the absolute maximum value.

25. $f(x) = 4x + \dfrac{1}{x}$ is continuous on [1,3] and differentiable in (1,3). To find the critical points of f, we solve $f'(x) = 4 - \frac{1}{x^2} = 0$, obtaining $x = \pm\frac{1}{2}$. Since these critical points lie outside the interval [1,3], they are not candidates for the absolute extrema of f. Evaluating f at the endpoints of the interval [1,3], we find that the absolute maximum value of f is $f(3) = \frac{37}{3}$, and the absolute minimum value of f is $f(1) = 5$.

27. $f(x) = \frac{1}{2}x^2 - 2\sqrt{x} = \frac{1}{2}x^2 - 2x^{1/2}$. To find the critical points of f, we solve
$$f'(x) = x - x^{-1/2} = 0, \quad \text{or} \quad x^{3/2} - 1 = 0,$$
obtaining $x = 1$. From the table

x	0	1	3
$f(x)$	0	$-\frac{3}{2}$	$\frac{9}{2} - 2\sqrt{3} \approx 1.04$

we conclude that $f(3) \approx 1.04$ is the absolute maximum value and $f(1) = -3/2$ is the absolute minimum value.

29. The graph of $f(x) = 1/x$ over the interval $(0,\infty)$ follows.

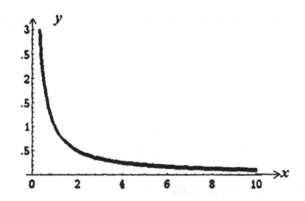

From the graph of f, we conclude that f has no absolute extrema.

31. $f(x) = 3x^{2/3} - 2x$. The function f is continuous on $[0,3]$ and differentiable on $(0,3)$. To find the critical points of f, we solve
$$f'(x) = 2x^{-1/3} - 2 = 0$$
obtaining $x = 1$ as the critical point. From the table,

x	0	1	3
$f(x)$	0	1	$3^{5/3} - 6 \approx 0.24$

4 Applications of the Derivative

we conclude that the absolute maximum value is $f(1) = 1$ and the absolute minimum value is $f(0) = 0$.

33. $f(x) = x^{2/3}(x^2 - 4)$. $f'(x) = x^{2/3}(2x) + \frac{2}{3}x^{-1/3}(x^2 - 4) = \frac{2}{3}x^{-1/3}[3x^2 + (x^2 - 4)]$
$$= \frac{8(x^2 - 1)}{3x^{1/3}} = 0.$$

Observe that f' is not defined at $x = 0$. Furthermore, $f'(x) = 0$ at $x \pm 1$. So the critical points of f are -1, 0, 1. From the following table,

x	-1	0	1	2
$f(x)$	-3	0	-3	0

we see that f has an absolute minimum at (-1,-3) and (1,-3) and absolute maxima at (0,0) and (2,0).

35. $f(x) = \dfrac{x}{x^2 + 2}$. To find the critical points of f, we solve
$$f'(x) = \frac{(x^2 + 2) - x(2x)}{(x^2 + 2)^2} = \frac{2 - x^2}{(x^2 + 2)^2} = 0$$
obtaining $x = \pm\sqrt{2}$. Since $x = -\sqrt{2}$ lies outside [-1,2], $x = \sqrt{2}$ is the only critical point in the given interval. From the table

x	-1	$\sqrt{2}$	2
$f(x)$	$-\frac{1}{3}$	$\sqrt{2}/4 \approx 0.35$	$\frac{1}{3}$

we conclude that $f(\sqrt{2})) = \sqrt{2}/4 \approx 0.35$ is the absolute maximum value and $f(-1) = -1/3$ is the absolute minimum value.

37. The function $f(x) = \dfrac{x}{\sqrt{x^2 + 1}} = \dfrac{x}{(x^2 + 1)^{1/2}}$ is continuous and defined on the

closed interval $[-1,1]$ and differentiable on $(-1,1)$. To find the critical points of f, we first compute

$$f'(x) = \frac{(x^2+1)^{1/2}(1) - x(\tfrac{1}{2})(x^2+1)^{-1/2}(2x)}{[(x^2+1)^{1/2}]^2}$$

$$= \frac{(x^2+1)^{-1/2}[x^2+1-x^2]}{x^2+1} = \frac{1}{(x^2+1)^{3/2}}$$

which is never equal to zero. Next, we compute the values of f shown in the following table.

x	-1	1
$f(x)$	$-\sqrt{2}/2$	$\sqrt{2}/2$

We conclude that $f(-1) = -\sqrt{2}/2$ is the absolute minimum value and $f(1) = \sqrt{2}/2$ is the absolute maximum value.

39. $h(t) = -16t^2 + 64t + 80$. To find the maximum value of h, we solve
$$h'(t) = -32t + 64 = -32(t - 2) = 0$$
giving $t = 2$ as the critical point of h. Furthermore, this value of t gives rise to the absolute maximum value of h since the graph of h is parabola that opens downward. The maximum height is given by
$$h(2) = -16(4) + 64(2) + 80 = 144, \text{ or } 144 \text{ feet.}$$

41. $P(x) = -0.04x^2 + 240x - 10,000$. We compute $P'(x) = -0.08x + 240$. Setting $P'(x) = 0$ gives $x = 3000$. The graph of P is a parabola that opens downward and so $x = 3000$ gives rise to the absolute maximum of P. Thus, to maximize profits, the company should produce 3000 cameras per month.

43. $N(t) = 0.81t - 1.14\sqrt{t} + 1.53$. $N'(t) = 0.81 - 1.14(\tfrac{1}{2}t^{-1/2}) = 0.81 - \dfrac{0.57}{t^{1/2}}$. Setting $N'(t) = 0$ gives $t^{1/2} = \dfrac{0.57}{0.81}$, or $t = 0.4952$ as a critical point of N. Evaluating $N(t)$ at the endpoints $t = 0$ and $t = 6$ as well as at the critical point, we have

t	0	0.4952	6
$N(t)$	1.53	1.13	3.60

From the table, we see that the absolute maximum of N occurs at $t = 6$ and the absolute minimum occurs at $t \approx 0.5$. Our results tell us that the number of nonfarm full-time self-employed women over the time interval from 1963 to 1993 was the highest in 1993 and stood at approximately 3.6 million.

45. $P(x) = -0.000002x^3 + 6x - 400$. $P'(x) = -0.000006x^2 + 6 = 0$ if $x = \pm 1000$. We reject the negative root. Next, we compute $P''(x) = -0.000012x$. Since $P''(1000) = -0.012 < 0$, the Second Derivative Test shows that $x = 1000$ affords a relative maximum of f. From physical considerations, or from a sketch of the graph of f, we see that the maximum profit is realized if 1000 cases are produced per day. The profit is $P(1000) = -0.000002(1000)^3 + 6(1000) - 400$, or $3600/day.

47. The revenue is $R(x) = px = -0.0004x^2 + 10x$, and the profit is
$$P(x) = R(x) - C(x) = -0.0004x^2 + 10x - (400 + 4x + 0.0001x^2)$$
$$= -0.0005x^2 + 6x - 400.$$
$$P'(x) = -0.001x + 6 = 0$$
if $x = 6000$, a critical point. Since $P''(x) = -0.001 < 0$ for all x, we see that the graph of P is a parabola that opens downward. Therefore, a level of production of 6000 rackets/day will yield a maximum profit.

49. The total cost function is given by
$$C(x) = V(x) + 20,000$$
$$= 0.000001x^3 - 0.01x^2 + 50x + 20,000$$
The profit function is
$$P(x) = R(x) - C(x)$$
$$= -0.02x^2 + 150x - 0.000001x^3 + 0.01x^2 - 50x + 20,000$$
$$= -0.000001x^3 - 0.01x^2 + 100x - 20,000$$
We want to maximize P on $[0, 7000]$.
$$P'(x) = -0.000003x^2 - 0.02x + 100$$
Setting $P'(x) = 0$ gives $3x^2 + 20,000x - 100,000,000 = 0$

or $x = \dfrac{-20,000 \pm \sqrt{20,000^2 + 1,200,000,000}}{6} = -10,000$ or $3,333.3$.

So $x = 3,333.30$ is a critical point in the interval $[0, 7000]$.

x	0	3,333	7,000
$P(x)$	-20,000	165,185	-519,700

From the table, we see that a level of production of 3,333 pagers per week will yield a maximum profit of \$165,185 per week.

51. a. $\overline{C}(x) = \dfrac{C(x)}{x} = 0.0025x + 80 + \dfrac{10,000}{x}$.

   b. $\overline{C}'(x) = 0.0025 - \dfrac{10,000}{x^2} = 0$ if $0.0025x^2 = 10,000$, or $x = 2000$.

   Since $\overline{C}''(x) = \dfrac{20,000}{x^3}$, we see that $\overline{C}''(x) > 0$ for $x > 0$ and so $\overline{C}$ is concave upward on $(0, \infty)$. Therefore, $x = 2000$ yields a minimum.

   c. We solve $\overline{C}(x) = C'(x)$. $0.0025x + 80 + \dfrac{10,000}{x} = 0.005x + 80$,

   $0.0025x^2 = 10,000$, or $x = 2000$.

   d. It appears that we can solve the problem in two ways.
   REMARK This can be proved.

53. The demand equation is $p = \sqrt{800 - x} = (800 - x)^{1/2}$. The revenue function is $R(x) = xp = x(800 - x)^{1/2}$. To find the maximum of R, we compute
$$R'(x) = \tfrac{1}{2}(800 - x)^{-1/2}(-1)(x) + (800 - x)^{1/2}$$
$$= \tfrac{1}{2}(800 - x)^{-1/2}[-x + 2(800 - x)]$$
$$= \tfrac{1}{2}(800 - x)^{-1/2}(1600 - 3x).$$

Next, $R'(x) = 0$ implies $x = 800$ or $x = 1600/3$ are critical points of R. Next, we compute the values of R given in the following table.

x	0	800	1600/3
$R(x)$	0	0	8709

We conclude that $R(\frac{1600}{3}) = 8709$ is the absolute maximum value. Therefore, the revenue is maximized by producing $1600/3 \approx 533$ dresses.

55. $f(t) = 100 \left[\dfrac{t^2 - 4t + 4}{t^2 + 4} \right].$

    a. $f'(t) = 100 \left[\dfrac{(t^2 + 4)(2t - 4) - (t^2 - 4t + 4)(2t)}{(t^2 + 4)^2} \right] = \dfrac{400(t^2 - 4)}{(t^2 + 4)^2}$

$$= \frac{400(t - 2)(t + 2)}{(t^2 + 4)^2}.$$

From the sign diagram for f'

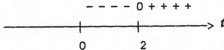

we see that $t = 2$ gives a relative minimum, and we conclude that the oxygen content is the lowest 2 days after the organic waste has been dumped into the pond.

    b.

$$f''(t) = 400 \left[\frac{(t^2 + 4)^2 (2t) - (t^2 - 4)2(t^2 + 4)(2t)}{(t + 4)^4} \right] = 400 \left[\frac{(2t)(t^2 + 4)(t^2 + 4 - 2t^2 + 8)}{(t^2 + 4)^4} \right]$$

$$= -\frac{800t(t^2 - 12)}{(t^2 + 4)^3}$$

and $f''(t) = 0$ when $t = 0$ and $t = \pm 2\sqrt{3}$. We reject $t = 0$ and $t = -2\sqrt{3}$. From the sign diagram for f'',

$$0\ +\ +\ +\ 0\ \ -\ -\ -\ -$$

$$\begin{array}{cc} & \phantom{2\sqrt{3}} \\ 0 & 2\sqrt{3} \end{array} \longrightarrow t$$

we see that $f'(2\sqrt{3})$ gives an inflection point of f and we conclude that this is an absolute maximum. Therefore, the rate of oxygen regeneration is greatest 3.5 days after the organic waste has been dumped into the pond.

57. We compute $\overline{R}'(x) = \dfrac{xR'(x) - R(x)}{x^2}$. Setting $\overline{R}'(x) = 0$ gives $xR'(x) - R(x) = 0$

    or $R'(x) = \dfrac{R(x)}{x} = \overline{R}(x)$, so a critical point of $\overline{R}$ occurs when $\overline{R}(x) = R'(x)$.

Next, we compute

$$\bar{R}''(x) = \frac{x^2[R'(x) + xR''(x) - R'(x)] - [xR'(x) - R(x)](2x)}{x^4} = \frac{R''(x)}{x} < 0.$$

So, by the Second Derivative Test, the critical point does give a maximum revenue.

59. The growth rate is $G'(t) = -0.6t^2 + 4.8t$. To find the maximum growth rate, we compute $G''(t) = -1.2t + 4.8$. Setting $G''(t) = 0$ gives $t = 4$ as a critical point.

t	0	4	8
$G'(t)$	0	9.6	0

From the table, we see that G is maximal at $t = 4$; that is, the growth rate is greatest in 1997.

61. $f(t) = -0.0129t^4 + 0.3087t^3 + 2.1760t^2 + 62.8466t + 506.2955$.
To find the maximum of $f(t)$, we first compute
$$f'(t) = -0.0516t^3 + 0.92611t^2 + 4.352t + 62.8466.$$
Then
$$f'(23.6811) = -0.0516(23.6811)^3 + 0.9261(23.6811)^2 + 4.352(23.6811) + 62.8466$$
$$\approx 0$$
Next, we compute $f'(23) \approx 25.03$ and $f'(24) = -15.18$. Since f is a polynomial function it is continuous. We conclude that $f(t)$ is maximized when $t = 23.6811$ since f' changes sign from positive to negative as we move across the critical point $t = 23.6811$. These results may be confirmed by graphing the derivative function f' on your graphing calculator.

63. $R = D^2\left(\frac{k}{2} - \frac{D}{3}\right) = \frac{kD^2}{2} - \frac{D^3}{3}$. $\frac{dR}{dD} = \frac{2kD}{2} - \frac{3D^2}{3} = kD - D^2 = D(k - D)$

Setting $\frac{dR}{dD} = 0$, we have $D = 0$ or $k = D$. We only consider $k = D$

(since $D > 0$). If $k > 0$, $\frac{dR}{dD} > 0$ and if $k < 0$, $\frac{dR}{dD} < 0$. Therefore $k = D$ provides a relative maximum. The nature of the problem suggests that $k = D$ gives the absolute maximum of R. We can also verify this by graphing R.

65. False. Let $f(x) = \begin{cases} |x| & \textit{if } x \neq 0 \\ 1 & \textit{if } x = 0 \end{cases}$ on [-1, 1].

67. False. Let $f(x) = \begin{cases} -x & \text{if } -1 \leq x < 0 \\ \dfrac{1}{2} & \text{if } 0 \leq x < 1 \end{cases}$. Then f is discontinuous at $x = 0$. But f

has an absolute maximum value of 1 attained at $x = -1$.

69. Since $f(x) = c$ for all x, the function f satisfies $f(x) \leq c$ for all x and so f has an absolute maximum at all points of x. Similarly, f has an absolute minimum at all points of x.

71. a. f is not continuous at $x = 0$ because $\lim\limits_{x \to 0} f(x)$ does not exist.

b. $\lim\limits_{x \to 0} f(x) = \lim\limits_{x \to 0^-} \dfrac{1}{x} = -\infty$ and $\lim\limits_{x \to 0^+} f(x) = \lim\limits_{x \to 0^+} \dfrac{1}{x} = \infty$

c.

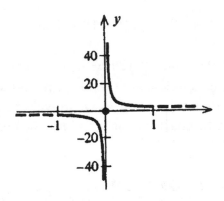

USING TECHNOLOGY EXERCISES 4.4, page 314

1. Absolute maximum value: 145.8985; absolute minimum value: -4.3834

3. Absolute maximum value: 16; absolute minimum value: -0.1257

5. Absolute maximum value: 2.8889; absolute minimum value: 0

7. a. 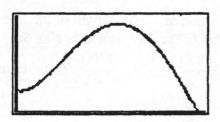   b. 200.1410 banks/yr

9. a. 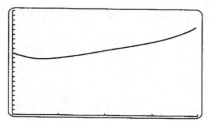   11. a.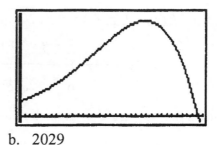

   b. 21.51%   b. 2029

EXERCISES 4.5, page 325

1. Refer to the following figure.

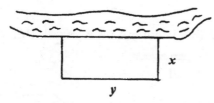

We have $2x + y = 3000$ and we want to maximize the function
$$A = f(x) = xy = x(3000 - 2x) = 3000x - 2x^2$$
on the interval $[0,1500]$. The critical point of A is obtained by solving
$f'(x) = 3000 - 4x = 0$, giving $x = 750$. From the table of values

x	0	750	1500
$f(x)$	0	1,125,000	0

we conclude that $x = 750$ yields the absolute maximum value of A. Thus, the required dimensions are 750×1500 yards. The maximum area is $1,125,000$ sq yd.

3. Let x denote the length of the side made of wood and y the length of the side made of steel. The cost of construction will be $C = 6(2x) + 3y$. But $xy = 800$. So $y = 800/x$ and therefore $C = f(x) = 12x + 3\left(\dfrac{800}{x}\right) = 12x + \dfrac{2400}{x}$. To minimize C,

we compute
$$f'(x) = 12 - \frac{2400}{x^2} = \frac{12x^2 - 2400}{x^2} = \frac{12(x^2 - 200)}{x^2}.$$

Setting $f'(x) = 0$ gives $x = \pm\sqrt{200}$ as critical points of f. The sign diagram of f'

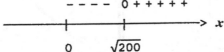

shows that $x = \pm\sqrt{200}$ gives a relative minimum of f. $f''(x) = \dfrac{4800}{x^3} > 0$

if $x > 0$ and so f is concave upward for $x > 0$. Therefore $x = \sqrt{200} = 10\sqrt{2}$ actually

yields the absolute minimum. So the dimensions of the enclosure should be
$$10\sqrt{2} \text{ ft} \times \frac{800}{10\sqrt{2}} \text{ ft, or } 14.1 \text{ ft} \times 56.6 \text{ ft.}$$

5. Let the dimensions of each square that is cut out be $x'' \times x''$. Refer to the following

diagram.

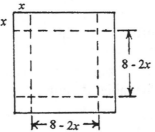 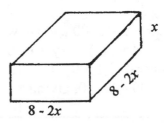

Then the dimensions of the box will be $(8 - 2x)''$ by $(8 - 2x)''$ by x''. Its volume will be $V = f(x) = x(8 - 2x)^2$. We want to maximize f on $[0,4]$.

$$f'(x) = (8 - 2x)^2 + x(2)(8 - 2x)(-2) \qquad \text{[Using the Product Rule.]}$$
$$= (8 - 2x)[(8 - 2x) - 4x] = (8 - 2x)(8 - 6x) = 0$$

if $x = 4$ or $4/3$. The latter is a critical point in $(0,4)$.

x	0	4/3	4
$f(x)$	0	1024/27	0

We see that $x = 4/3$ yields an absolute maximum for f. So the dimensions of the box should be $\frac{16}{3}" \times \frac{16}{3}" \times \frac{4}{3}"$.

7. Let x denote the length of the sides of the box and y denote its height. Referring to the following figure, we see that the volume of the box is given by $x^2 y = 128$. The

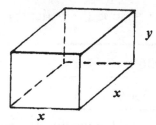

amount of material used is given by
$$S = f(x) = 2x^2 + 4xy$$
$$= 2x^2 + 4x\left(\frac{128}{x^2}\right)$$
$$= 2x^2 + \frac{512}{x} \text{ square inches.}$$

We want to minimize f subject to the condition that $x > 0$. Now
$$f'(x) = 4x - \frac{512}{x^2} = \frac{4x^3 - 512}{x^2} = \frac{4(x^3 - 128)}{x^2}.$$

Setting $f'(x) = 0$ yields $x = 5.04$, a critical point of f. Next,
$$f''(x) = 4 + \frac{1024}{x^3} > 0$$

for all $x > 0$. Thus, the graph of f is concave upward and so $x = 5.04$ yields an absolute minimum of f. Thus, the required dimensions are 5.04" $\times$ 5.04" $\times$ 5.04".

9. The length plus the girth of the box is $4x + h = 108$ and $h = 108 - 4x$. Then
$$V = x^2h = x^2(108 - 4x) = 108x^2 - 4x^3$$
and $V' = 216x - 12x^2$. We want to maximize V on the interval $[0,27]$. Setting $V'(x) = 0$ and solving for x, we obtain $x = 18$ and $x = 0$. Evaluating $V(x)$ at $x = 0$, $x = 18$, and $x = 27$, we obtain
$$V(0) = 0, \ V(18) = 11{,}664, \text{ and } V(27) = 0$$
Thus, the dimensions of the box are $18" \times 18" \times 36"$ and its maximum volume is approximately 11,664 cu in.

11. We take $2\pi r + \ell = 108$. We want to maximize
$$V = \pi r^2 \ell = \pi r^2 (-2\pi r + 108) = -2\pi^2 r^3 + 108\pi r^2$$
subject to the condition that $0 \le r \le \frac{54}{\pi}$. Now
$$V'(r) = -6\pi^2 r^2 + 216\pi r = -6\pi r(\pi r - 36).$$
Since $V' = 0$, we find $r = 0$ or $r = 36/\pi$, the critical points of V. From the table

r	0	36/π	54/π
V	0	46,656/π	0

we conclude that the maximum volume occurs when $r = 36/\pi \approx 11.5$ inches and $\ell = 108 - 2\pi\left(\frac{36}{\pi}\right) = 36$ inches and its volume is $46{,}656/\pi$ cu in .

13. Let y denote the height and x the width of the cabinet. Then $y = (3/2)x$. Since the volume is to be 2.4 cu ft, we have $xyd = 2.4$, where d is the depth of the cabinet.
We have $\quad x\left(\frac{3}{2}x\right)d = 2.4 \ $ or $\ d = \dfrac{2.4(2)}{3x^2} = \dfrac{1.6}{x^2}$.
The cost for constructing the cabinet is
$$C = 40(2xd + 2yd) + 20(2xy) = 80\left[\frac{1.6}{x} + \left(\frac{3}{2}x\right)\left(\frac{1.6}{x^2}\right)\right] + 40x\left(\frac{3}{2}x\right)$$
$$= \frac{320}{x} + 60x^2.$$
$$C'(x) = -\frac{320}{x^2} + 120x = \frac{120x^3 - 320}{x^2} = 0 \ \text{ if } x = \sqrt[3]{\frac{8}{3}} = \frac{2}{\sqrt[3]{3}} = \frac{2}{3}\sqrt[3]{9}$$
Therefore, $x = \frac{2}{3}\sqrt[3]{9}$ is a critical point of C. The sign diagram

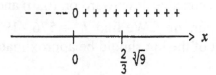

shows that $x = \frac{2}{3}\sqrt[3]{9}$ gives a relative minimum. Next, $C''(x) = \dfrac{640}{x^3} + 120 > 0$

for all $x > 0$ tells us that the graph of C is concave upward. So $x = \frac{2}{3}\sqrt[3]{9}$ yields an

absolute minimum. The required dimensions are $\frac{2}{3}\sqrt[3]{9}\,' \times \sqrt[3]{9}\,' \times \frac{2}{3}\sqrt[3]{9}\,'$.

15. We want to maximize the function
$$R(x) = (200 + x)(300 - x) = -x^2 + 100x + 60000.$$
$$R'(x) = -2x + 100 = 0$$
gives $x = 50$ and this is a critical point of R. Since $R''(x) = -2 < 0$, we see that $x = 50$ gives an absolute maximum of R. Therefore, the number of passengers should be 250. The fare will then be \$250/passenger and the revenue will be \$62,500.

17. Let x denote the number of people beyond 20 who sign up for the cruise. Then the revenue is $R(x) = (20 + x)(600 - 4x) = -4x^2 + 520x + 12,000$. We want to maximize R on the closed bounded interval $[0, 70]$.
$$R'(x) = -8x + 520 = 0 \text{ implies } x = 65,$$
a critical point of R. Evaluating R at this critical point and the endpoints, we have

x	0	65	70
$R(x)$	12,000	28,900	28,800

From this table, we see that R is maximized if $x = 65$. Therefore, 85 passengers will result in a maximum revenue of \$28,900. The fare would be \$340/passenger.

19. We want to maximize $S = kh^2w$. But $h^2 + w^2 = 24^2$ or $h^2 = 576 - w^2$. So
$S = f(w) = kw(576 - w^2) = k(576w - w^3)$. Now, setting
$$f'(w) = k(576 - 3w^2) = 0$$
gives $w = \pm\sqrt{192} \approx \pm 13.86$. Only the positive root is a critical point of interest.
Next, we find $f''(w) = -6kw$, and in particular,
$$f''(\sqrt{192}) = -6\sqrt{192}\,k < 0,$$

so that $w = \pm\sqrt{192} \approx \pm 13.86$ gives a relative maximum of f. Since $f''(w) < 0$ for $w > 0$, we see that the graph of f is concave downward on $(0,\infty)$ and so, $w = \sqrt{192}$ gives an absolute maximum of f. We find $h^2 = 576 - 192 = 384$ or $h \approx 19.60$. So the width and height of the log should be approximately 13.86 inches and 19.60 inches, respectively.

21. We want to minimize $C(x) = 1.50(10,000 - x) + 2.50\sqrt{3000^2 + x^2}$ subject to $0 \leq x \leq 10,000$. Now

$$C'(x) = -1.50 + 2.5(\tfrac{1}{2})(9,000,000 + x^2)^{-1/2}(2x) = -1.50 + \frac{2.50x}{\sqrt{9,000,000 + x^2}}$$

$$C'(x) = 0 \Rightarrow 2.5x = 1.50\sqrt{9,000,000 + x^2}$$

$$6.25x^2 = 2.25(9,000,000 + x^2) \quad \text{or} \quad 4x^2 = 20250000, \; x = 2250.$$

x	0	2250	10000
$f(x)$	22500	21000	26101

From the table, we see that $x = 2250$ gives the absolute minimum.

23. The time taken for the flight is $T = f(x) = \dfrac{12 - x}{6} + \dfrac{\sqrt{x^2 + 9}}{4}$.

$$f'(x) = -\frac{1}{6} + \frac{1}{4}\left(\frac{1}{2}\right)(x^2 + 9)^{-1/2}(2x) = -\frac{1}{6} + \frac{x}{4\sqrt{x^2 + 9}} = \frac{3x - 2\sqrt{x^2 + 9}}{12\sqrt{x^2 + 9}}.$$

Setting $f'(x) = 0$ gives $3x = 2\sqrt{x^2 + 9}$, $9x^2 = 4(x^2 + 9)$ or $5x^2 = 36$. Therefore, $x = \pm 6/\sqrt{5} = \pm 6\sqrt{5}/5$. Only the critical point $x = 6\sqrt{5}/5$ is of interest. The nature of the problem suggests $x \approx 2.68$ gives an absolute minimum for T.

25. Let x denote the number of motorcycle tires in each order. We want to minimize

$$C(x) = 400\left(\frac{40,000}{x}\right) + x = \frac{16,000,000}{x} + x.$$

We compute $C'(x) = -\dfrac{16,000,000}{x^2} + 1 = \dfrac{x^2 - 16,000,000}{x^2}.$

Setting $C'(x) = 0$ gives $x = 4000$, a critical point of C. Since

$$C''(x) = \frac{32{,}000{,}000}{x^3} > 0 \text{ for all } x > 0,$$

we see that the graph of C is concave upward and so $x = 4000$ gives an absolute minimum of C. So there should be 10 orders per year, each order of 4000 tires.

27. We want to minimize the function $C(x) = \dfrac{500{,}000{,}000}{x} + 0.2x + 500{,}000$ on the

interval $(0, 1{,}000{,}000)$. Differentiating $C(x)$, we have $C'(x) = -\dfrac{500{,}000{,}000}{x^2} + 0.2$.

Setting $C'(x) = 0$ and solving the resulting equation, we find $0.2x^2 = 500{,}000{,}000$

and $x = \sqrt{2{,}500{,}000{,}000}$ or $x = 50{,}000$. Next, we find

$$C''(x) = \frac{1{,}000{,}000{,}000}{x^3} > 0 \text{ for all } x \text{ and so the graph of } C \text{ is concave upward on}$$

$(0,\infty)$. Thus, $x = 50{,}000$ gives rise to the absolute minimum of C. So, the company should produce 50,000 containers of cookies per production run.

CHAPTER 4 REVIEW, page 330

1. a. $f(x) = \frac{1}{3}x^3 - x^2 + x - 6$. $f'(x) = x^2 - 2x + 1 = (x-1)^2$. $f'(x) = 0$ gives $x = 1$, the critical point of f. Now, $f'(x) > 0$ for all $x \neq 1$. Thus, f is increasing on $(-\infty,1) \cup (1,\infty)$.
    b. Since $f'(x)$ does not change sign as we move across the critical point $x = 1$, the First Derivative Test implies that $x = 1$ does not give rise to a relative extremum of f.
    c. $f''(x) = 2(x-1)$. Setting $f''(x) = 0$ gives $x = 1$ as a candidate for an inflection point of f. Since $f''(x) < 0$ for $x < 1$, and $f''(x) > 0$ for $x > 1$, we see that f is concave downward on $(-\infty,1)$ and concave upward on $(1,\infty)$.
    d. The results of (c) imply that $(1, -\frac{17}{3})$ is an inflection point.

3. a. $f(x) = x^4 - 2x^2$. $f'(x) = 4x^3 - 4x = 4x(x^2 - 1) = 4x(x+1)(x-1)$. The sign diagram of f' shows that f is decreasing on $(-\infty,-1) \cup (0,1)$ and increasing on $(-1,0) \cup (1,\infty)$.

```
      - -  -0 + + 0 - - - 0 + + +
    ————+————+————+————> x
       -1    0    1
```

    b. The results of (a) and the First Derivative Test show that $(-1,-1)$ and $(1,-1)$ are relative minima and (0,0) is a relative maximum.

c. $f''(x) = 12x^2 - 4 = 4(3x^2 - 1) = 0$ if $x = \pm\sqrt{3}/3$. The sign diagram

$$+\ +\ +\ +\ 0\ -\ -\ -\ -\ -\ -\ -\ 0\ +\ +\ +\ +$$

$$\xrightarrow{\hspace{3cm}} x$$

$$-\frac{\sqrt{3}}{3} \qquad 0 \qquad \frac{\sqrt{3}}{3}$$

shows that f is concave upward on $(-\infty, -\sqrt{3}/3) \cup (\sqrt{3}/3, \infty)$ and concave downward on $(-\sqrt{3}/3, \sqrt{3}/3)$.

d. The results of (c) show that $(-\sqrt{3}/3, -5/9)$ and $(\sqrt{3}/3, -5/9)$ are inflection points.

5. a. $f(x) = \dfrac{x^2}{x-1}$. $f'(x) = \dfrac{(x-1)(2x) - x^2(1)}{(x-1)^2} = \dfrac{x^2 - 2x}{(x-1)^2} = \dfrac{x(x-2)}{(x-1)^2}$.

The sign diagram of f'

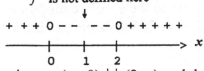

f' is not defined here

$$+\ +\ +\ 0\ -\ -\ \ -\ -\ 0\ +\ +\ +\ +\ +$$

$$\xrightarrow{\hspace{3cm}} x$$

$$0 \qquad 1 \qquad 2$$

shows that f is increasing on $(-\infty, 0) \cup (2, \infty)$ and decreasing on $(0,1) \cup (1,2)$.

b. The results of (a) show that $(0,0)$ is a relative maximum and $(2,4)$ is a relative minimum.

c. $f''(x) = \dfrac{(x-1)^2(2x-2) - x(x-2)2(x-1)}{(x-1)^4} = \dfrac{2(x-1)[(x-1)^2 - x(x-2)]}{(x-1)^4}$

$$= \dfrac{2}{(x-1)^3}.$$

Since $f''(x) < 0$ if $x < 1$ and $f''(x) > 0$ if $x > 1$, we see that f is concave downward on $(-\infty, 1)$ and concave upward on $(1, \infty)$.

d. Since $x = 1$ is not in the domain of f, there are no inflection points.

7. $f(x) = (1-x)^{1/3}$. $f'(x) = -\dfrac{1}{3}(1-x)^{-2/3} = -\dfrac{1}{3(1-x)^{2/3}}$.

The sign diagram for f' is

f' not defined here

$$-\ -\ -\ -\ -\ -\ -\ -\ -\ -\ \downarrow\ -\ -\ -\ -\ -$$

$$\xrightarrow{\hspace{3cm}} x$$

$$0 \qquad 1$$

a. f is decreasing on $(-\infty,1) \cup (1,\infty)$.

b. There are no relative extrema.

c. Next, we compute $f''(x) = -\dfrac{2}{9}(1-x)^{-5/3} = -\dfrac{2}{9(1-x)^{5/3}}$.

The sign diagram for f'' is

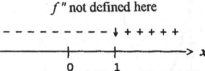

f'' not defined here

We find f is concave downward on $(-\infty,1)$ and concave upward on $(1,\infty)$.

d. $x = 1$ is a candidate for an inflection point of f. Referring to the sign diagram for f'', we see that $(1,0)$ is an inflection point.

9. a. $f(x) = \dfrac{2x}{x+1}$. $f'(x) = \dfrac{(x+1)(2)-2x(1)}{(x+1)^2} = \dfrac{2}{(x+1)^2} > 0$ if $x \neq -1$.

Therefore f is increasing on $(-\infty,-1) \cup (-1,\infty)$.

b. Since there are no critical points, f has no relative extrema.

c. $f''(x) = -4(x+1)^{-3} = -\dfrac{4}{(x+1)^3}$. Since $f''(x) > 0$ if $x < -1$ and $f''(x) < 0$ if $x > -1$,

we see that f is concave upward on $(-\infty,-1)$ and concave downward on $(-1,\infty)$.

d. There are no inflection points since $f''(x) \neq 0$ for all x in the domain of f.

11. $f(x) = x^2 - 5x + 5$

1. The domain of f is $(-\infty, \infty)$.

2. Setting $x = 0$ gives 5 as the y-intercept.

3. $\lim\limits_{x \to -\infty}(x^2 - 5x + 5) = \lim\limits_{x \to \infty}(x^2 - 5x + 5) = \infty$.

4. There are no asymptotes because f is a quadratic function.

5. $f'(x) = 2x - 5 = 0$ if $x = 5/2$. The sign diagram

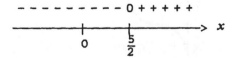

shows that f is increasing on $(\frac{5}{2},\infty)$ and decreasing on $(-\infty,\frac{5}{2})$.

6. The First Derivative Test implies that $(\frac{5}{2},-\frac{5}{4})$ is a relative minimum.

7. $f''(x) = 2 > 0$ and so f is concave upward on $(-\infty, \infty)$.
8. There are no inflection points.
The graph of f follows.

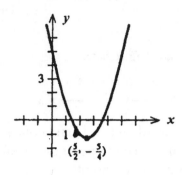

13. $g(x) = 2x^3 - 6x^2 + 6x + 1$.
    1. The domain of g is $(-\infty, \infty)$.
    2. Setting $x = 0$ gives 1 as the y-intercept.
    3. $\lim\limits_{x \to -\infty} g(x) = -\infty, \quad \lim\limits_{x \to \infty} g(x) = \infty$.
    4. There are no vertical or horizontal asymptotes.
    5. $g'(x) = 6x^2 - 12x + 6 = 6(x^2 - 2x + 1) = 6(x - 1)^2$. Since $g'(x) > 0$ for all $x \neq 1$, we
    see that g is increasing on $(-\infty, 1) \cup (1, \infty)$.
    6. $g'(x)$ does not change sign as we move across the critical point $x = 1$, so there is
    no extremum.

    7. $g''(x) = 12x - 12 = 12(x - 1)$. Since $g''(x) < 0$ if $x < 1$ and $g''(x) > 0$ if $x > 1$, we
    see that g is concave upward on $(1, \infty)$ and concave downward on $(-\infty, 1)$.
    8. The point $x = 1$ gives rise to the inflection point $(1, 3)$.
    9. The graph of g follows.

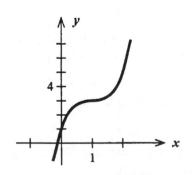

15. $h(x) = x\sqrt{x-2}$.

1. The domain of h is $[2,\infty)$.

2. There are no y-intercepts. Next, setting $y = 0$ gives 2 as the x-intercept.

3. $\lim\limits_{x \to \infty} x\sqrt{x-2} = \infty$.

4. There are no asymptotes.

5. $h'(x) = (x-2)^{1/2} + x(\tfrac{1}{2})(x-2)^{-1/2} = \tfrac{1}{2}(x-2)^{-1/2}[2(x-2)+x]$

$$= \frac{3x-4}{2\sqrt{x-2}} > 0 \quad \text{on } [2,\infty)$$

and so h is increasing on $[2,\infty)$.

6. Since h has no critical points in $(2,\infty)$, there are no relative extrema.

7. $h''(x) = \dfrac{1}{2}\left[\dfrac{(x-2)^{1/2}(3)-(3x-4)\tfrac{1}{2}(x-2)^{-1/2}}{x-2}\right]$

$$= \frac{(x-2)^{-1/2}[6(x-2)-(3x-4)]}{4(x-2)} = \frac{3x-8}{4(x-2)^{3/2}}.$$

The sign diagram for h''

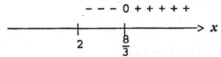

shows that h is concave downward on $(2,\tfrac{8}{3})$ and concave upward on $(\tfrac{8}{3},\infty)$.

8. The results of (7) tell us that $(\tfrac{8}{3},\tfrac{8\sqrt{6}}{9})$ is an inflection point.

The graph of h follows.

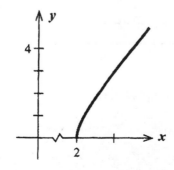

17. $f(x) = \dfrac{x-2}{x+2}$.

1. The domain of f is $(-\infty, -2) \cup (-2, \infty)$.

2. Setting $x = 0$ gives -1 as the y-intercept. Setting $y = 0$ gives 2 as the x-intercept.

3. $\displaystyle\lim_{x \to -\infty} \frac{x-2}{x+2} = \lim_{x \to \infty} \frac{x-2}{x+2} = 1$.

4. The results of (3) tell us that $y = 1$ is a horizontal asymptote. Next, observe that the denominator of $f(x)$ is equal to zero at $x = -2$, but its numerator is not equal to zero there. Therefore, $x = -2$ is a vertical asymptote.

5. $\qquad f'(x) = \dfrac{(x+2)(1) - (x-2)(1)}{(x+2)^2} = \dfrac{4}{(x+2)^2}$.

The sign diagram of f'

f' is not defined here

$\downarrow$

$+\ +\ + \qquad +\ +\ +\ +\ \ +\ +$

```
————————+———————+————————> x
        -2       0
```

tells us that f is increasing on $(-\infty, -2) \cup (-2, \infty)$.

6. The results of (5) tells us that there are no relative extrema.

7. $f''(x) = -\dfrac{8}{(x+2)^3}$. The sign diagram of f'' follows

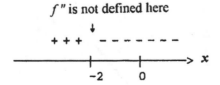

f'' is not defined here

$\downarrow$

$+\ +\ + \qquad -\ -\ -\ -\ -\ -\ -$

```
————————+———————+————————> x
        -2       0
```

and it shows that f is concave upward on $(-\infty, -2)$ and concave downward on $(-2, \infty)$.

8. There are no inflection points.

The graph of f follows.

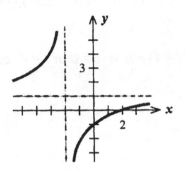

19. $\lim\limits_{x\to-\infty}\dfrac{1}{2x+3}=\lim\limits_{x\to\infty}\dfrac{1}{2x+3}=0$ and so $y=0$ is a horizontal asymptote. Since the denominator is equal to zero at $x=-3/2$, but the numerator is not equal to zero there, we see that $x=-3/2$ is a vertical asymptote.

21. $\lim\limits_{x\to-\infty}\dfrac{5x}{x^2-2x-8}=\lim\limits_{x\to\infty}\dfrac{5x}{x^2-2x-8}=0$ and so $y=0$ is a horizontal asymptote. Next, note that the denominator is zero if $x^2-2x-8=(x-4)(x+2)=0$, or $x=-2$ or $x=4$. Since the numerator is not equal to zero at these points, we see that $x=-2$ and $x=4$ are vertical asymptotes.

23. $f(x)=2x^2+3x-2$; $f'(x)=4x+3$. Setting $f'(x)=0$ gives $x=-3/4$ as a critical point of f. Next, $f''(x)=4>0$ for all x, so f is concave upward on $(-\infty,\infty)$. Therefore, $f(-\tfrac{3}{4})=-\tfrac{25}{8}$ is an absolute minimum of f. There is no absolute maximum.

25. $g(t)=\sqrt{25-t^2}=(25-t^2)^{1/2}$. Differentiating $g(t)$, we have
$$g'(t)=\tfrac{1}{2}(25-t^2)^{-1/2}(-2t)=-\dfrac{t}{\sqrt{25-t^2}}.$$
Setting $g'(t)=0$ gives $t=0$ as a critical point of g. The domain of g is given by solving the inequality $25-t^2\geq0$ or $(5-t)(5+t)\geq0$ which implies that $t\in[-5,5]$. From the table

t	-5	0	5
$g(t)$	0	5	0

we conclude that $g(0) = 5$ is the absolute maximum of g and $g(-5) = 0$ and $g(5) = 0$ is the absolute minimum value of g.

27. $h(t) = t^3 - 6t^2$. $h'(t) = 3t^2 - 12t = 3t(t-4) = 0$ if $t = 0$ or $t = 4$, critical points of h. But only $t = 4$ lies in $(2,5)$.

t	2	4	5
$h(t)$	-16	-32	-25

From the table, we see that there is an absolute minimum at $(4,-32)$ and an absolute maximum at $(2,-16)$.

29. $f(x) = x - \dfrac{1}{x}$ on $[1,3]$. $f'(x) = 1 + \dfrac{1}{x^2}$. Since $f'(x)$ is never zero, f has no critical point.

x	1	3
$f(x)$	0	$\frac{8}{3}$

We see that $f(1) = 0$ is the absolute minimum value and $f(3) = 8/3$ is the absolute maximum value.

31. $f(s) = s\sqrt{1-s^2}$ on $[-1,1]$. The function f is continuous on $[-1,1]$ and differentiable on $(-1,1)$. Next,

$$f'(s) = (1-s^2)^{1/2} + s(\tfrac{1}{2})(1-s^2)^{-1/2}(-2s) = \frac{1-2s^2}{\sqrt{1-s^2}}.$$

Setting $f'(s) = 0$, we have $s = \pm\sqrt{2}/2$, giving the critical points of f. From the table

x	-1	$-\sqrt{2}/2$	$\sqrt{2}/2$	1
$f(x)$	0	-1/2	1/2	0

we see that $f(-\sqrt{2}/2) = -1/2$ is the absolute minimum value and

$f(\sqrt{2}/2) = 1/2$ is the absolute maximum value of f.

33. We want to maximize $P(x) = -x^2 + 8x + 20$. Now, $P'(x) = -2x + 8 = 0$ if $x = 4$, a critical point of P. Since $P''(x) = -2 < 0$, the graph of P is concave downward. Therefore, the critical point $x = 4$ yields an absolute maximum. So, to maximize profit, the company should spend $4000 on advertising per month.

35. a. $I(t) = \dfrac{50t^2 + 600}{t^2 + 10}$.

$I'(t) = \dfrac{(t^2 + 10)(100t) - (50t^2 + 600)(2t)}{(t^2 + 10)^2} = -\dfrac{200t}{(t^2 + 10)^2} < 0$ on $(0,10)$ and so I is

decreasing on $(0,10)$.

b. $I''(t) = -200\left[\dfrac{(t^2 + 10)^2(1) - t(2)(t^2 + 10)(2t)}{(t^2 + 10)^4}\right]$

$= \dfrac{-200(t^2 + 10)[(t^2 + 10) - 4t^2]}{(t^2 + 10)^4} = -\dfrac{200(10 - 3t^2)}{(t^2 + 10)^3}$.

The sign diagram of I'' (for $t > 0$)

```
        - - - -  0 + + + + + +
   ─────────────┼─────────────────→ t
              0     √10/3 ≈ 1.8
```

shows that I is concave downward on $(0, \sqrt{10/3})$ and concave upward on $(\sqrt{10/3}, \infty)$.

c.

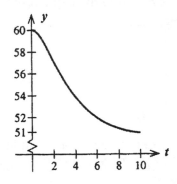

d. The rate of decline in the environmental quality of the wildlife was increasing the first 1.8 years. After that time the rate of decline decreased.

     4 Applications of the Derivative

37. a. $C(x) = 0.001x^2 + 100x + 4000$.

$$\overline{C}(x) = \frac{C(x)}{x} = \frac{0.001x^2 + 100x + 4000}{x} = 0.001x + 100 + \frac{4000}{x}.$$

b. $\overline{C}'(x) = 0.001 - \dfrac{4000}{x^2} = \dfrac{0.001x^2 - 4000}{x^2} = \dfrac{0.001(x^2 - 4,000,000)}{x^2}.$

Setting $\overline{C}'(x) = 0$ gives $x = \pm 2000$. We reject the negative root.

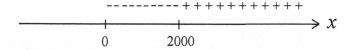

The sign diagram of $\overline{C}'$ shows that $x = 2000$ gives rise to a relative minimum of $\overline{C}$. Since $\overline{C}''(x) = \dfrac{8000}{x^3} > 0$ if $x > 0$, we see that $\overline{C}$ is concave upward on $(0, \infty)$. So $x = 2000$ yields an absolute minimum. So the required production level is 2000 units.

39. $R'(x) = k\dfrac{d}{dx}x(M - x) = k[(M - x) + x(-1)] = k(M - 2x)$

Setting $R'(x) = 0$ gives $M - 2x = 0$, or $x = \frac{M}{2}$, a critical point of R. Since $R''(x) = -2k < 0$, we see that $x = M/2$ affords a maximum; that is R is greatest when half the population is infected.

41. Suppose the radius is r and the height is h. Then the capacity is $\pi r^2 h$ and we want

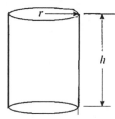

it to be 32π cu ft; that is, $\pi r^2 h = 32\pi$. Let the cost for the side by $c/sq ft. Then the cost of construction is $C = 2\pi rhc + 2(\pi r^2)(2c) = 2\pi crh + 4\pi cr^2$. But $h = \dfrac{32\pi}{\pi r^2} = \dfrac{32}{r^2}$. Therefore,

$$C = f(r) = -\frac{64\pi c}{r^2} + 8\pi cr = \frac{-64\pi c + 8\pi cr^3}{r^2} = \frac{8\pi c(-8 + r^3)}{r^2}$$

Setting $f'(r) = 0$ gives $r^3 = 8$ or $r = 2$. Next, $f''(r) = \frac{128\pi c}{r^3} + 8\pi c$ and so

$f''(2) > 0$. Therefore, $r = 2$ minimizes f. The required dimensions are $r = 2$ and

$h = \frac{32}{4} = 8$. That is, its radius is 2 ft and its height is 8 ft.

43. Let x denote the number of cases in each order. Then the average number of cases of beer in storage during the year is $x/2$. The storage cost is $2(x/2)$, or x dollars. Next, we see that the number of orders required is $800,000/x$, and so the ordering cost is $\frac{500(800,000)}{x} = \frac{400,000,000}{x}$ dollars. Thus, the total cost incurred by the company per year is given by

$$C(x) = x + \frac{400,000,000}{x}.$$

We want to minimize C in the interval $(0, \infty)$. Now

$$C'(x) = 1 - \frac{400,000,000}{x^2}.$$

Setting $C'(x) = 0$ gives $x^2 = 400,000,000$, or $x = 20,000$ (we reject $x = -20,000$).

Next, $C''(x) = \frac{800,000,000}{x^3} > 0$ for all x, so C is concave upward. Thus,

$x = 20,000$ gives rise to the absolute minimum of C. Thus, the company should order 20,000 cases of beer per order.

CHAPTER 5

EXERCISES 5.1, page 337

1. a. $4^{-3} \times 4^5 = 4^{-3+5} = 4^2 = 16$

   b. $3^{-3} \times 3^6 = 3^{6-3} = 3^3 = 27$.

3. a. $9(9)^{-1/2} = \dfrac{9}{9^{1/2}} = \dfrac{9}{3} = 3$.

   b. $5(5)^{-1/2} = 5^{1/2} = \sqrt{5}$.

5. a. $\dfrac{(-3)^4(-3)^5}{(-3)^8} = (-3)^{4+5-8} = (-3)^1 = -3$.

   b. $\dfrac{(2^{-4})(2^6)}{2^{-1}} = 2^{-4+6+1} = 2^3 = 8$.

7. a. $\dfrac{5^{3.3} \cdot 5^{-1.6}}{5^{-0.3}} = \dfrac{5^{3.3-1.6}}{5^{-0.3}} = 5^{1.7+(0.3)} = 5^2 = 25$.

   b. $\dfrac{4^{2.7} \cdot 4^{-1.3}}{4^{-0.4}} = 4^{2.7-1.3+0.4} = 4^{1.8} \approx 12.1257$.

9. a. $(64x^9)^{1/3} = 64^{1/3}(x^{9/3}) = 4x^3$.

   b. $(25x^3y^4)^{1/2} = 25^{1/2}(x^{3/2})(y^{4/2}) = 5x^{3/2}y^2 = 5xy^2\sqrt{x}$.

11. a. $\dfrac{6a^{-5}}{3a^{-3}} = 2a^{-5+3} = 2a^{-2} = \dfrac{2}{a^2}$.

    b. $\dfrac{4b^{-4}}{12b^{-6}} = \dfrac{1}{3}b^{-4+6} = \dfrac{1}{3}b^2$.

13. a. $(2x^3y^2)^3 = 2^3 \times x^{3(3)} \times y^{2(3)} = 8x^9y^6$.

    b. $(4x^2y^2z^3)^2 = 4^2 \times x^{2(2)} \times y^{2(2)} \times z^{3(2)} = 16x^4y^4z^6$.

15. a. $\dfrac{5^0}{(2^{-3}x^{-3}y^2)^2} = \dfrac{1}{2^{-3(2)}x^{-3(2)}y^{2(2)}} = \dfrac{2^6x^6}{y^4} = \dfrac{64x^6}{y^4}$.

    b. $\dfrac{(x+y)(x-y)}{(x-y)^0} = (x+y)(x-y)$.

17. $6^{2x} = 6^4$ if and only if $2x = 4$ or $x = 2$.

19. $3^{3x-4} = 3^5$ if and only if $3x - 4 = 5$, $3x = 9$, or $x = 3$.

21. $(2.1)^{x+2} = (2.1)^5$ if and only if $x + 2 = 5$, or $x = 3$.

23. $8^x = (\frac{1}{32})^{x-2}$, $(2^3)^x = (32)^{2-x} = (2^5)^{2-x}$, so $2^{3x} = 2^{5(2-x)}$, $3x = 10 - 5x$, $8x = 10$,

   or $x = 5/4$.

25. Let $y = 3^x$, then the given equation is equivalent to

$$y^2 - 12y + 27 = 0$$
$$(y-9)(y-3) = 0$$

giving $y = 3$ or 9. So $3^x = 3$ or $3^x = 9$, and therefore, $x = 1$ or $x = 2$.

27. $y = 2^x$, $y = 3^x$, and $y = 4^x$

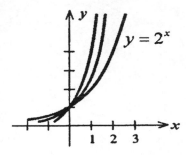

29. $y = 2^{-x}$, $y = 3^{-x}$, and $y = 4^{-x}$

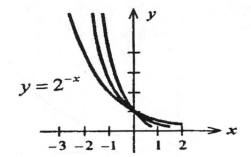

31. $y = 4^{0.5x}$, $y = 4x$, and $y = 4^{2x}$

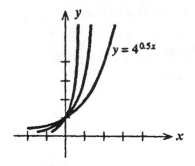

33. $y = e^{0.5x}$, $y = e^x$, $y = e^{1.5x}$

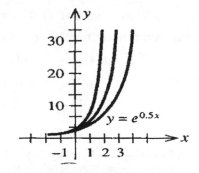

35. $y = 0.5e^{-x}$, $y = e^{-x}$, and $y = 2e^{-x}$

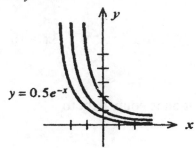

37. a.

Year	0	1	2	3	4	5
Number (billions)	0.45	0.80	1.41	2.49	4.39	7.76

b.

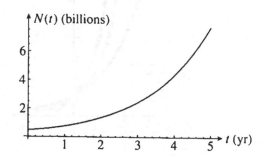

39. a. The initial concentration is given by $C(0) = 0.3(0) - 18(1 - e^{-(0)/60})$, or 0 g/cm^3

   b. The concentration after 10 seconds is given by
   $$C(10) = 0.3(10) - 18(1 - e^{-(10)/60}) = 0.23667, \text{ or } 0.2367 \text{ g/cm}^3.$$

   c. The concentration after 30 seconds is given by
   $$C(30) = 18e^{-(30)/60} - 12e^{-(30-20)/60} = 0.75977, \text{ or } 0.7598 \text{ g/cm}^3.$$

   d. The concentration of the drug in the long run is given by
   $$\lim_{t \to \infty} C(t) = \lim_{t \to \infty}[18e^{-t/60} - 12e^{-\frac{t-20}{60}}] = 0$$

41. a. The concentration initially is given by
   $$N(0) = 0.08 + 0.12(1 - e^{-0.02(0)}) = 0.08, \text{ or } 0.08 \text{ g/cm}^3.$$

   b. The concentration after 20 seconds is given by
   $$N(20) = 0.08 + 0.12(1 - e^{-0.02(20)}) = 0.11956, \text{ or } 0.1196 \text{ g/cm}^3.$$

   c. The concentration in the long run is given by

$$\lim_{t\to\infty} x(t) = \lim_{t\to\infty}[0.08 + 0.12(1 - e^{-0.02t})] = 0.2, \text{ or } 0.2 \text{ g/cm}^3.$$

d.

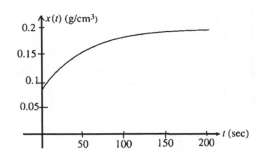

43. False. $(x^2 + 1)^3 = x^6 + 3x^4 + 3x^2 + 1$.

45. True. $f(x) = e^x$ is an increasing function and so if $x < y$, then $f(x) < f(y)$, or $e^x < e^y$.

USING TECHNOLOGY EXERCISES 5.1, page 338

1.

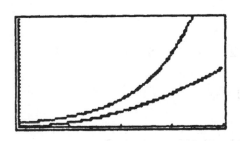

3.

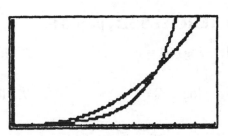

5..

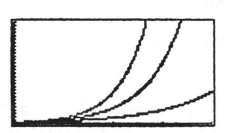

7.

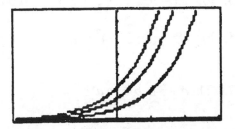

9.

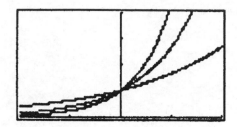

11. a.

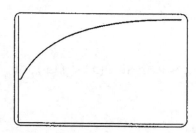

b. 0.08 g/cm^3 c. 0.12 g/cm^3

d. 0.2 g/cm^3

13. a.

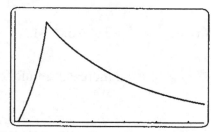

b. 20 sec.   c. 35.1 sec

EXERCISES 5.2 , page 346

1. $\log_2 64 = 6$

3. $\log_3 \dfrac{1}{9} = -2$

5. $\log_{1/3} \dfrac{1}{3} = 1$

7. $\log_{32} 8 = \dfrac{3}{5}$

9. $\log_{10} 0.001 = -3$

11. $\log 12 = \log 4 \times 3 = \log 4 + \log 3 = 0.6021 + 0.4771 = 1.0792.$

13. $\log 16 = \log 4^2 = 2 \log 4 = 2(0.6021) = 1.2042.$

15. $\log 48 = \log 3 \times 4^2 = \log 3 + 2 \log 4 = 0.4771 + 2(0.6021) = 1.6813.$

17. $2 \ln a + 3 \ln b = \ln a^2 b^3.$

19. $\ln 3 + \dfrac{1}{2} \ln x + \ln y - \dfrac{1}{3} \ln z = \ln \dfrac{3\sqrt{x}\, y}{\sqrt[3]{z}}$

21. $\log x(x + 1)^4 = \log x + \log (x + 1)^4 = \log x + 4 \log (x + 1).$

23. $\log \dfrac{\sqrt{x+1}}{x^2+1} = \log (x + 1)^{1/2} - \log(x^2 + 1) = \frac{1}{2} \log (x + 1) - \log (x^2 + 1)$

25. $\ln xe^{-x^2} = \ln x - x^2.$

27. $\ln \left(\dfrac{x^{1/2}}{x^2 \sqrt{1+x^2}} \right) = \ln x^{1/2} - \ln x^2 - \ln (1 + x^2)^{1/2}$

$\qquad = \frac{1}{2} \ln x - 2 \ln x - \frac{1}{2} \ln (1 + x^2) = -\frac{3}{2} \ln x - \frac{1}{2} \ln (1 + x^2).$

29. $y = \log_3 x$

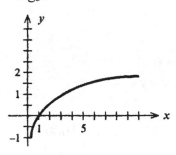

31. $y = \ln 2x$

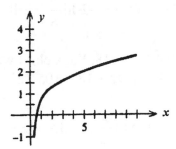

33. $y = 2^x$ and $y = \log_2 x$

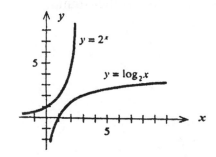

35. $e^{0.4t} = 8$, $0.4t \ln e = \ln 8$, and $0.4t = \ln 8$ ($\ln e = 1.$) So, $t = \dfrac{\ln 8}{0.4} = 5.1986.$

37. $5e^{-2t} = 6$, $e^{-2t} = \frac{6}{5} = 1.2$. Taking the logarithm, we have

$\qquad -2t \ln e = \ln 1.2$, or $t = -\dfrac{\ln 1.2}{2} \approx -0.0912.$

39. $2e^{-0.2t} - 4 = 6$, $2e^{-0.2t} = 10$. Taking the logarithm on both sides of this last equation, we have $\ln e^{-0.2t} = \ln 5$; $-0.2t \ln e = \ln 5$; $-0.2t = \ln 5$;

and $t = -\dfrac{\ln 5}{0.2} \approx -8.0472.$

41. $\dfrac{50}{1+4e^{0.2t}} = 20, \quad 1+4e^{0.2t} = \dfrac{50}{20} = 2.5, \quad 4e^{0.2t} = 1.5,$

$e^{0.2t} = \dfrac{1.5}{4} = 0.375, \; \ln e^{0.2t} = \ln 0.375, \; 0.2t = \ln 0.375.$ So $t = \dfrac{\ln 0.375}{0.2} \approx -4.9041.$

43. Taking the logarithm on both sides, we obtain

$\ln A = \ln Be^{-t/2}, \;\; \ln A = \ln B + \ln e^{-t/2}, \; \ln A - \ln B = -t/2 \ln e,$

$\ln \dfrac{A}{B} = -\dfrac{t}{2}$ or $t = -2 \ln \dfrac{A}{B} = 2 \ln \dfrac{B}{A}$

45. $p(x) = 19.4 \ln x + 18.$ For a child weighing 92 lb, we find
$p(92) = 19.4 \ln 92 + 18 = 105.72$ millimeters of mercury.

47. a. $\quad 30 = 10 \log \dfrac{I}{I_0}; \quad 3 = \log \dfrac{I}{I_0}; \quad \dfrac{I}{I_0} = 10^3 = 1000.$ So $I = 1000 \, I_0.$

b. When $D = 80$, $I = 10^8 I_0$ and when $D = 30$, $I = 10^3 I_0$. Therefore, an 80–decibel sound is $10^8/10^3$ or $10^5 = 100,000$ times louder than a 30–decibel sound.

c. It is $10^{15}/10^8 = 10^7$, or $10,000,000$, times louder.

49. We solve the following equation for t. Thus,

$$\dfrac{160}{1+240e^{-0.2t}} = 80; \quad 1+240e^{-0.2t} = \dfrac{160}{80},$$

$$240e^{-0.2t} = 2-1 = 1; \quad e^{-0.2t} = \dfrac{1}{240}; \quad -0.2t = \ln\dfrac{1}{240}$$

$$t = -\dfrac{1}{0.2}\ln\dfrac{1}{240} \approx 27.40, \text{ or approximately 27.4 years old.}$$

51. We solve the following equation for t:

$$200(1-0.956e^{-0.18t}) = 140; \quad 1-0.956e^{-0.18t} = \dfrac{140}{200} = 0.7$$

$$-0.956e^{-0.18t} = 0.7-1 = -0.3 \; ; \quad e^{-0.18t} = \tfrac{0.3}{0.956}$$

$$-0.18t = \ln\left(\tfrac{0.3}{0.956}\right) \text{ and } t = -\dfrac{\ln\left(\tfrac{0.3}{0.956}\right)}{0.18} \approx 6.43875.$$

So, its approximate age is 6.44 years.

53. a. We solve the equation $0.08 + 0.12e^{-0.02t} = 0.18$.

$$0.12e^{-0.02t} = 0.1; \ e^{-0.02t} = \frac{0.1}{0.12} = \frac{1}{1.2}$$

$$\ln e^{-0.02t} = \ln \frac{1}{1.2} = \ln 1 - \ln 1.2 = -\ln 1.2$$

$$-0.02t = -\ln 1.2 \quad \text{and} \quad t = \frac{\ln 1.2}{0.02} \approx 9.116, \quad \text{or 9.12 sec.}$$

b. We solve the equation $0.08 + 0.12e^{-0.02t} = 0.16$.

$$0.12e^{-0.02t} = 0.08; \ e^{-0.02t} = \frac{0.08}{0.12} = \frac{2}{3}; \ -0.02t = \ln \frac{2}{3}$$

$$t = -\frac{\ln\left(\frac{2}{3}\right)}{0.02} \approx 20.2733, \quad \text{or 20.27 sec.}$$

55. False. Take $x = e$. Then $(\ln e)^3 = 1^3 = 1 \neq 3 \ln e = 3$.

57. True. $g(x) = \ln x$ is continuous and greater than zero on $(1, \infty)$. Therefore,

$f(x) = \dfrac{1}{\ln x}$ is continuous on $(1, \infty)$.

59. a. Taking the logarithm on both sides gives $\ln 2^x = \ln e^{kx}$, $x \ln 2 = kx(\ln e) = kx$.
So, $x(\ln 2 - k) = 0$ for all x and this implies that $k = \ln 2$.
b. Tracing the same steps as done in (a), we find that $k = \ln b$.

61. Let $\log_b m = p$, then $m = b^p$. Therefore, $m^n = (b^p)^n = b^{np}$. Therefore,
$$\log_b m^n = \log_b b^{np} = np \log_b b = np \qquad (\text{Since } \log_b b = 1.)$$
$$= n \log_b m,$$
as was to be shown.

EXERCISES 5.3, page 359

1. $A = 2500\left(1 + \dfrac{0.07}{2}\right)^{20} = 4974.47$, or $\$4974.47$.

3. $A = 150,000 \left(1 + \dfrac{0.1}{12}\right)^{48} = 223,403.11,$ or $\$223,403.11$

5. a. Using the formula $r_{eff} = \left(1 + \dfrac{r}{m}\right)^{m} - 1$ with $r = 0.10$ and $m = 2$, we have

$$r_{eff} = \left(1 + \dfrac{0.10}{2}\right)^{2} - 1 = 0.1025, \quad \text{or } 10.25 \text{ percent/yr}$$

   b. Using the formula $r_{eff} = \left(1 + \dfrac{r}{m}\right)^{m} - 1$ with $r = 0.09$ and $m = 4$, we have

$$r_{eff} = \left(1 + \dfrac{0.09}{4}\right)^{4} - 1 = 0.09308, \text{ or } 9.308 \text{ percent/yr.}$$

7. a. The present value is given by $P = 40,000 \left(1 + \dfrac{0.08}{2}\right)^{-8} = 29,227.61,$

   or $\$29,227.61$.

   b. The present value is given by $P = 40,000 \left(1 + \dfrac{0.08}{4}\right)^{-16} = 29,137.83,$ or

   $\$29,137.83$.

9. $A = 5000 e^{0.08(4)} \approx 6885.64,$ or $\$6,885.64$.

11. We use formula (6) with $A = 7500$, $P = 5000$, $m = 12$, and $t = 3$. Thus
$$7500 = 5000 \left(1 + \tfrac{r}{12}\right)^{36};$$
$$\left(1 + \tfrac{r}{12}\right)^{36} = \tfrac{7500}{5000} = \tfrac{3}{2}, \quad \ln\left(1 + \tfrac{r}{12}\right)^{36} = \ln 1.5;$$
$$36\left(1 + \tfrac{r}{12}\right) = \ln 1.5$$
$$\left(1 + \tfrac{r}{12}\right) = \tfrac{\ln 1.5}{36} = 0.0112629$$
$$1 + \tfrac{r}{12} = e^{0.0112629} = 1.011327; \tfrac{r}{12} = 0.011327;$$
$$r = 0.13592$$
So the interest rate is 13.59% per year.

13. We use formula (6) with $A = 8000$, $P = 4000$, $m = 2$, and $t = 4$. Thus
$$8000 = 4000 \left(1 + \tfrac{r}{2}\right)^{8}$$
$$\left(1 + \tfrac{r}{2}\right)^{8} = \tfrac{8000}{5000} = 1.6$$

$$\ln(1+\tfrac{r}{2})^8 = \ln 1.6$$
$$8\ln\left(1+\tfrac{r}{2}\right) = \ln 1.6$$
$$\ln\left(1+\tfrac{r}{2}\right) = \tfrac{\ln 1.6}{8} = 0.05875$$
$$1+\tfrac{r}{2} = e^{0.05875} = 1.06051; \ \tfrac{r}{2} = 0.06051$$
$$r = 0.1210$$

So the required interest rate is 12.1% per year.

15. We use formula (6) with $A = 4000$, $P = 2000$, $m = 1$, and $t = 5$. Thus
$$4000 = 2000(1+r)^5; \quad (1+r)^5 = 2 ; 5\ln(1+r) = \ln 2; \ \ln(1+r) = \tfrac{\ln 2}{5} = 0.138629$$
$$1+r = e^{0.138629} = 1.148698; \ r = 0.1487$$

So the required interst rate is 14.87% per year.

17. We use formula (6) with $A = 6500$, $P = 5000$, $m = 12$, and $r = 0.12$. Thus
$$6500 = 5000\left(1+\frac{0.12}{12}\right)^{12t}; \quad (1.01)^{12t} = \frac{6500}{5000} = 1.3; \ 12t\ln(1.01) = \ln 1.3$$
$$t = \frac{\ln 1.3}{12\ln 1.01} \approx 2.197$$

So, it will take approximately 2.2 years.

19. We use formula (6) with $A = 4000$, $P = 2000$, $m = 12$, and $r = 0.09$. Thus,
$$4000 = 2000\left(1+\tfrac{0.09}{12}\right)^{12t}$$
$$\left(1+\tfrac{0.09}{12}\right)^{12t} = 2$$
$$12t\ln\left(1+\tfrac{0.09}{12}\right) = \ln 2 \ \text{ and } \ t = \frac{\ln 2}{12\ln\left(1+\tfrac{0.09}{12}\right)} \approx 7.73.$$

So it will take approximately 7.7 years.

21. We use formula (10) with $A = 6000$, $P = 5000$, and $t = 3$. Thus,
$$6000 = 5000e^{3r}$$
$$e^{3r} = \frac{6000}{5000} = 1.2; \quad 3r = \ln 1.2$$
$$r = \frac{\ln 1.2}{3} \approx 0.6077$$

So the interest rate is 6.08% per year.

23. We use formula (10) formula (6) with $A = 7000$, $P = 6000$, and $r = 0.075$. Thus

$$7000 = 6000e^{0.075t}; \quad e^{0.075t} = \tfrac{7000}{6000} = \tfrac{7}{6}$$

$$0.075t \ln e = \ln \tfrac{7}{6} \quad \text{and} \quad t = \frac{\ln \tfrac{7}{6}}{0.075} \approx 2.055.$$

So, it will take 2.06 years.

25. The Estradas can expect to pay $80,000(1+0.09)^4$, or approximately \$112,926.52.

27. The investment will be worth

$$A = 1.5\left(1+\frac{0.095}{2}\right)^{20} = 3.794651 \text{ , or approximately \$3.8 million dollars.}$$

29. The present value of the \$8000 loan due in 3 years is given by

$$P = 8000\left(1+\frac{0.10}{2}\right)^{-6} = 5969.72, \text{ or } \$5969.72.$$

The present value of the \$15,000 loan due in 6 years is given by

$$P = 15,000\left(1+\frac{0.10}{2}\right)^{-12} = 8352.56, \text{ or } \$8352.56.$$

Therefore, the amount the proprietors of the inn will be required to pay at the end

of 5 years is given by $A = 14,322.28\left(1+\frac{0.10}{2}\right)^{10} = 23,329.48$, or \$23,329.48.

31. We solve the equation $2 = 1(1+0.075)^t$ for t. Taking the logarithm on both sides,

we have $\ln 2 = \ln(1.075)^t \approx t \ln 1.075$. So $t = \frac{\ln 2}{\ln 1.075} \approx 9.58$, or 9.6 years.

33. The effective annual rate of return on his investment is found by solving the

equation $(1+r)^2 = \dfrac{32100}{25250}$

$$1+r = \left(\frac{32100}{25250}\right)^{1/2}$$

$1+r \approx 1.1275$ and $r = 0.1275$, or 12.75 percent.

35. $P = Ae^{-rt} = 59673e^{-(0.08)5} \approx 40{,}000.008$, or approximately $40,000.

37. a. If they invest the money at 10.5 percent compounded quarterly, they should set aside $P = 70{,}000\left(1 + \frac{0.105}{4}\right)^{-28} \approx 33{,}885.14$, or $33,885.14.

b. If they invest the money at 10.5 percent compounded continuously, they should set aside $P = 70{,}000e^{-0.735} = 33{,}565.38$, or $33,565.38.

39. a. If inflation over the next 15 years is 6 percent, then Eleni's first year's pension will be worth $P = 40{,}000e^{-0.9} = 16{,}262.79$, or $16,262.79.

b. If inflation over the next 15 years is 8 percent, then Eleni's first year's pension will be worth $P = 40{,}000e^{-1.2} = 12{,}047.77$, or $12,047.77.

c. If inflation over the next 15 years is 12 percent, then Eleni's first year's pension will be worth $P = 40{,}000e^{-1.8} = 6611.96$, or $6,611.96.

41. $r_{eff} = \lim\limits_{m \to \infty}\left(1 + \dfrac{r}{m}\right)^{m} - 1 = e^r - 1$.

43. The effective rate of interest at Bank A is given by
$$R = \left(1 + \tfrac{0.07}{4}\right)^4 - 1 = 0.07186,$$
or 7.186 percent. The effective rate at Bank B is given by
$$R = e^r - 1 = e^{0.07125} - 1 = 0.07385$$
or 7.385 percent. We conclude that Bank B has the higher effective rate of interest.

45. The nominal rate of interest that, when compounded continuously, yields an effective rate of interest of 10 percent per year is found by solving the equation
$$R = e^r - 1, \ 0.10 = e^r - 1, \ 1.10 = e^r, \quad \ln 1.10 = r \ln e, \ r = \ln 1.10 \approx 0.09531,$$
or 9.531 percent.

EXERCISES 5.4 , page 368

1. $f(x) = e^{3x}; \ f'(x) = 3e^{3x}$

3. $g(t) = e^{-t}; g'(t) = -e^{-t}$

5. $f(x) = e^x + x; \ f'(x) = e^x + 1$

7. $f(x) = x^3 e^x, \ f'(x) = x^3 e^x + e^x(3x^2) = x^2 e^x(x + 3)$.

9. $f(x) = \dfrac{2e^x}{x}$, $f'(x) = \dfrac{x(2e^x) - 2e^x(1)}{x^2} = \dfrac{2e^x(x-1)}{x^2}$.

11. $f(x) = 3(e^x + e^{-x})$; $f'(x) = 3(e^x - e^{-x})$.

13. $f(w) = \dfrac{e^w + 1}{e^w} = 1 + \dfrac{1}{e^w} = 1 + e^{-w}$. $f'(w) = -e^{-w} = -\dfrac{1}{e^w}$.

15. $f(x) = 2e^{3x-1}$, $f'(x) = 2e^{3x-1}(3) = 6e^{3x-1}$.

17. $h(x) = e^{-x^2}$; $h'(x) = e^{-x^2}(-2x) = -2xe^{-x^2}$.

19. $f(x) = 3e^{-1/x}$; $f'(x) = 3e^{-1/x} \cdot \dfrac{d}{dx}\left(-\dfrac{1}{x}\right) = 3e^{-1/x}\left(\dfrac{1}{x^2}\right) = \dfrac{3e^{-1/x}}{x^2}$.

21. $f(x) = (e^x + 1)^{25}$, $f'(x) = 25(e^x + 1)^{24}e^x = 25e^x(e^x + 1)^{24}$.

23. $f(x) = e^{\sqrt{x}}$; $f'(x) = e^{\sqrt{x}}\dfrac{d}{dx}x^{1/2} = e^{\sqrt{x}}\dfrac{1}{2}x^{-1/2} = \dfrac{e^{\sqrt{x}}}{2\sqrt{x}}$.

25. $f(x) = (x-1)e^{3x+2}$; $f'(x) = (x-1)(3)e^{3x+2} + e^{3x+2} = e^{3x+2}(3x - 3 + 1) = e^{3x+2}(3x - 2)$.

27. $f(x) = \dfrac{e^x - 1}{e^x + 1}$; $f'(x) = \dfrac{(e^x + 1)(e^x) - (e^x - 1)(e^x)}{(e^x + 1)^2} = \dfrac{e^x(e^x + 1 - e^x + 1)}{(e^x + 1)^2} = \dfrac{2e^x}{(e^x + 1)^2}$.

29. $f(x) = e^{-4x} + 2e^{3x}$; $f'(x) = -4e^{-4x} + 6e^{3x}$ and
    $f''(x) = 16e^{-4x} + 18e^{3x} = 2(8e^{-4x} + 9e^{3x})$.

31. $f(x) = 2xe^{3x}$; $f'(x) = 2e^{3x} + 2xe^{3x}(3) = 2(3x + 1)e^{3x}$.
    $f''(x) = 6e^{3x} + 2(3x + 1)e^{3x}(3) = 6(3x + 2)e^{3x}$.

33. $y = f(x) = e^{2x - 3}$. $f'(x) = 2e^{2x-3}$. To find the slope of the tangent line to the graph
    of f at $x = 3/2$, we compute $f'(\tfrac{3}{2}) = 2e^{3-3} = 2$. Next, using the point–slope form of
    the equation of a line, we find that
    $$y - 1 = 2(x - \tfrac{3}{2})$$

$$= 2x - 3, \quad \text{or} \quad y = 2x - 2.$$

35. $f(x) = e^{-x^2/2}$, $f'(x) = e^{-x^2/2}(-x) = -xe^{-x^2/2}$. Setting $f'(x) = 0$, gives $x = 0$ as the only critical point of f. From the sign diagram,

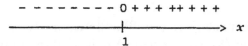

we conclude that f is increasing on $(-\infty, 0)$ and decreasing on $(0, \infty)$.

37. $f(x) = \frac{1}{2}e^x - \frac{1}{2}e^{-x}$, $f'(x) = \frac{1}{2}(e^x + e^{-x})$, $f''(x) = \frac{1}{2}(e^x - e^{-x})$. Setting $f''(x) = 0$, gives $e^x = e^{-x}$ or $e^{2x} = 1$, and $x = 0$. From the sign diagram for f'',

we conclude that f is concave upward on $(0, \infty)$ and concave downward on $(-\infty, 0)$.

39. $f(x) = xe^{-2x}$. $f'(x) = e^{-2x} + xe^{-2x}(-2) = (1 - 2x)e^{-2x}$.
$f''(x) = -2e^{-2x} + (1 - 2x)e^{-2x}(-2) = 4(x - 1)e^{-2x}$.
Observe that $f''(x) = 0$ if $x = 1$. The sign diagram of f''

shows that $(1, e^{-2})$ is an inflection point.

1. $f(x) = e^{-x^2}$. $f'(x) = -2xe^{-x^2} = 0$ if $x = 0$, the only critical point of f.

x	-1	0
	1	
$f(x)$	e^{-1}	1
	e^{-1}	

From the table, we see that f has an absolute minimum value of e^{-1} attained at $x = -1$ and $x = 1$. It has an absolute maximum at $(0, 1)$.

3. $g(x) = (2x - 1)e^{-x}$; $g'(x) = 2e^{-x} + (2x - 1)e^{-x}(-1) = (3 - 2x)e^{-x} = 0$, if $x = 3/2$. The

graph of g shows that $(\frac{3}{2}, 2e^{-3/2})$ is an absolute maximum, and $(0,-1)$ is an absolute minimum.

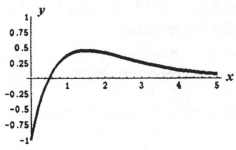

45. $f(t) = e^t - t$;

We first gather the following information on f.
1. The domain of f is $(-\infty,\infty)$.
2. Setting $t = 0$ gives 1 as the y-intercept.
3. $\lim\limits_{t\to-\infty} (e^t - t) = \infty$ and $\lim\limits_{t\to\infty} (e^t - t) = \infty$.
4. There are no asymptotes.
5. $f'(t) = e^t - 1$ if $t = 0$, a critical point of f. From the sign diagram for f'

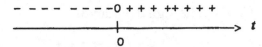

we see that f is decreasing on $(-\infty,0)$ and increasing on $(0,\infty)$.
6. From the results of (5), we see that $(0,1)$ is a relative minimum of f.
7. $f''(t) = e^t > 0$ for all t in $(-\infty,\infty)$. So the graph of f is concave upward on $(-\infty,\infty)$.
8. There are no inflection points.
The graph of f follows.

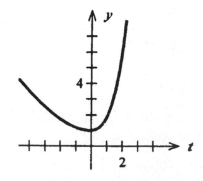

47. $f(x) = 2 - e^{-x}$.
We first gather the following information on f.
1. The domain of f is $(-\infty, \infty)$.
2. Setting $x = 0$ gives 1 as the y-intercept.
3. $\lim\limits_{x \to -\infty} (2 - e^{-x}) = -\infty$ and $\lim\limits_{x \to \infty} (2 - e^{-x}) = 2$,
4. From the results of (3), we see that $y = 2$ is a horizontal asymptote of f.
5. $f'(x) = e^{-x}$. Observe that $f'(x) > 0$ for all x in $(-\infty, \infty)$ and so f is increasing on $(-\infty, \infty)$.
6. Since there are no critical points, f has no relative extrema.
7. $f''(x) = -e^{-x} < 0$ for all x in $(-\infty, \infty)$ and so the graph of f is concave downward on $(-\infty, \infty)$.
8. There are no inflection points
The graph of f follows.

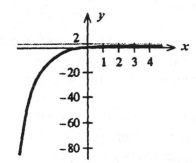

49. $P'(t) = 20.6(-0.009)e^{-0.009t} = -0.1854e^{-0.009t}$
$P'(10) = -0.1694$, $P'(20) = -0.1549$, and $P'(30) = -0.1415$,
and this tells us that the percentage of the total population relocating was decreasing at the rate of 0.17% in 1970, 0.15% in 1980, and 0.14% in 1990.

51. a. The number of air passengers in 2000 is $N(0) = 666$, or 666 million. The number in 2005 is $N(5) = 818,759$, or 819 million.
   b. $f'(t) = 666(0.0413)e^{0.0413t} = 27.5058e^{0.0413t}$. The required rate is $f'(5) = 33.8147$, or 33.8 million/yr.

53. a. $S(t) = 20,000(1 + e^{-0.5t})$
   $S'(t) = 20,000(-0.5e^{-0.5t}) = -10,000e^{-0.5t}$;
   $S'(1) = -10,000e^{-0.5} = -6065$, or $-\$6065$/day.
   $S'(2) = -10,000e^{-1} = -3679$, or $-\$3679$/day.

$$S'(3) = -10{,}000(e^{-1.5}) = -2231, \text{ or } -\$2231/\text{day}.$$
$$S'(4) = -10{,}000e^{-2} = -1353, \text{ or } -\$1353/\text{day}.$$

b. $\qquad S(t) = 20{,}000(1 + e^{-0.5t}) = 27{,}400$

$$1 + e^{-0.5t} = \frac{27{,}400}{20{,}000}$$

$$e^{-0.5t} = \frac{274}{200} - 1$$

$$-0.5t = \ln\left(\frac{274}{200} - 1\right)$$

$$t = \frac{\ln\left(\dfrac{274}{200} - 1\right)}{-0.5} \approx 2$$

55. $N(t) = 5.3e^{0.095t^2 - 0.85t}$.

a. $N'(t) = 5.3e^{0.095t^2 - 0.85t}(0.19t - 0.85)$. Since $N'(t)$ is negative for $(0 \le t \le 4)$, we see that $N(t)$ is decreasing over that interval.

b. To find the rate at which the number of polio cases was decreasing at the beginning of 1959, we compute
$$N'(0) = 5.3e^{0.095(0^2) - 0.85(0)}(0.85) = 5.3(-0.85) = -4.505$$
(t is measured in thousands), or 4,505 cases per year. To find the rate at which the number of polio cases was decreasing at the beginning of 1962, we compute
$$N'(3) = 5.3e^{0.095(9) - 0.85(3)}(0.57 - 0.85)$$
$$= (-0.28)(0.9731) \approx -0.273, \text{ or } 273 \text{ cases per year.}$$

57. From the results of Exercise 56, we see that $R'(x) = 100(1 - 0.0001x)e^{-0.0001x}$.
Setting $R'(x) = 0$ gives $x = 10{,}000$, a critical point of R. From the graph of R

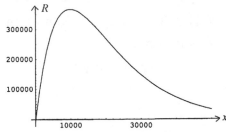

we see that the revenue is maximized when $x = 10{,}000$. So 10,000 pairs must be sold, yielding a maximum revenue of $R(10{,}000) = 367{,}879.44$, or \$367,879.

59. $p = 240\left(1 - \dfrac{3}{3 + e^{-0.0005x}}\right) = 240[1 - 3(3 + e^{-0.0005x})^{-1}].$

$p' = 720(3 + e^{-0.0005x})^{-2}(-0.0005e^{-0.0005x})$

$p'(1000) = 720(3 + e^{-0.0005(1000)})^{-2}(-0.0005e^{-0.0005(1000)})$

$\qquad = -\dfrac{0.36(0.606531)}{(3 + 0.606531)^2} \approx -0.0168,$ or -1.68 cents per case.

$p(1000) = 240(1 - \dfrac{3}{3.606531}) \approx 40.36,$ or $\$40.36$/case.

61. a. $W = 2.4e^{1.84h}$; $W = 2.4e^{1.84(16)} \approx 45.58,$ or approximately 45.6 kg.

  b. $\Delta W \approx dW = (2.4)(1.84)e^{1.84h}dh$. With $h = 1.6$ and $dh = \Delta h = 1.65 - 1.6 = 0.05$,

  we find $\Delta W \approx (2.4)(1.84)e^{1.84(1.6)} \cdot (0.05) \approx 4.19$, or approximately 4.2 kg.

63. $P(t) = 80{,}000\, e^{\sqrt{t}/2 - 0.09t} = 80{,}000\, e^{\frac{1}{2}t^{1/2} - 0.09t}$.

  $P'(t) = 80{,}000(\frac{1}{4}t^{-1/2} - 0.09)e^{\frac{1}{2}t^{1/2} - 0.09t}$.

  Setting $P'(t) = 0$, we have

  $$\tfrac{1}{4}t^{-1/2} = 0.09, \quad t^{-1/2} = 0.36, \quad \frac{1}{\sqrt{t}} = 0.36, \quad t = \left(\frac{1}{0.36}\right)^2 \approx 7.72.$$

  Evaluating $P(t)$ at each of its endpoints and at the point $t = 7.72$, we find

t	$P(t)$
0	80,000
7.72	160,207.69
8	160,170.71

  We conclude that P is optimized at $t = 7.72$. The optimal price is $\$160{,}207.69$.

65. $f(t) = 1.5 + 1.8te^{-1.2t}$

  $f'(t) = 1.8\dfrac{d}{dt}(te^{-1.2t}) = 1.8[e^{-1.2t} + te^{-1.2t}(-1.2)] = 1.8e^{-1.2t}(1 - 1.2t).$

  $f'(0) = 1.8,\ f'(1) = -0.11,\ f'(2) = -0.23,$ and $f'(3) = -0.13,$

and this tells us that the rate of change of the amount of oil used is 1.8 barrels per $1000 of output per decade in 1965; it is decreasing at the rate of 0.11 barrels per $1000 of output per decade in 1966, and so on.

67. a. The price at $t = 0$ is $8 + 4$, or 12, dollars per unit.

b. $\dfrac{dp}{dt} = -8e^{-2t} + e^{-2t} - 2te^{-2t}$.

$\dfrac{dp}{dt}\bigg|_{t=0} = -8e^{-2t} + e^{-2t} - 2te^{-2t}\bigg|_{t=0} = -8 + 1 = -7$.

That is, the price is decreasing at the rate of $7/week.

c. The equilibrium price is $\lim\limits_{t\to\infty}(8 + 4e^{-2t} + te^{-2t}) = 8 + 0 + 0$, or $8 per unit.

69. We are given that
$$c(1 - e^{-at/V}) < m$$

$$1 - e^{-at/V} < \frac{m}{c}$$

$$-e^{-at/V} < \frac{m}{c} - 1 \quad \text{and} \quad e^{-at/V} > 1 - \frac{m}{c}.$$

Taking the log of both sides of the inequality, we have

$$-\frac{at}{V}\ln e > \ln\frac{c - m}{c}$$

$$-\frac{at}{V} > \ln\frac{c - m}{c}$$

$$-t > \frac{V}{a}\ln\frac{c - m}{c} \quad \text{or} \quad t < \frac{V}{a}\left(-\ln\frac{c - m}{c}\right) = \frac{V}{a}\ln\left(\frac{c}{c - m}\right).$$

Therefore the liquid must not be allowed to enter the organ for a time longer than
$$t = \frac{V}{a}\ln\left(\frac{c}{c - m}\right) \text{ minutes.}$$

71. $C'(t) = \begin{cases} 0.3 + 18e^{-t/60}(-\frac{1}{60}) & 0 \le t \le 20 \\ -\frac{18}{60}e^{-t/60} + \frac{12}{60}e^{-(t-20)/60} & t > 20 \end{cases} = \begin{cases} 0.3(1 - e^{-t/60}) & 0 \le t \le 20 \\ -0.3e^{-t/60} + 0.2e^{-(t-20)/60} & t > 20 \end{cases}$

a. $C'(10) = 0.3\left(1 - e^{-10/60}\right) \approx 0.05$ or 0.05 g/cm^3/sec.

b. $C'(30) = -0.3e^{-30/60} + 0.2e^{-10/60} \approx -0.01$, or decreasing at the rate of 0.01 g/ cm^3/sec.

c. On the interval $(0, 20)$, $C'(t) = 0$ implies $1 - e^{-t/60} = 0$, or $t = 0$.
Therefore, C attains its absolute maximum value at an endpoint. In this
case, at $t = 20$. On the interval $[20, \infty)$, $C'(t) = 0$ implies

$$-0.3e^{-t/60} = -0.2e^{-(t-20)/60}$$

$$\frac{e^{-\left(\frac{t-20}{60}\right)}}{e^{-t/60}} = \frac{3}{2}; \text{ or } e^{1/3} = \frac{3}{2},$$

which is not possible. Therefore $C'(t) \neq 0$ on $[20, \infty)$. Since $C(t) \to 0$ as
$t \to \infty$, the absolute maximum of c occurs at $t = 20$. Thus, the
concentration of the drug reaches a maximum at $t = 20$.

d. The maximum concentration is $C(20) = 0.90$ g/cm^3.

73. False. $f(x) = 3^x = e^{x\ln 3}$ and so $f'(x) = e^{x\ln 3} \cdot \dfrac{d}{dx}(x\ln 3) = (\ln 3)e^{x\ln 3} = (\ln 3)3^x$.

75. False. $f'(x) = (\ln \pi)\pi^x$..

USING TECHNOLOGY EXERCISES 5.4, page 414

1. 5.4366   3. 12.3929   5. 0.1861

7. a. The initial population of crocodiles is $P(0) = \frac{300}{6} = 50$.

   b. $\lim\limits_{t \to 0} P(t) = \lim\limits_{t \to 0} \dfrac{300e^{-0.024t}}{5e^{-0.024t} + 1} = \dfrac{0}{0+1} = 0.$

   c.

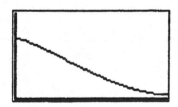

9. a.   b. 4.2720 billion/half century

11. a. Using the function evaluation capabilities of a graphing utility, we find
$$f(11) = 153.024 \text{ and } g(11) = 235.180977624$$
and this tells us that the number of violent-crime arrests will be 153,024 at the beginning of the year 2000, but if trends like inner-city drug use and wider availability of guns continue, then the number of arrests will be 235,181.

b. Using the differentiation capability of a graphing utility, we find
$$f'(11) = -0.634 \text{ and } g'(11) = 18.4005596893$$
and this tells us that the number of violent-crime arrests will be decreasing at the rate of 634 per year at the beginning of the year 2000. But if the trends like inner-city drug use and wider availability of guns continues, then the number of arrests will be increasing at the rate of 18,401 per year at the beginning of the year 2000.

13. a. $P(10) = \dfrac{74}{1 + 2.6e^{-0.166(10) + 0.04536(10)^2 - 0.0066(10)^3}} \approx 69.63$ percent.

b. $P'(10) = 5.09361$, or $5.09361\%/\text{decade}$

EXERCISES 5.5, page 379

1. $f(x) = 5 \ln x; f'(x) = 5\left(\dfrac{1}{x}\right) = \dfrac{5}{x}.$ 3. $f(x) = \ln (x + 1); f'(x) = \dfrac{1}{x+1}.$

5. $f(x) = \ln x^8; f'(x) = \dfrac{8x^7}{x^8} = \dfrac{8}{x}.$

7. $f(x) = \ln x^{1/2}; \quad f'(x) = \dfrac{\frac{1}{2}x^{-1/2}}{x^{1/2}} = \dfrac{1}{2x}.$

9. $f(x) = \ln\left(\dfrac{1}{x^2}\right) = \ln x^{-2} = -2 \ln x; \quad f'(x) = -\dfrac{2}{x}.$

11. $f(x) = \ln (4x^2 - 6x + 3); \quad f'(x) = \dfrac{8x - 6}{4x^2 - 6x + 3} = \dfrac{2(4x - 3)}{4x^2 - 6x + 3}.$

13. $f(x) = \ln\left(\dfrac{2x}{x+1}\right) = \ln 2x - \ln (x + 1).$

$$f'(x) = \frac{2}{2x} - \frac{1}{x+1} = \frac{2(x+1) - 2x}{2x(x+1)} = \frac{2x + 2 - 2x}{2x(x+1)} = \frac{2}{2x(x+1)} = \frac{1}{x(x+1)}.$$

5. $f(x) = x^2 \ln x$; $f'(x) = x^2\left(\frac{1}{x}\right) + (\ln x)(2x) = x + 2x \ln x = x(1 + 2 \ln x)$

7. $f(x) = \dfrac{2 \ln x}{x}$. $f'(x) = \dfrac{x\left(\frac{2}{x}\right) - 2 \ln x}{x^2} = \dfrac{2(1 - \ln x)}{x^2}$.

9. $f(u) = \ln (u-2)^3$; $f'(u) = \dfrac{3(u-2)^2}{(u-2)^3} = \dfrac{3}{u-2}$.

21. $f(x) = (\ln x)^{1/2}$ and $f'(x) = \dfrac{1}{2}(\ln x)^{-1/2}\left(\dfrac{1}{x}\right) = \dfrac{1}{2x\sqrt{\ln x}}$.

23. $f(x) = (\ln x)^3$; $f'(x) = 3(\ln x)^2\left(\dfrac{1}{x}\right) = \dfrac{3(\ln x)^2}{x}$.

25. $f(x) = \ln (x^3 + 1)$; $f'(x) = \dfrac{3x^2}{x^3 + 1}$.

27. $f(x) = e^x \ln x$. $f'(x) = e^x \ln x + e^x\left(\dfrac{1}{x}\right) = \dfrac{e^x(x \ln x + 1)}{x}$.

29. $f(t) = e^{2t} \ln (t+1)$

$f'(t) = e^{2t}\left(\dfrac{1}{t+1}\right) + \ln(t+1)\cdot(2e^{2t}) = \dfrac{[2(t+1)\ln(t+1) + 1]e^{2t}}{t+1}$.

31. $f(x)$ $\dfrac{\ln x}{x}$. $f'(x) = \dfrac{x\left(\frac{1}{x}\right) - \ln x}{x^2} = \dfrac{1 - \ln x}{x^2}$.

33. $f(x) = \ln 2 + \ln x$; So $f'(x) = \dfrac{1}{x}$ and $f''(x) = -\dfrac{1}{x^2}$.

35. $f(x) = \ln (x^2 + 2)$; $f'(x) = \dfrac{2x}{(x^2 + 2)}$ and

$$f''(x) = \frac{(x^2+2)(2) - 2x(2x)}{(x^2+2)^2} = \frac{2(2-x^2)}{(x^2+2)^2}.$$

37. $y = (x+1)^2(x+2)^3$

$\ln y = \ln (x+1)^2(x+2)^3 = \ln (x+1)^2 + \ln (x+2)^3$

$\quad = 2 \ln (x+1) + 3 \ln (x+2).$

$$\frac{y'}{y} = \frac{2}{x+1} + \frac{3}{x+2} = \frac{2(x+2) + 3(x+1)}{(x+1)(x+2)} = \frac{5x+7}{(x+1)(x+2)}$$

$$y' = \frac{(5x+7)(x+1)^2(x+2)^3}{(x+1)(x+2)} = (5x+7)(x+1)(x+2)^2.$$

39. $y = (x-1)^2(x+1)^3(x+3)^4$

$\ln y = 2 \ln (x-1) + 3 \ln (x+1) + 4 \ln (x+3)$

$$\frac{y'}{y} = \frac{2}{x-1} + \frac{3}{x+1} + \frac{4}{x+3}$$

$$= \frac{2(x+1)(x+3) + 3(x-1)(x+3) + 4(x-1)(x+1)}{(x-1)(x+1)(x+3)}$$

$$= \frac{2x^2 + 8x + 6 + 3x^2 + 6x - 9 + 4x^2 - 4}{(x-1)(x+1)(x+3)} = \frac{9x^2 + 14x - 7}{(x-1)(x+1)(x+3)}.$$

Therefore,

$$y' = \frac{9x^2 + 14x - 7}{(x-1)(x+1)(x+3)} \cdot y$$

$$= \frac{(9x^2 + 14x - 7)(x-1)^2(x+1)^3(x+3)^4}{(x-1)(x+1)(x+3)}$$

$$= (9x^2 + 14x - 7)(x-1)(x+1)^2(x+3)^3.$$

41. $y = \dfrac{(2x^2-1)^5}{\sqrt{x+1}}.$

$$\ln y = \ln \frac{(2x^2-1)^5}{(x+1)^{1/2}} = 5 \ln(2x^2-1) - \frac{1}{2}\ln(x+1)$$

So $\dfrac{y'}{y} = \dfrac{20x}{2x^2-1} - \dfrac{1}{2(x+1)} = \dfrac{40x(x+1) - (2x^2-1)}{2(2x^2-1)(x+1)}$

$$= \frac{38x^2 + 40x + 1}{2(2x^2-1)(x+1)}.$$

$$y' = \frac{38x^2 + 40x + 1}{2(2x^2 - 1)(x+1)} \cdot \frac{(2x^2 - 1)^5}{\sqrt{x+1}} = \frac{(38x^2 + 40x + 1)(2x^2 - 1)^4}{2(x+1)^{3/2}}.$$

43. $y = 3^x$; $\quad \ln y = x \ln 3$; $\quad \frac{1}{y} \cdot \frac{dy}{dx} = \ln 3$; $\quad \frac{dy}{dx} = y \ln 3 = 3^x \ln 3$.

45. $y = (x^2 + 1)^x$; $\ln y = \ln (x^2 + 1)^x = x \ln (x^2 + 1)$. So

$$\frac{y'}{y} = \ln(x^2 + 1) + x\left(\frac{2x}{x^2 + 1}\right) = \frac{(x^2 + 1)\ln(x^2 + 1) + 2x^2}{x^2 + 1}.$$

$$y' = \frac{[(x^2 + 1)\ln(x^2 + 1) + 2x^2](x^2 + 1)^x}{x^2 + 1}$$

47. $y = x \ln x$. The slope of the tangent line at any point is
$$y' = \ln x + x\left(\tfrac{1}{x}\right) = \ln x + 1.$$

In particular, the slope of the tangent line at $(1,0)$ where $x = 1$ is $m = \ln 1 + 1 = 1$. So, an equation of the tangent line is $y - 0 = 1(x - 1)$ or $y = x - 1$.

49. $f(x) = \ln x^2 = 2 \ln x$ and so $f'(x) = 2/x$. Since $f'(x) < 0$ if $x < 0$, and $f'(x) > 0$ if $x > 0$, we see that f is decreasing on $(-\infty,0)$ and increasing on $(0,\infty)$.

51. $f(x) = x^2 + \ln x^2$; $\quad f'(x) = 2x + \frac{2x}{x^2} = 2x + \frac{2}{x}$; $\quad f''(x) = 2 - \frac{2}{x^2}$.

To find the intervals of concavity for f, we first set $f''(x) = 0$ giving
$$2 - \frac{2}{x^2} = 0, \quad 2 = \frac{2}{x^2}, \quad 2x^2 = 2$$

or $\qquad\qquad x^2 = 1$ and $x = \pm 1$.

Next, we construct the sign diagram for f''

f'' is not defined here

and conclude that f is concave upward on $(-\infty,-1) \cup (1,\infty)$ and concave downward on $(-1,0) \cup (0,1)$.

53. $f(x) = \ln(x^2 + 1)$. $f'(x) = \dfrac{2x}{x^2 + 1}$; $f''(x) = \dfrac{(x^2 + 1)(2) - (2x)(2x)}{(x^2 + 1)^2} = -\dfrac{2(x^2 - 1)}{(x^2 + 1)^2}$.

Setting $f''(x) = 0$ gives $x = \pm 1$ as candidates for inflection points of f.
From the sign diagram for f''

$$- \; - \; - \; - \; 0 + + + 0 + + + 0 - \; - \; -$$

we see that $(-1, \ln 2)$ and $(1, \ln 2)$ are inflection points of f.

we see that $(e^{-3/2}, -\tfrac{3}{2}e^{-3})$ is an inflection point of f.

55. $f(x) = x - \ln x$; $f'(x) = 1 - \dfrac{1}{x} = \dfrac{x-1}{x} = 0$ if $x = 1$, a critical point of f.

x	1/2	1	3
$f(x)$	$1/2 + \ln 2$	1	$3 -$
	$\ln 3$		

From the table, we see that f has an absolute minimum at $(1,1)$ and an absolute maximum at $(3, 3 - \ln 3)$.

57. $f(x) = 7.2956 \ln(0.0645012 x^{0.95} + 1)$

$$f'(x) = 7.2956 \cdot \dfrac{\frac{d}{dx}(0.0645012 x^{0.95} + 1)}{0.0645012 x^{0.95} + 1} = \dfrac{7.2956(0.0645012)(0.95 x^{-0.05})}{0.0645012 x^{0.95} + 1}$$

$$= \dfrac{0.4470462}{x^{0.05}(0.0645012 x^{0.95} + 1)}$$

So $f'(100) = 0.05799$, or approximately 0.0580 percent/kg and
$f'(500) = 0.01329$, or approximately 0.0133 percent/kg.

59. a. If $0 < r < 100$, then $c = 1 - \tfrac{r}{100}$ satisfies $0 < c < 1$. It suffices to show that
$A_1(n) = -(1 - \tfrac{r}{100})^n$ is increasing, (why?), or equivalently $A_2(n) = -A_1(n) = \left(1 - \tfrac{r}{100}\right)^n$
is decreasing. Let $y = \left(1 - \tfrac{r}{100}\right)^n$. Then $\ln y = \ln\left(1 - \tfrac{r}{100}\right)^n = \ln c^n = n \ln c$.

Differentiating both sides with respect to n, we find

$$\frac{y'}{y} = \ln c \quad \text{and so} \quad y' = (\ln c)\left(1 - \tfrac{r}{100}\right)^n < 0$$

since $\ln c < 0$ and $\left(1 - \tfrac{r}{m}\right)^n > 0$ for $0 < r < 100$. Therefore, A is an increasing function of n on $(0, \infty)$.

b.

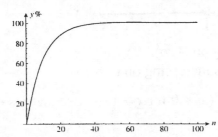

c. $\displaystyle \lim_{n \to \infty} A(n) = \lim_{n \to \infty} 100\left[1 - \left(1 - \tfrac{r}{100}\right)^n\right] = 100$

61. a. $\displaystyle R = \log \frac{10^6 I_0}{I_0} = \log 10^6 = 6.$

b. $I = I_0 10^R$ by definition. Taking the natural logarithm on both sides, we find

$$\ln I = \ln I_0 10^R = \ln I_0 + \ln 10^R = \ln I_0 + R \ln 10.$$

Differentiating implicitly with respect to R, we obtain

$$\frac{I'}{I} = \ln 10 \quad \text{or} \quad \frac{dI}{dR} = (\ln 10)I.$$

Therefore, $\Delta I \approx dI = \dfrac{dI}{dR} \Delta R = (\ln 10)I\Delta R.$ With $|\Delta R| \leq (0.02)(6) = 0.12$ and $I = 1{,}000{,}000 I_0$, (see part a), we have

$$|\Delta I| \leq (\ln 10)(1{,}000{,}000 I_0)(0.12) = 276310.21 I_0$$

So the error is at most 276,310 times the standard reference intensity.

63. $f(x) = \ln(x - 1).$

1. The domain of f is obtained by requiring that $x - 1 > 0$. We find the domain to be $(1, \infty)$.
2. Since $x \neq 0$, there are no y-intercepts. Next, setting $y = 0$ gives $x - 1 = 1$ or $x = 2$ as the x-intercept.
3. $\displaystyle \lim_{x \to 1^+} \ln(x - 1) = -\infty.$

4. There are no horizontal asymptotes. Observe that $\lim_{x \to 1^+} \ln(x-1) = -\infty$ so $x = 1$ is a vertical asymptote.

5. $f'(x) = \dfrac{1}{x-1}$.

The sign diagram for f' is

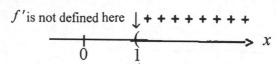

We conclude that f is increasing on $(1,\infty)$.

6. The results of (5) show that f is increasing on $(1,\infty)$.

7. $f''(x) = -\dfrac{1}{(x-1)^2}$. Since $f''(x) < 0$ for $x > 1$, we see that f is concave downward on $(1,\infty)$.

8. From the results of (7), we see that f has no inflection points. The graph of f follows.

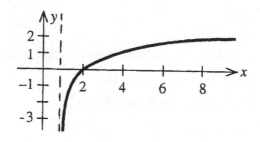

65. False. ln 5 is a constant function and $f'(x) = 0$.

67. If $x \le 0$, then $|x| = -x$. Therefore, $\ln |x| = \ln(-x)$. Writing $f(x) = \ln |x|$ we have $|x| = -x = e^{f(x)}$. Differentiating both sides with respect to x and using the Chain Rule, we have $-1 = e^{f(x)} \cdot f'(x)$ or $f'(x) = -\dfrac{1}{e^{f(x)}} = -\dfrac{1}{-x} = \dfrac{1}{x}$.

EXERCISES 5.6 , page 388

1. a. The growth constant is $k = 0.05$. b. Initially, the quantity present is 400 units.

c.

t	0	10	20	100	1000
Q	400	660	1087	59365	2.07×10^{24}

3. a. $Q(t) = Q_0 e^{kt}$. Here $Q_0 = 100$ and so $Q(t) = 100e^{kt}$. Since the number of cells doubles in 20 minutes, we have
$$Q(20) = 100e^{20k} = 200, \; e^{20k} = 2, \; 20k = \ln 2, \text{ or } k = \tfrac{1}{20} \ln 2 \approx 0.03466.$$
$$Q(t) = 100e^{0.03466t}$$
b. We solve the equation $100e^{0.03466t} = 1{,}000{,}000$. We obtain
$$e^{0.03466t} = 10000 \text{ or } 0.03466t = \ln 10000,$$
$$t = \frac{\ln 10{,}000}{0.03466} \approx 266, \quad \text{or } 266 \text{ minutes.}$$
c. $Q(t) = 1000e^{0.03466t}$.

5. a. We solve the equation
$$5.3e^{0.0198t} = 3(5.3) \text{ or } e^{0.0198t} = 3,$$
or
$$0.0198t = \ln 3 \text{ and } t = \frac{\ln 3}{0.0198} \approx 55.5.$$
So the world population will triple in approximately 55.5 years.
b. If the growth rate is 1.8 percent, then proceeding as before, we find
$$1.018(5.3) = 5.3e^k, \text{ and } k = \ln 1.018 \approx 0.0178.$$
So $N(t) = 5.3e^{0.0178t}$. If $t = 55.5$, the population would be
$$N(55.5) = 5.3e^{0.0178(55.5)} \approx 14.23, \text{ or approximately } 14.23 \text{ billion.}$$

7. $P(h) = p_0 e^{-kh}$, $P(0) = 15$, therefore, $p_0 = 15$.
$$P(4000) = 15e^{-4000k} = 12.5; \; e^{-4000k} = \frac{12.5}{15},$$
$$-4000k = \ln\left(\frac{12.5}{15}\right) \text{ and } k = 0.00004558.$$
Therefore, $P(12{,}000) = 15e^{-0.00004558(12{,}000)} = 8.68$, or 8.7 lb/sq in.
The rate of change of the atmospheric pressure with respect to altitude is given by
$$P'(h) = \frac{d}{dh}(15e^{-0.00004558h}) = -0.0006837e^{-0.00004558h}.$$
So, the rate of change of the atmospheric pressure with respect to altitude when the altitude is 12,000 feet is $P'(12{,}000) = -0.0006837e^{-0.00004558(12{,}000)} \approx -0.00039566$.

That is, it is dropping at the rate of approximately 0.0004 lbs per square inch/foot.

9. Suppose the amount of phosphorus 32 at time t is given by
$$Q(t) = Q_0e^{-kt}$$

where Q_0 is the amount present initially and k is the decay constant. Since this element has a half–life of 14.2 days, we have
$$\tfrac{1}{2}Q_0 = Q_0e^{-14.2k}, \quad e^{-14.2k} = \tfrac{1}{2}, \; -14.2k = \ln\tfrac{1}{2}, \; k = -\frac{\ln\tfrac{1}{2}}{14.2} \approx 0.0488.$$
Therefore, the amount of phosphorus 32 present at any time t is given by
$$Q(t) = 100e^{-0.0488t}$$
The amount left after 7.1 days is given by
$$Q(7.1) = 100e^{-0.0488(7.1)} = 100e^{-0.3465}$$
$$= 70.617, \text{ or } 70.617 \text{ grams.}$$
The rate at which the phosphorus 32 is decaying when $t = 7.1$ is given by
$$Q'(t) = \frac{d}{dt}[100e^{-0.0488t}] = 100(-0.0488)e^{-0.0488t} = -4.88e^{-0.0488t}.$$

Therefore, $Q'(7.1) = -4.88e^{-0.0488(7.1)} \approx -3.451$; that is, it is changing at the rate of 3.451 gms/day.

11. We solve the equation $0.2Q_0 = Q_0e^{-0.00012t}$

obtaining
$$t = \frac{\ln 0.2}{-0.00012} \approx 13{,}412, \text{ or approximately } 13{,}412 \text{ years.}$$

13. The graph of $Q(t)$ follows.

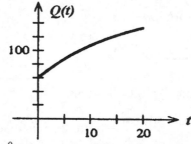

a. $Q(0) = 120(1 - e^0) + 60 = 60$, or 60 w.p.m.
b. $Q(10) = 120(1 - e^{-0.5}) + 60 = 107.22$, or approximately 107 w.p.m.
c. $Q(20) = 120(1 - e^{-1}) + 60 = 135.65$, or approximately 136 w.p.m.

15. The graph of $D(t)$ follows.

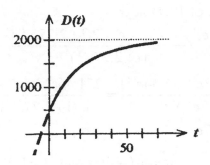

a. After one month, the demand is $D(1) = 2000 - 1500e^{-0.05} \approx 573$.
After twelve months, the demand is $D(12) = 2000 - 1500e^{-0.6} \approx 1177$.
After twenty-four months the demand is $D(24) = 2000 - 1500e^{-1.2} \approx 1548$.
After sixty months, the demand is $D(60) = 2000 - 1500e^{-3} \approx 1925$.

b. $\qquad \lim_{t \to \infty} D(t) = \lim_{t \to \infty} 2000 - 1500e^{-0.05t} = 2000$

and we conclude that the demand is expected to stabilize at 2000 computers per month.

c. $D'(t) = -1500e^{-0.05t}(-0.05) = 75e^{-0.05t}$. Therefore, the rate of growth after ten months is given by $D'(10) = 75e^{-0.5} \approx 45.49$, or approximately 46 computers per month.

17. a. The length is given by $f(5) = 200(1 - 0.956e^{-0.18(5)}) \approx 122.26$, or approximately 122.3 cm.

b. $f'(t) = 200(-0.956)e^{-0.18t}(-0.18) = 34.416e^{-0.18t}$. So, a 5-yr old is growing at the rate of $f'(5) = 34.416e^{-0.18(5)} \approx 13.9925$, or approximately 14 cm/yr.

c. The maximum length is given by $\lim_{t \to \infty} 200(1 - 0.956e^{-0.18t}) = 200$, or 200 cm.

19. a. The percent of lay teachers is $f(3) = \dfrac{98}{1 + 2.77e^{-3}} \approx 86.1228$, or 86.12%.

b. $f'(t) = \dfrac{d}{dt}[98(1 + 2.77e^{-t})^{-1}] = 98(-1)(1 + 2.77e^{-t})^{-2}(2.77e^{-t})(-1)$

$\qquad = \dfrac{271.46e^{-t}}{(1 + 2.77e^{-t})^2}$

$$f'(3) = \frac{271.46e^{-3}}{(1+2.77e^{-3})^2} \approx 10.4377.$$

So it is increasing at the rate of 10.44%/yr.

c. $f''(t) = 271.46\left[\dfrac{(1+2.77e^{-t})^2(-e^{-t}) - e^{-t}\cdot 2(1+2.77e^{-t})(-2.77e^{-t})}{(1+2.77e^{-t})^4}\right]$

$$= \frac{271.46[-(1+2.77e^{-t}+5.54e^{-t}]}{e^{t}(1+2.77e^{-t})^3} = \frac{271.46(2.77e^{-t}-1)}{e^{t}(1+2.77e^{-t})^3}.$$

Setting $f''(t) = 0$ gives $2.77e^{-t} = 1$

$$e^{-t} = \frac{1}{2.77}; \quad -t = \ln\left(\tfrac{1}{2.77}\right), \quad \text{and} \quad t = 1.0188.$$

The sign diagram of f'' shows that $t = 1.02$ gives an inflection point of P. So, the

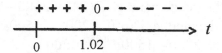

percent of lay teachers was increasing most rapidly in 1970.

21. $P(t) = \dfrac{68}{1+21.67e^{-0.62t}}.$

The percentage of households that owned VCRs at the beginning of 1985 is given

by $P(0) = \dfrac{68}{1+21.67e^{-0.62(0)}} = \dfrac{68}{22.67} \approx 3$, or approximately 3 percent.

The percentage of households that owned VCRs at the beginning of 1995 is given

by $P(10) = \dfrac{68}{1+21.67e^{-0.62(10)}} \approx 65.14$, or approximately 65.14 percent.

23. The first of the given conditions implies that $f(0) = 300$, that is,

$$300 = \frac{3000}{1+Be^{0}} = \frac{3000}{1+B}.$$

So $1+B = 10$, or $B = 9$. Therefore, $f(t) = \dfrac{3000}{1+9e^{-kt}}$. Next, the condition

$f(2) = 600$ gives the equation

$$600 = \frac{3000}{1+9e^{-2k}}, \quad 1+9e^{-2k} = 5, \quad e^{-2k} = \frac{4}{9}, \quad \text{or} \quad k = -\frac{1}{2}\ln\left(\frac{4}{9}\right).$$

Therefore, $f(t) = \dfrac{3000}{1+9e^{(1/2)t \cdot \ln(4/9)}} = \dfrac{3000}{1+9\left(\frac{4}{9}\right)^{t/2}}$.

The number of students who had heard about the policy four hours later is given by

$$f(4) = \dfrac{3000}{1+9\left(\frac{4}{9}\right)^2} = 1080, \quad \text{or 1080 students.}$$

To find the rate at which the rumor was spreading at any time time, we compute

$$f'(t) = \dfrac{d}{dt}\left[3000(1+9e^{-0.405465t})^{-1}\right]$$

$$= (3000)(-1)(1+9e^{-0.405465})^{-2}\dfrac{d}{dt}(9e^{-0.405465t})$$

$$= -3000(9)(-0.405465)e^{-0.405465t}(1+9e^{-0.405465t})^{-2}$$

$$= \dfrac{10947.555\,e^{-0.405465t}}{(1+9e^{-0.405465t})^2}$$

In particular, the rate at which the rumor was spreading 4 hours after the ceremony

is given by $f'(4) = \dfrac{10947.555e^{-0.405465(4)}}{(1+9e^{-0.405465(4)})^2} \approx 280.25737$.

So , the rumor is spreading at the rate of 280 students per hour.

25. $x(t) = \dfrac{15\left(1-\left(\frac{2}{3}\right)^{3t}\right)}{1-\frac{1}{4}\left(\frac{2}{3}\right)^{3t}}$; $\displaystyle\lim_{t\to\infty} x(t) = \lim_{t\to\infty}\dfrac{15\left(1-\left(\frac{2}{3}\right)^{3t}\right)}{1-\frac{1}{4}\left(\frac{2}{3}\right)^{3t}} = \dfrac{15(1-0)}{1-0} = 15$

or 15 lbs.

27. a. $C(t) = \dfrac{k}{b-a}\left(e^{-at} - e^{-bt}\right)$;

$$C'(t) = \dfrac{k}{b-a}(-ae^{-at} + be^{-bt}) = \dfrac{kb}{b-a}\left[e^{-bt} - \left(\dfrac{a}{b}\right)e^{-at}\right]$$

$$= \dfrac{kb}{b-a}e^{-bt}\left[1 - \dfrac{a}{b}e^{(b-a)t}\right]$$

$C'(t) = 0$ implies that $1 = \dfrac{a}{b}e^{(b-a)t}$ or $t = \dfrac{\ln\left(\frac{b}{a}\right)}{b-a}$.

The sign diagram of C'

shows that this value of t gives a minimum.

b. $\displaystyle \lim_{t \to \infty} C(t) = \frac{k}{b-a}$.

29. a. We solve $Q_0 e^{-kt} = \dfrac{1}{2} Q_0$ for t. Proceeding, we have

$$e^{-kt} = \frac{1}{2}, \quad \ln e^{-kt} = \ln \frac{1}{2} = \ln 1 = \ln 2 = -\ln 2;$$

$$-kt = -\ln 2;$$

So $\qquad \overline{t} = \dfrac{\ln 2}{k}$

b. $\qquad \overline{t} = \dfrac{\ln 2}{0.0001238} \approx 5598.927$, or approximately 5599 years.

USING TECHNOLOGY EXERCISES 5.6, page 391

1. a.

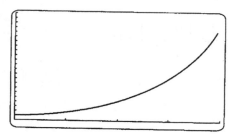

b. 12.146%/yr c. 9.474%/yr

3. a.

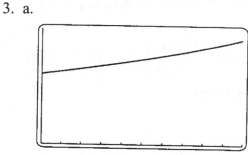

b. 666 million, 818.8 million
c. 33.8 million/yr

5. a.

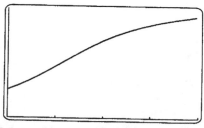

b. 86.12%/yr c. 10.44%/yr

7. a.

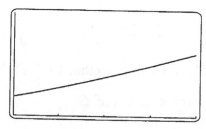

b. 325 million

d. 1970   c. 76.84 million/decade

9.   a.

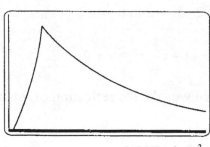

b. 0   c. 0.237 g/cm^3 d. 0.760 g/cm^3 e. 0

CHAPTER 5 REVIEW EXERCISES, page 395

1.   a-b

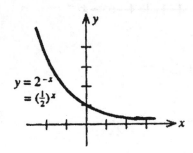

Since $y = \left(\dfrac{1}{2}\right)^x = \dfrac{1}{2^x} = 2^{-x}$, it has the same graph as that of $y = 2^{-x}$.

3. $16^{-3/4} = 0.125$ is equivalent to $-\dfrac{3}{4} = \log_{16} 0.125$.

5. $\ln(x-1) + \ln 4 = \ln(2x+4) - \ln 2$
       $\ln(x-1) - \ln(2x+4) = -\ln 2 - \ln 4 = -(\ln 2 + \ln 4)$

$$\ln\left(\frac{x-1}{2x+4}\right) = -\ln 8 = \ln \tfrac{1}{8}.$$

$$\left(\frac{x-1}{2x+4}\right) = \frac{1}{8}$$

$$8x - 8 = 2x + 4$$
$$6x = 12, \text{ or } x = 2.$$

CHECK: l.h.s. $\ln(2-1) + \ln 4 = \ln 4$

r.h.s $\ln(4+4) - \ln 2 = \ln 8 - \ln 2 = \ln \frac{8}{2} = \ln 4.$

7. $\ln 3.6 = \ln \frac{36}{10} = \ln 36 - \ln 10 = \ln 6^2 - \ln 2 \cdot 5 = 2\ln 6 - \ln 2 - \ln 5$

$= 2(\ln 2 + \ln 3) - \ln 2 - \ln 5 = 2(x+y) - x - z = x + 2y - z.$

9. We first sketch the graph of $y = 2^{x-3}$. Then we take the reflection of this graph with respect to the line $y = x$.

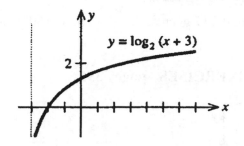

11. $f(x) = xe^{2x}; f'(x) = e^{2x} + xe^{2x}(2) = (1 + 2x)e^{2x}.$

13. $g(t) = \sqrt{t}e^{-2t}; g'(t) = \frac{1}{2}t^{-1/2}e^{-2t} + \sqrt{t}e^{-2t}(-2) = \frac{1-4t}{2\sqrt{t}e^{2t}}.$

15. $y = \dfrac{e^{2x}}{1+e^{-2x}}; \; y' = \dfrac{(1+e^{-2x})e^{2x}(2) - e^{2x} \cdot e^{-2x}(-2)}{(1+e^{-2x})^2} = \dfrac{2(e^{2x}+2)}{(1+e^{-2x})^2}.$

17. $f(x) = xe^{-x^2}; f'(x) = e^{-x^2} + xe^{-x^2}(-2x) = (1 - 2x^2)e^{-x^2}.$

19. $f(x) = x^2 e^x + e^x;$
$f'(x) = 2xe^x + x^2 e^x + e^x = (x^2 + 2x + 1)e^x = (x+1)^2 e^x.$

21. $f(x) = \ln(e^{x^2} + 1); \; f'(x) = \dfrac{e^{x^2}(2x)}{e^{x^2}+1} = \dfrac{2xe^{x^2}}{e^{x^2}+1}.$

23. $f(x) = \dfrac{\ln x}{x+1}$. $f'(x) = \dfrac{(x+1)\left(\dfrac{1}{x}\right) - \ln x}{(x+1)^2} = \dfrac{1 + \dfrac{1}{x} - \ln x}{(x+1)^2} = \dfrac{x - x\ln x + 1}{x(x+1)^2}.$

25. $y = \ln(e^{4x} + 3)$; $y' = \dfrac{e^{4x}(4)}{e^{4x} + 3} = \dfrac{4e^{4x}}{e^{4x} + 3}.$

27. $f(x) = \dfrac{\ln x}{1 + e^x}$;

$$f'(x) = \frac{(1 + e^x)\dfrac{d}{dx}\ln x - \ln x \dfrac{d}{dx}(1 + e^x)}{(1 + e^x)^2} = \frac{(1 + e^x)\left(\dfrac{1}{x}\right) - (\ln x)e^x}{(1 + e^x)^2}$$

$$= \frac{1 + e^x - xe^x \ln x}{x(1 + e^x)^2} = \frac{1 + e^x(1 - x\ln x)}{x(1 + e^x)^2}.$$

29. $y = \ln(3x + 1)$; $y' = \dfrac{3}{3x + 1}$;

$$y'' = 3\frac{d}{dx}(3x + 1)^{-1} = -3(3x + 1)^{-2}(3) = -\frac{9}{(3x + 1)^2}.$$

31. $h'(x) = g'(f(x))f'(x)$. But $g'(x) = 1 - \dfrac{1}{x^2}$ and $f'(x) = e^x$.

So $f(0) = e^0 = 1$ and $f'(0) = e^0 = 1$. Therefore,

$h'(0) = g'(f(0))f'(0) = g'(1)f'(0) = 0 \cdot 1 = 0.$

33. $y = (2x^3 + 1)(x^2 + 2)^3$. $\ln y = \ln(2x^3 + 1) + 3\ln(x^2 + 2)$.

$$\frac{y'}{y} = \frac{6x^2}{2x^3 + 1} + \frac{3(2x)}{x^2 + 2} = \frac{6x^2(x^2 + 2) + 6x(2x^3 + 1)}{(2x^3 + 1)(x^2 + 2)}$$

$$= \frac{6x^4 + 12x^2 + 12x^4 + 6x}{(2x^3 + 1)(x^2 + 2)} = \frac{18x^4 + 12x^2 + 6x}{(2x^3 + 1)(x^2 + 2)}.$$

Therefore, $y' = 6x(3x^3 + 2x + 1)(x^2 + 2)^2.$

35. $y = e^{-2x}$. $y' = -2e^{-2x}$ and this gives the slope of the tangent line to the graph of

$y = e^{-2x}$ at any point (x, y). In particular, the slope of the tangent line at $(1, e^{-2})$ is

$y'(1) = -2e^{-2}$. The required equation is $y - e^{-2} = -2e^{-2}(x - 1)$ or $y = \dfrac{1}{e^2}(-2x + 3)$.

37. $f(x) = xe^{-2x}$.

We first gather the following information on f.
1. The domain of f is $(-\infty, \infty)$.
2. Setting $x = 0$ gives 0 as the y-intercept.
3. $\lim\limits_{x \to -\infty} xe^{-2x} = -\infty$ and $\lim\limits_{x \to \infty} xe^{-2x} = 0$.
4. The results of (3) show that $y = 0$ is a horizontal asymptote.
5. $f'(x) = e^{-2x} + xe^{-2x}(-2) = (1 - 2x)e^{-2x}$. Observe that $f'(x) = 0$ if $x = 1/2$, a critical point of f. The sign diagram of f'

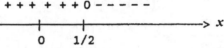

shows that f is increasing on $(-\infty, \frac{1}{2})$ and decreasing on $(\frac{1}{2}, \infty)$.

6. The results of (5) show that $(\frac{1}{2}, \frac{1}{2}e^{-1})$ is a relative maximum.

7. $f''(x) = -2e^{-2x} + (1 - 2x)e^{-2x}(-2) = 4(x - 1)e^{-2x}$ and is equal to zero if $x = 1$. The sign diagram of f''

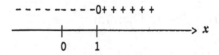

shows that the graph of f is concave downward on $(-\infty, 1)$ and concave upward on $(1, \infty)$.
The graph of f follows.

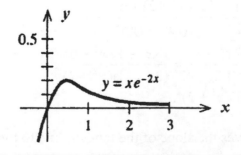

39. $f(t) = te^{-t}$. $f'(t) = e^{-t} + t(-e^{-t}) = e^{-t}(1 - t)$. Setting $f'(t) = 0$ gives $t = 1$ as the only critical point of f. From the sign diagram of f'

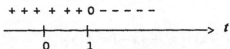

we see that $f(1) = e^{-1} = 1/e$ is the absolute maximum value of f.

41. We want to find r where r satisfies the equation $8.2 = 4.5 \, e^{r(5)}$. We have

$$e^{5r} = \frac{8.2}{4.5} \quad \text{or} \quad r = \frac{1}{5} \ln\left(\frac{8.2}{4.5}\right) \approx 0.12$$

and so the annual rate of return is 12 percent per year.

43. a. $Q(t) = 2000e^{kt}$. Now $Q(120) = 18{,}000$ gives $2000e^{120k} = 18{,}000$, $e^{120k} = 9$, or $120k = \ln 9$. So $k = \frac{1}{120} \ln 9 \approx 0.01831$ and $Q(t) = 2000e^{0.01831t}$.

 b. $Q(4) = 2000e^{0.01831(240)} \approx 161{,}992$, or approximately 162,000.

45.

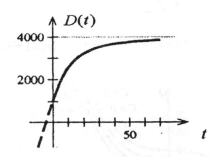

 a. $D(1) = 4000 - 3000 \, e^{-0.06} = 1175$, $D(12) = 4000 - 3000 \, e^{-0.72} = 2540$, and $D(24) = 4000 - 3000 \, e^{-1.44} = 3289$.

 b. $\lim_{t \to \infty} D(t) = \lim_{t \to \infty} (4000 - 3000e^{-0.06t}) = 4000$.

CHAPTER 6

EXERCISES 6.1, page 407

1. $F(x) = \frac{1}{3}x^3 + 2x^2 - x + 2;\ F'(x) = x^2 + 4x - 1 = f(x).$

3. $F(x) = (2x^2 - 1)^{1/2};\ F'(x) = \frac{1}{2}(2x^2 - 1)^{-1/2}(4x) = 2x(2x^2 - 1)^{-1/2} = f(x).$

5. a. $G'(x) = \dfrac{d}{dx}(2x) = 2 = f(x)$ b. $F(x) = G(x) + C = 2x + C$

    c.

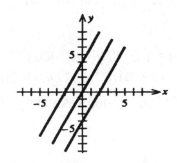

7. a. $G'(x) = \dfrac{d}{dx}(\frac{1}{3}x^3) = x^2 = f(x)$ b. $F(x) = G(x) + C = \frac{1}{3}x^3 + C$

    c.

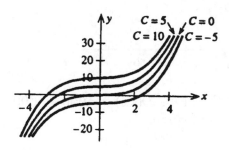

9. $\displaystyle\int 6\, dx = 6x + C.$ 11. $\displaystyle\int x^3 dx = \frac{1}{4}x^4 + C$

13. $\displaystyle\int x^{-4} dx = -\frac{1}{3}x^{-3} + C$ 15. $\displaystyle\int x^{2/3} dx = \frac{3}{5}x^{5/3} + C$

17. $\displaystyle\int x^{-5/4} dx = -4x^{-1/4} + C$

19. $\displaystyle\int \frac{2}{x^2}\,dx = 2\int x^{-2}dx = 2(-1x^{-1}) + C = -\frac{2}{x} + C$

21. $\displaystyle\int \pi\sqrt{t}\,dt = \pi\int t^{1/2}dt = \pi(\tfrac{2}{3}t^{3/2}) + C = \frac{2\pi}{3}t^{3/2} + C$

23. $\displaystyle\int (3-2x)\,dx = \int 3\,dx - 2\int x\,dx = 3x - x^2 + C$

25. $\displaystyle\int (x^2 + x + x^{-3})\,dx = \int x^2\,dx + \int x\,dx + \int x^{-3}\,dx = \tfrac{1}{3}x^3 + \tfrac{1}{2}x^2 - \tfrac{1}{2}x^{-2} + C$

27. $\displaystyle\int 4e^x\,dx = 4e^x + C$

29. $\displaystyle\int (1 + x + e^x)\,dx = x + \tfrac{1}{2}x^2 + e^x + C$

31. $\displaystyle\int (4x^3 - \frac{2}{x^2} - 1)\,dx = \int (4x^3 - 2x^{-2} - 1)\,dx = x^4 + 2x^{-1} - x + C = x^4 + \frac{2}{x} - x + C$

33. $\displaystyle\int (x^{5/2} + 2x^{3/2} - x)\,dx = \tfrac{2}{7}x^{7/2} + \tfrac{4}{5}x^{5/2} - \tfrac{1}{2}x^2 + C$

35. $\displaystyle\int (x^{1/2} + 3x^{-1/2})\,dx = \tfrac{2}{3}x^{3/2} + 6x^{1/2} + C$

37. $\displaystyle\int \left(\frac{u^3 + 2u^2 - u}{3u}\right)du = \frac{1}{3}\int (u^2 + 2u - 1)\,du = \frac{1}{9}u^3 + \frac{1}{3}u^2 - \frac{1}{3}u + C$

39. $\displaystyle\int (2t+1)(t-2)\,dt = \int (2t^2 - 3t - 2)\,dt = \tfrac{2}{3}t^3 - \tfrac{3}{2}t^2 - 2t + C$

41. $\displaystyle\int \frac{1}{x^2}(x^4 - 2x^2 + 1)\,dx = \int (x^2 - 2 + x^{-2})\,dx = \frac{1}{3}x^3 - 2x - x^{-1} + C$

$$= \frac{1}{3}x^3 - 2x - \frac{1}{x} + C$$

43. $\displaystyle\int \frac{ds}{(s+1)^{-2}} = \int (s+1)^2\, ds = \int (s^2 + 2s + 1)\, ds = \tfrac{1}{3}s^3 + s^2 + s + C$

45. $\displaystyle\int (e^t + t^e)\, dt = e^t + \frac{1}{e+1}t^{e+1} + C$

47. $\displaystyle\int \left(\frac{x^3 + x^2 - x + 1}{x^2}\right) dx = \int \left(x + 1 - \frac{1}{x} + \frac{1}{x^2}\right) dx = \frac{1}{2}x^2 + x - \ln|x| - x^{-1} + C$

49. $\displaystyle\int \left(\frac{(x^{1/2}-1)^2}{x^2}\right) dx = \int \left(\frac{x - 2x^{1/2} + 1}{x^2}\right) dx = \int (x^{-1} - 2x^{-3/2} + x^{-2})\, dx$

$\displaystyle\qquad = \ln|x| + 4x^{-1/2} - x^{-1} + C = \ln|x| + \frac{4}{\sqrt{x}} - \frac{1}{x} + C$

51. $\displaystyle\int f'(x)\, dx = \int (2x + 1)\, dx = x^2 + x + C.$ The condition $f(1) = 3$ gives
$f(1) = 1 + 1 + C = 3$, or $C = 1$. Therefore, $f(x) = x^2 + x + 1$.

53. $f'(x) = 3x^2 + 4x - 1$; $f(x) = x^3 + 2x^2 - x + C$. Using the given initial condition,
we have $f(2) = 8 + 2(4) - 2 + C = 9$, so $16 - 2 + C = 9$, or $C = -5$. Therefore,
$f(x) = x^3 + 2x^2 - x - 5$.

55. $\displaystyle f(x) = \int f'(x)\, dx = \int \left(1 + \frac{1}{x^2}\right) dx = \int (1 + x^{-2})\, dx = x - \frac{1}{x} + C.$

Using the given initial condition, we have $f(1) = 1 - 1 + C = 2$, or $C = 2$.

Therefore, $\displaystyle f(x) = x - \frac{1}{x} + 2.$

57. $\displaystyle f(x) = \int \frac{x+1}{x}\, dx = \int \left(1 + \frac{1}{x}\right) dx = x + \ln|x| + C.$ Using the initial condition, we

have $f(1) = 1 + \ln 1 + C = 1 + C = 1$, or $C = 0$. So $f(x) = x + \ln|x|$.

59. $f(x) = \int f'(x)\,dx = \int \frac{1}{2}x^{-1/2}\,dx = \frac{1}{2}(2x^{1/2}) + C = x^{1/2} + C;\ f(2) = \sqrt{2} + C = \sqrt{2}$

implies $C = 0$. So $f(x) = \sqrt{x}$.

61. $f'(x) = e^x + x;\ f(x) = e^x + \frac{1}{2}x^2 + C;\ f(0) = e^0 + \frac{1}{2}(0) + C = 1 + C$

So $3 = 1 + C$ or $2 = C$. Therefore, $f(x) = e^x + \frac{1}{2}x^2 + 2$.

63. The position of the car is

$s(t) = \int f(t)\,dt = \int 2\sqrt{t}\,dt = \int 2t^{1/2}\,dt = 2(\frac{2}{3}t^{3/2}) + C = \frac{4}{3}t^{3/2} + C.$

$s(0) = 0$ implies $s(0) = C = 0$. So $\ s(t) = \frac{4}{3}t^{3/2}.$

65. $C(x) = \int C'(x)\,dx = \int (0.000009x^2 - 0.009x + 8)\,dx$

$= 0.000003x^3 - 0.0045x^2 + 8x + k.$

$C(0) = k = 120\ $ and so $\ C(x) = 0.000003x^3 - 0.0045x^2 + 8x + 120.$

$C(500) = 0.000003(500)^3 - 0.0045(500)^2 + 8(500) + 120,\ $ or $\$3370$.

67. $P'(x) = -0.004x + 20,\ P(x) = -0.002x^2 + 20x + C$. Since $C = -16{,}000$, we find
that $P(x) = -0.002x^2 + 20x - 16{,}000$. The company realizes a maximum profit
when $P'(x) = 0$, that is, when $x = 5000$ units. Next,
$$P(5000) = -0.002(5000)^2 + 20(5000) - 16{,}000 = 34{,}000.$$
Thus, a maximum profit of $\$34{,}000$ is realized at a production level of 5000 units.

69. a. $N(t) = \int N'(t)\,dt = \int (-3t^2 + 12t + 45)\,dt = -t^3 + 6t^2 + 45t + C$. But $N(0) = C = 0$

and so $N(t) = -t^3 + 6t^2 + 45t.$
b. The number is $N(4) = -4^3 + 6(4)^2 + 45(4) = 212.$

71. a. We have the initial-value problem:
$$C'(t) = 12.288t^2 - 150.5594t + 695.23$$
$$C(0) = 3142$$
Integrating, we find
$$C(t) = \int C'(t)\,dt = \int (12.288t^2 - 150.5594t + 695.23)\,dt$$

$$= 4.096t^3 - 75.2797t^2 + 695.23t + k$$

Using the initial condition, we find

$$C(0) = 0 + k = 3142, \quad \text{and so } k = 3142.$$

Therefore, $C(t) = 4.096t^3 - 75.2797t^2 + 695.23t + 3142.$

The projected average out-of-pocket costs for beneficiaries is 2010 is

$$C(2) = 4.096(8) - 75.2797(4) + 695.23(2) + 3142 = 4264.1092$$

or $4264.11.

73. The number of new subscribers at any time is

$$N(t) = \int (100 + 210t^{3/4}) \, dt = 100t + 120t^{7/4} + C.$$

The given condition implies that $N(0) = 5000$. Using this condition, we find $C = 5000$. Therefore, $N(t) = 100t + 120t^{7/4} + 5000$. The number of subscribers 16 months from now is

$$N(16) = 100(16) + 120(16)^{7/4} + 5000, \quad \text{or } 21{,}960.$$

75. a. We have the initial-value problem

$$S'(t) = R(t) = -0.033t^2 + 0.3428t + 0.07 \quad \text{and} \quad S(0) = 2.9$$

where $S(t)$ denotes the share of online advertisement, worldwide, as a precent of the total ad market. Integrating, we find

$$S(t) = \int S'(t)\,dt = \int (-0.033t^2 + 0.3428t + 0.07)\,dt$$

$$= -0.011t^3 + 0.1714t^2 + 0.07t + C$$

Using the initial condition, we find $S(0) = 0 + C = 2.9$ and so $C = 2.9$.
Therefore, $S(t) = -0.011t^3 + 0.1714t^2 + 0.07t + 2.9$.

b. The projected online ad market share at the beginning of 2005 is

$$S(5) = -0.011(125) + 0.1714(25) + 0.07(5) + 2.9 = 6.16, \quad \text{or } 6.16\%.$$

77. $A(t) = \int A'(t)\,dt = \int (3.2922t^2 - 0.366t^3)\,dt = 1.0974t^3 - 0.0915t^4 + C$

Now, $A(0) = C = 0$. So, $A(t) = 1.0974t^3 - 0.0915t^4$.

79. $h(t) = \int h'(t)\,dt = \int (-3t^2 + 192t + 120)\,dt = -t^3 + 96t^2 + 120t + C$

$$= -t^3 + 96t^2 + 120t + C.$$

$h(0) = C = 0$ implies $h(t) = -t^3 + 96t^2 + 120t$.
The altitude 30seconds after lift-off is
$h(30) = -30^3 + 96(30)^2 + 120(30) = 63{,}000$ ft.

81. $v(r) = \int v'(r)\,dr = \int -kr\,dr = -\frac{1}{2}kr^2 + C$.

But $v(R) = 0$ and so $v(R) = -\frac{1}{2}kR^2 + C = 0$, or $C = \frac{1}{2}kR^2$. Therefore,
$v(R) = -\frac{1}{2}kr^2 + \frac{1}{2}kR^2 = \frac{1}{2}k(R^2 - r^2)$.

83. Denote the constant deceleration by k (ft/sec^2). Then $f''(t) = -k$,so
$f'(t) = v(t) = -kt + C_1$. Next, the given condition implies that $v(0) = 88$. This gives
$C_1 = 88$,or $f'(t) = -kt + 88$.
$$s = f(t) = \int f'(t)\,dt = \int (-kt + 88)\,dt = -\frac{1}{2}kt^2 + 88t + C_2.$$
Also, $f(0) = 0$ gives $s = f(t) = -\frac{1}{2}kt^2 + 88t$. Since the car is brought to rest in 9
seconds, we have $v(9) = -9k + 88 = 0$, or $k = \frac{88}{9}$, or $9\frac{7}{9}$. So the deceleration is
$9\frac{7}{9}$ ft/sec^2. The distance covered is
$$s = f(9) = -\frac{1}{2}\left(\frac{88}{9}\right)(81) + 88(9) = 396.$$
So the stopping distance is 396 ft.

85. a. We have the initial-value problem $R'(t) = \dfrac{8}{(t+4)^2}$ and $R(0) = 0$

Integrating, we find $R(t) = \int \dfrac{8}{(t+4)^2}\,dt = 8\int (t+4)^{-2}\,dt = -\dfrac{8}{t+4} + C$

$R(0) = 0$ implies $-\dfrac{8}{4} + C = 0$ or $C = 2$.

Therefore, $R(t) = -\dfrac{8}{t+4} + 2 = \dfrac{-8 + 2t + 8}{t+4} = \dfrac{2t}{t+4}$.

b. After 1 hr, $R(1) = \dfrac{2}{5} = 0.4$, or 0.4" had fallen. After 2 hr, $R(2) = \dfrac{4}{6} = \dfrac{2}{3}$, or

$\dfrac{2}{3}$" had fallen.

87. The net amount on deposit in Branch A is given by the area under the graph of f from $t = 0$ to $t = 180$. On the other hand, the net amount on deposit in Branch B is given by the area under the graph of g over the same interval. Evidently the first area is longer than the second. Therefore, we see that Branch A has the larger net deposit.

89. True. See proof in Section 6.1 in the text.

91. True. Use the Sum Rule followed by the Constant Multiple Rule.

EXERCISES 6.2, page 419

1. Put $u = 4x + 3$ so that $du = 4\, dx$, or $dx = \frac{1}{4} du$. Then
$$\int 4(4x+3)^4\, dx = \int u^4\, du = \tfrac{1}{5} u^5 + C = \tfrac{1}{5}(4x+3)^5 + C.$$

3. Let $u = x^3 - 2x$ so that $du = (3x^2 - 2)\, dx$. Then
$$\int (x^3 - 2x)^2 (3x^2 - 2)\, dx = \int u^2\, du = \tfrac{1}{3} u^3 + C = \tfrac{1}{3}(x^3 - 2x)^3 + C.$$

5. Let $u = 2x^2 + 3$ so that $du = 4x\, dx$. Then
$$\int \frac{4x}{(2x^2+3)^3}\, dx = \int \frac{1}{u^3}\, du = \int u^{-3}\, du = -\tfrac{1}{2} u^{-2} + C = -\frac{1}{2(2x^2+3)^2} + C.$$

7. Put $u = t^3 + 2$ so that $du = 3t^2\, dt$ or $t^2\, dt = \frac{1}{3} du$. Then
$$\int 3t^2 \sqrt{t^3 + 2}\, dt = \int u^{1/2}\, du = \tfrac{2}{3} u^{3/2} + C = \tfrac{2}{3}(t^3 + 2)^{3/2} + C$$

9. Let $u = x^2 - 1$ so that $du = 2x\, dx$ and $x\, dx = \frac{1}{2} du$. Then,
$$\int (x^2 - 1)^9 x\, dx = \int \tfrac{1}{2} u^9\, du = \tfrac{1}{20} u^{10} + C = \tfrac{1}{20}(x^2 - 1)^{10} + C.$$

11. Let $u = 1 - x^5$ so that $du = -5x^4\, dx$ or $x^4\, dx = -\frac{1}{5} du$. Then
$$\int \frac{x^4}{1 - x^5}\, dx = -\frac{1}{5} \int \frac{du}{u} = -\frac{1}{5} \ln |u| + C = -\frac{1}{5} \ln \left| 1 - x^5 \right| + C.$$

13. Let $u = x - 2$ so that $du = dx$. Then

$$\int \frac{2}{x-2} dx = 2 \int \frac{du}{u} = 2 \ln|u| + C = \ln u^2 + C = \ln(x-2)^2 + C$$

15. Let $u = 0.3x^2 - 0.4x + 2$. Then $du = (0.6x - 0.4) dx = 2(0.3x - 0.2) dx$.

$$\int \frac{0.3x - 0.2}{0.3x^2 - 0.4x + 2} dx = \int \frac{1}{2u} du = \frac{1}{2} \ln|u| + C = \frac{1}{2} \ln(0.3x^2 - 0.4x + 2) + C.$$

17. Let $u = 3x^2 - 1$ so that $du = 6x \, dx$, or $x \, dx = \frac{1}{6} du$. Then

$$\int \frac{x}{3x^2 - 1} dx = \frac{1}{6} \int \frac{du}{u} = \frac{1}{6} \ln|u| + C = \frac{1}{6} \ln|3x^2 - 1| + C.$$

19. Let $u = -2x$ so that $du = -2 \, dx$ or $dx = -\frac{1}{2} du$. Then

$$\int e^{-2x} dx = -\frac{1}{2} \int e^u du = -\frac{1}{2} e^u + C = -\frac{1}{2} e^{-2x} + C.$$

21. Let $u = 2 - x$ so that $du = - dx$ or $dx = - du$. Then

$$\int e^{2-x} dx = -\int e^u du = -e^u + C = -e^{2-x} + C.$$

23. Let $u = -x^2$, then $du = -2x \, dx$ or $x \, dx = -\frac{1}{2} du$.

$$\int xe^{-x^2} dx = \int -\frac{1}{2} e^u du = -\frac{1}{2} e^u + C = -\frac{1}{2} e^{-x^2} + C.$$

25. $\int (e^x - e^{-x}) dx = \int e^x dx - \int e^{-x} dx = e^x - \int e^{-x} dx.$

To evaluate the second integral on the right, let $u = -x$ so that $du = -dx$ or $dx = -du$. Therefore,

$$\int (e^x - e^{-x}) dx = e^x + \int e^u du = e^x + e^u + C = e^x + e^{-x} + C.$$

27. Let $u = 1 + e^x$ so that $du = e^x dx$. Then

$$\int \frac{e^x}{1+e^x} dx = \int \frac{du}{u} = \ln|u| + C = \ln(1 + e^x) + C.$$

29. Let $u = \sqrt{x} = x^{1/2}$. Then $du = \frac{1}{2}x^{-1/2}\,dx$ or $2\,du = x^{-1/2}\,dx$.

$$\int \frac{e^{\sqrt{x}}}{\sqrt{x}}\,dx = \int 2e^{u}\,du = 2e^{u} + C = 2e^{\sqrt{x}} + C.$$

31. Let $u = e^{3x} + x^3$ so that $du = (3e^{3x} + 3x^2)\,dx = 3(e^{3x} + x^2)\,dx$ or $(e^{3x} + x^2)\,dx = \frac{1}{3}du$.
Then

$$\int \frac{e^{3x} + x^2}{(e^{3x} + x^3)^3}\,dx = \frac{1}{3}\int \frac{du}{u^3} = \frac{1}{3}\int u^{-3}\,du = -\frac{1}{6}u^{-2} + C = -\frac{1}{6(e^{3x} + x^3)^2} + C.$$

33. Let $u = e^{2x} + 1$, so that $du = 2e^{2x}\,dx$, or $\frac{1}{2}du = e^{2x}\,dx$.

$$\int e^{2x}(e^{2x} + 1)^3\,dx = \int \frac{1}{2}u^3\,du = \frac{1}{8}u^4 + C = \frac{1}{8}(e^{2x} + 1)^4 + C.$$

35. Let $u = \ln 5x$ so that $du = \frac{1}{x}\,dx$. Then

$$\int \frac{\ln 5x}{x}\,dx = \int u\,du = \frac{1}{2}u^2 + C = \frac{1}{2}(\ln 5x)^2 + C.$$

37. Let $u = \ln x$ so that $du = \frac{1}{x}\,dx$. Then

$$\int \frac{1}{x\,\ln x}\,dx = \int \frac{du}{u} = \ln|u| + C = \ln|\ln x| + C.$$

39. Let $u = \ln x$ so that $du = \frac{1}{x}\,dx$. Then

$$\int \frac{\sqrt{\ln x}}{x}\,dx = \int \sqrt{u}\,du = \frac{2}{3}u^{3/2} + C = \frac{2}{3}(\ln x)^{3/2} + C$$

41. $\displaystyle \int \left(xe^{x^2} - \frac{x}{x^2 + 2} \right)\,dx = \int xe^{x^2} - \int \frac{x}{x^2 + 2}\,dx.$

To evaluate the first integral, let $u = x^2$ so that $du = 2x\,dx$, or $x\,dx = \frac{1}{2}du$. Then

$$\int xe^{x^2}\,dx = \frac{1}{2}\int e^{u}\,du + C_1 = \frac{1}{2}e^{u} + C_1 = \frac{1}{2}e^{x^2} + C_1.$$

To evaluate the second integral, let $u = x^2 + 2$ so that $du = 2x\,dx$, or $x\,dx = \frac{1}{2}\,du$. Then

$$\int \frac{x}{x^2+2}\,dx = \frac{1}{2}\int \frac{du}{u} = \frac{1}{2}\ln|u| + C_2 = \frac{1}{2}\ln(x^2+2) + C_2.$$

Therefore, $\displaystyle\int \left(xe^{x^2} - \frac{x}{x^2+2} \right) dx = \frac{1}{2}e^{x^2} - \frac{1}{2}\ln(x^2+2) + C.$

43. Let $u = \sqrt{x} - 1$ so that $du = \frac{1}{2}x^{-1/2}\,dx = \dfrac{1}{2\sqrt{x}}\,dx$ or $dx = 2\sqrt{x}\,du$.

Also, we have $\sqrt{x} = u + 1$, so that $x = (u+1)^2 = u^2 + 2u + 1$ and $dx = 2(u+1)\,du$. So

$$\int \frac{x+1}{\sqrt{x}-1}\,dx = \int \frac{u^2+2u+2}{u}\cdot 2(u+1)\,du = 2\int \frac{(u^3+3u^2+4u+2)}{u}\,du$$

$$= 2\int \left(u^2 + 3u + 4 + \frac{2}{u} \right) du = 2\left(\frac{1}{3}u^3 + \frac{3}{2}u^2 + 4u + 2\ln|u| \right) + C$$

$$= 2\left[\frac{1}{3}(\sqrt{x}-1)^3 + \frac{3}{2}(\sqrt{x}-1)^2 + 4(\sqrt{x}-1) + 2\ln\left|\sqrt{x}-1\right| \right] + C.$$

45. Let $u = x - 1$ so that $du = dx$. Also, $x = u + 1$ and so

$$\int x(x-1)^5\,dx = \int (u+1)u^5\,du = \int (u^6 + u^5)\,du$$

$$= \frac{1}{7}u^7 + \frac{1}{6}u^6 + C = \frac{1}{7}(x-1)^7 + \frac{1}{6}(x-1)^6 + C$$

$$= \frac{(6x+1)(x-1)^6}{42} + C.$$

47. Let $u = 1 + \sqrt{x}$ so that $du = \frac{1}{2}x^{-1/2}\,dx$ and $dx = 2\sqrt{x} = 2(u-1)\,du$

$$\int \frac{1-\sqrt{x}}{1+\sqrt{x}}\,dx = \int \left(\frac{1-(u-1)}{u} \right) \cdot 2(u-1)\,du = 2\int \frac{(2-u)(u-1)}{u}\,du$$

$$= 2\int \frac{-u^2+3u-2}{u}\,du = 2\int \left(-u+3-\frac{2}{u} \right) du = -u^2 + 6u - 4\ln|u| + C$$

6 Integration

$$= -(1+\sqrt{x})^2 + 6(1+\sqrt{x}) - 4\ln(1+\sqrt{x}) + C$$
$$= -1 - 2\sqrt{x} - x + 6 + 6\sqrt{x} - 4\ln(1+\sqrt{x}) + C$$
$$= -x + 4\sqrt{x} + 5 - 4\ln(1+\sqrt{x}) + C.$$

49. $I = \int v^2 (1-v)^6 \, dv.$ Let $u = 1 - v$, then $du = -dv$. Also, $1 - u = v$, and $(1-u)^2 = v^2$. Therefore,

$$I = \int -(1 - 2u + u^2)u^6 \, du = \int -(u^6 - 2u^7 + u^8) \, du = -\left(\frac{u^7}{7} - \frac{2u^8}{8} + \frac{u^9}{9}\right) + C$$

$$= -u^7\left(\frac{1}{7} - \frac{1}{4}u + \frac{1}{9}u^2\right) + C = -\frac{1}{252}(1-v)^7[36 - 63(1-v) + 28(1 - 2v + v^2)]$$

$$= -\frac{1}{252}(1-v)^7[36 - 63 + 63v + 28 - 56v + 28v^2]$$

$$= -\frac{1}{252}(1-v)^7(28v^2 + 7v + 1) + C.$$

51. $f(x) = \int f'(x) \, dx = 5\int (2x-1)^4 \, dx.$ Let $u = 2x - 1$ so that $du = 2x-1$ so that $du = 2 \, dx$, or $dx = \frac{1}{2} du.$ Then

$$f(x) = \frac{5}{2}\int u^4 \, du = \frac{1}{2}u^5 + C = \frac{1}{2}(2x-1)^5 + C.$$

Next, $f(1) = 3$ implies $\frac{1}{2} + C = 3$ or $C = \frac{5}{2}.$ Therefore,

$$f(x) = \frac{1}{2}(2x-1)^5 + \frac{5}{2}.$$

53. $f(x) = \int -2xe^{-x^2+1} \, dx.$ Let $u = -x^2 + 1$ so that $du = -2x \, dx.$ Then

$f(x) = \int e^u \, du = e^u + C = e^{-x^2+1} + C.$ The condition $f(1) = 0$ implies

$f(1) = 1 + C = 0$, or $C = -1$. Therefore, $f(x) = e^{-x^2+1} - 1.$

55. $N'(t) = 2000(1 + 0.2t)^{-3/2}.$ Let $u = 1 + 0.2t.$ Then $du = 0.2 \, dt$ and $5 \, du = dt.$ Therefore, $N(t) = (5)(2000)$

$$\int u^{-3/2} \, du = -20{,}000u^{-1/2} + C = -20{,}000(1 + 0.2t)^{-1/2} + C.$$

Next, $N(0) = -20,000(1)^{-1/2} + C = 1000$. Therefore, $C = 21,000$ and
$N(t) = -\dfrac{20,000}{\sqrt{1+0.2t}} + 21,000$. In particular, $N(5) = -\dfrac{20,000}{\sqrt{2}} + 21,000 \approx 6,858$.

57. $p(x) = \displaystyle\int -\dfrac{250x}{(16+x^2)^{3/2}}\,dx = -250 \int \dfrac{x}{(16+x^2)^{3/2}}\,dx.$

Let $u = 16 + x^2$ so that $du = 2x\,dx$ and $x\,dx = \frac{1}{2}\,du$.

Then $p(x) = -\frac{250}{2} \displaystyle\int u^{-3/2}\,du = (-125)(-2)u^{-1/2} + C = \dfrac{250}{\sqrt{16+x^2}} + C.$

$p(3) = \dfrac{250}{\sqrt{16+9}} + C = 50$ implies $C = 0$ and $p(x) = \dfrac{250}{\sqrt{16+x^2}}.$

59. Let $u = 2t + 4$, so that $du = 2\,dt$. Then

$r(t) = \displaystyle\int \dfrac{30}{\sqrt{2t+4}}\,dt = 30 \int \dfrac{1}{2} u^{-1/2}\,du = 30u^{1/2} + C = 30\sqrt{2t+4} + C.$

$r(0) = 60 + C = 0$, and $C = -60$. Therefore, $r(t) = 30\left(\sqrt{2t+4} - 2\right)$. Then

$r(16) = 30\left(\sqrt{36} - 2\right) = 120$ ft. Therefore, the polluted area is

$\pi r^2 = \pi(120)^2 = 14,400\pi$, or $14,400\pi$ sq ft.

61. The population t years from now will be

$$P(t) = \int r(t)\,dt = \int 400\left(1 + \dfrac{2t}{24+t^2}\right)dt = \int 400\,dt + 800 \int \dfrac{t}{24+t^2}\,dt$$

In order to evaluate the second integral on the right, let
$u = 24 + t^2$, $du = 2t\,dt$, or $t\,dt = \frac{1}{2}\,du$

We obtain $P(t) = 400t + 800 \displaystyle\int \dfrac{\frac{1}{2}\,du}{u} = 400t + 400|\ln u| + C$

$= 400[t + \ln(24+t^2)] + C$

To find C, use the condition $P(0) = 60,000$ giving

$$400[0 + \ln 24] + C = 60,000 \quad \text{or} \quad C = 58728.78$$

So $P(t) = 400[t + \ln(24 + t^2)] + 58728.78$. Therefore, the population 5 years from now will be

$$400[5 + \ln(24 + 25)] + 58728.78 \approx 62,285.51, \text{ or approximately, } 62,286.$$

63. $N(t) = \int N'(t) \, dt = 6 \int e^{-0.05t} \, dt = \dfrac{6}{-0.05} e^{-0.05t}$ (Let u = -0.05t)

$$= -120e^{-0.05t} + C.$$

$N(0) = 60$ implies $-120 + C = 60$ or $C = 180$. Therefore, $N(t) = -120e^{-0.05t} + 180$.

65. $A(t) = \int A'(t) \, dt = r \int e^{-at} \, dt$. Let $u = -at$ so that $du = -a \, dt$, or $dt = -\frac{1}{a} du$.

$$A(t) = r(-\tfrac{1}{a}) \int e^u \, du = -\tfrac{r}{a} e^u + C = -\tfrac{r}{a} e^{-at} + C$$

$A(0)$ implies $-\dfrac{r}{a} + C = 0$, or $C = \dfrac{r}{a}$. So, $A(t) = -\dfrac{r}{a} e^{-at} + \dfrac{r}{a} = \dfrac{r}{a}(1 - e^{-at})$.

EXERCISES 6.3, page 430

1. $\frac{1}{3}(1.9 + 1.5 + 1.8 + 2.4 + 2.7 + 2.5) = \frac{12.8}{3} \approx 4.27$.

3. a. $A = \frac{1}{2}(2)(6) = 6$ sq units.

   b. $\Delta x = \frac{2}{4} = \frac{1}{2}$; $x_1 = 0$, $x_2 = \frac{1}{2}$, $x_3 = 1$, $x_4 = \frac{3}{2}$.

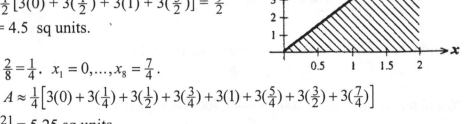

   $$A \approx \tfrac{1}{2}[3(0) + 3(\tfrac{1}{2}) + 3(1) + 3(\tfrac{3}{2})] = \tfrac{9}{2}$$
   $$= 4.5 \text{ sq units.}$$

   c. $\Delta x = \frac{2}{8} = \frac{1}{4}$. $x_1 = 0, \dots, x_8 = \frac{7}{4}$.

   $$A \approx \tfrac{1}{4}\left[3(0) + 3(\tfrac{1}{4}) + 3(\tfrac{1}{2}) + 3(\tfrac{3}{4}) + 3(1) + 3(\tfrac{5}{4}) + 3(\tfrac{3}{2}) + 3(\tfrac{7}{4})\right]$$
   $$= \tfrac{21}{4} = 5.25 \text{ sq units.}$$

   d. Yes.

5. a. $A = 4$

b. $\Delta x = \frac{2}{5} = 0.4$; $x_1 = 0$, $x_2 = 0.4$, $x_3 = 0.8$, $x_4 = 1.2$

$x_5 = 1.6$,

$A \approx 0.4\{[4 - 2(0)] + [4 - 2(0.4)] + [4 - 2(0.8)]$
$\qquad + [4 - 2(1.2)] + [4 - 2(1.6)]\}$

$\qquad = 4.8$

c. $\Delta x = \frac{2}{10} = 0.2$, $x_1 = 0$, $x_2 = 0.2$, $x_3 = 0.4$, ..., $x_{10} = 1.8$.

$A \approx 0.2\{[4 - 2(0)] + [4 - 2(0.2)] + [4 - 2(0.4)]$
$\qquad + \cdots + [4 - 2(1.8)]\} = 4.4$

d. Yes.

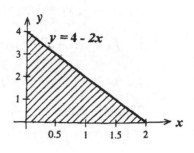

7. a. $\Delta x = \dfrac{4 - 2}{2} = 1$; $x_1 = 2.5$, $x_2 = 3.5$;   The Riemann sum is $[(2.5)^2 + (3.5)^2] = 18.5$.

b. $\Delta x = \dfrac{4 - 2}{5} = 0.4$; $x_1 = 2.2$, $x_2 = 2.6$, $x_3 = 3.0$, $x_4 = 3.4$, $x_5 = 3.8$

The Riemann sum is $0.4[2.2^2 + 2.6^2 + 3.0^2 + 3.4^2 + 3.8^2] = 18.64$.

c. $\Delta x = \dfrac{4 - 2}{10} = 0.2$; $x_1 = 2.1$, $x_2 = 2.3$, $x_2 = 2.5$, ..., $x_{10} = 3.9$

The Riemann sum is $0.2[2.1^2 + 2.3^2 + 2.5^2 + \cdots + 3.9^2] = 18.66$.

The area seems to be $18\frac{2}{3}$ sq units.

9. a. $\Delta x = \dfrac{4 - 2}{2} = 1$; $x_1 = 3$, $x_2 = 4$. The Riemann sum is $(1)[3^2 + 4^2] = 25$.

b. $\Delta x = \dfrac{4 - 2}{5} = 0.4$; $x_1 = 2.4$, $x_2 = 2.8$, $x_3 = 3.2$, $x_4 = 3.6$, $x_5 = 4$.

The Riemann sum is $0.4[2.4^2 + 2.8^2 + \cdots + 4^2] = 21.12$.

c. $\Delta x = \dfrac{4 - 2}{10} = 0.2$; $x_1 = 2.2$, $x_2 = 2.4$, $x_3 = 2.6$, ..., $x_{10} = 4$.

The Riemann sum is $0.2[2.2^2 + 2.4^2 + 2.6^2 + \cdots + 4^2] = 19.88$.

d. 19.9 sq units.

11. a. $\Delta x = \dfrac{1}{2}$, $x_1 = 0$, $x_2 = \dfrac{1}{2}$. The Riemann sum is

$f(x_1)\Delta x + f(x_2)\Delta x = \left[(0)^3 + (\tfrac{1}{2})^3\right]\tfrac{1}{2} = \tfrac{1}{16} = 0.0625.$

b. $\Delta x = \dfrac{1}{5}, x_1 = 0, x_2 = \dfrac{1}{5}, x_3 = \dfrac{2}{5}, x_4 = \dfrac{3}{5}, x_5 = \dfrac{4}{5}$. The Riemann sum

is $\ f(x_1)\Delta x + f(x_2)\Delta x + \cdots f(x_5)\Delta x = \left[(\tfrac{1}{5})^3 + (\tfrac{2}{5})^3 + \cdots + (\tfrac{4}{5})^3\right]\tfrac{1}{5} = \tfrac{100}{625} = 0.16.$

c. $\Delta x = \dfrac{1}{10}; x_1 = 0, x_2 = \dfrac{1}{10}, x_3 = \dfrac{2}{10}, \cdots, x_{10} = \dfrac{9}{10}.$

The Riemann sum is

$$f(x_1)\Delta x + f(x_2)\Delta x + \cdots + f(x_{10})\Delta x = \left[(\tfrac{1}{10})^3 + (\tfrac{2}{10})^3 + \cdots + (\tfrac{9}{10})^3\right]\tfrac{1}{10}$$
$$= \tfrac{2025}{10,000} = 0.2025 \approx 0.2 \text{ sq units.}$$

The Riemann sum seems to approach 0.2.

13. $\Delta x = \dfrac{2-0}{5} = \dfrac{2}{5}; \ x_1 = \dfrac{1}{5}, x_2 = \dfrac{3}{5}, x_3 = \dfrac{5}{5}, x_4 = \dfrac{7}{5}, x_5 = \dfrac{9}{5}.$

$A \approx \left\{\left[(\tfrac{1}{5})^2 + 1\right] + \left[(\tfrac{3}{5})^2 + 1\right] + \left[(\tfrac{5}{5})^2 + 1\right] + \left[(\tfrac{7}{5})^2 + 1\right] + \left[(\tfrac{9}{5})^2 + 1\right](\tfrac{2}{5})\right\}$

$\quad = \tfrac{580}{125} = 4.64 \text{ sq units.}$

15. $\Delta x = \dfrac{3-1}{4} = \dfrac{1}{2}; \ x_1 = \dfrac{3}{2}, x_2 = \dfrac{4}{2}, x_3 = \dfrac{5}{2}, x_4 = 3.$

$A \approx \left[\dfrac{1}{\frac{3}{2}} + \dfrac{1}{\frac{4}{2}} + \dfrac{1}{\frac{5}{2}} + \dfrac{1}{3}\right]\dfrac{1}{2} \approx 0.95 \text{ sq units.}$

17. $A = 20[f(10) + f(30) + f(50) + f(70) + f(90)]$
$\quad = 20(80 + 100 + 110 + 100 + 80) = 9400 \text{ sq ft.}$

EXERCISES 6.4, page 439

1. $A = \displaystyle\int_1^4 2\,dx = 2x\Big|_1^4 = 2(4-1) = 6$, or 6 square

units. The region is a rectangle whose area is

3 · 2, or 6, square units.

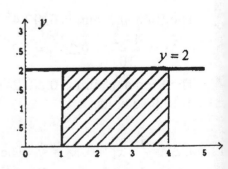

3. $A = \int_1^3 2x\,dx = x^2 \big|_1^3 = 9 - 1 = 8$, or 8 sq units.

The region is a parallelogram of area
$(1/2)(3 - 1)(2 + 6) = 8$ sq units.

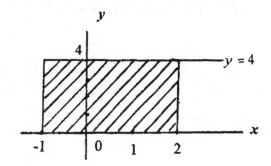

5. $A = \int_{-1}^2 (2x + 3)\,dx = x^2 + 3x \big|_{-1}^2 = (4 + 6) - (1 - 3) = 12$, or 12 sq. units.

7. $A = \int_{-1}^2 (-x^2 + 4)\,dx = -\dfrac{1}{3}x^3 + 4x \big|_{-1}^2 = \left(-\dfrac{8}{3} + 8\right) - \left(\dfrac{1}{3} - 4\right) = 9$, or 9 sq units.

9. $A = \int_1^2 \dfrac{1}{x}\,dx = \ln|x| \big|_1^2 = \ln 2 - \ln 1 = \ln 2$, or ln 2 sq units.

11. $A = \int_1^9 \sqrt{x}\,dx = \dfrac{2}{3}x^{3/2} \big|_1^9 = \dfrac{2}{3}(27 - 1) = \dfrac{52}{3}$, or $17\frac{1}{3}$ sq units.

13. $A = \int_{-8}^{-1} (1 - x^{1/3})\,dx = x - \frac{3}{4}x^{4/3} \big|_{-8}^{-1} = (-1 - \frac{3}{4}) - (-8 - 12) = 18\frac{1}{4}$, or $18\frac{1}{4}$ sq units.

15. $A = \int_0^2 e^x\,dx = e^x \big|_0^2 = (e^2 - 1)$, or approximately 6.39 sq units.

17. $\int_2^4 3\,dx = 3x \big|_2^4 = 3(4 - 2) = 6$.

19. $\int_1^3 (2x + 3)\,dx = x^2 + 3x \big|_1^3 = (9 + 9) - (1 + 3) = 14$.

21. $\int_{-1}^3 2x^2\,dx = \frac{2}{3}x^3 \big|_{-1}^3 = \frac{2}{3}(27) - \frac{2}{3}(-1) = \frac{56}{3}$.

23. $\int_{-2}^2 (x^2 - 1)\,dx = \frac{1}{3}x^3 - x \big|_{-2}^2 = \left(\frac{8}{3} - 2\right) - \left(-\frac{8}{3} + 2\right) = \frac{4}{3}$.

25. $\int_1^8 4x^{1/3}\,dx = (4)(\frac{3}{4})x^{4/3} \big|_1^8 = 3(16 - 1) = 45$.

27. $\int_0^1 (x^3 - 2x^2 + 1)\, dx = \frac{1}{4}x^4 - \frac{2}{3}x^3 + x\big|_0^1 = \frac{1}{4} - \frac{2}{3} + 1 = \frac{7}{12}$

29. $\int_2^4 \frac{1}{x}\, dx = \ln |x|\big|_2^4 = \ln 4 - \ln 2 = \ln(\frac{4}{2}) = \ln 2.$

31. $\int_0^4 x(x^2 - 1)\, dx = \int_0^4 (x^3 - x)\, dx = \frac{1}{4}x^4 - \frac{1}{2}x^2\big|_0^4 = 64 - 8 = 56.$

33. $\int_1^3 (t^2 - t)^2\, dt = \int_1^3 t^4 - 2t^3 + t^2)\, dt = \frac{1}{5}t^5 - \frac{1}{2}t^4 + \frac{1}{3}t^3\big|_1^3$

$$= \left(\frac{243}{5} - \frac{81}{2} + \frac{27}{3}\right) - \left(\frac{1}{5} - \frac{1}{2} + \frac{1}{3}\right) = \frac{512}{30} = \frac{256}{15}.$$

35. $\int_{-3}^{-1} x^{-2}\, dx = -\frac{1}{x}\Big|_{-3}^{-1} = 1 - \frac{1}{3} = \frac{2}{3}.$

37. $\int_1^4 \left(\sqrt{x} - \frac{1}{\sqrt{x}}\right) dx = \int_1^4 (x^{1/2} - x^{-1/2})\, dx = \frac{2}{3}x^{3/2} - 2x^{1/2}\Big|_1^4$

$$= \left(\frac{16}{3} - 4\right) - \left(\frac{2}{3} - 2\right) = \frac{8}{3}.$$

39. $\int_1^4 \frac{3x^3 - 2x^2 + 4}{x^2}\, dx = \int_1^4 (3x - 2 + 4x^{-2})\, dx = \frac{3}{2}x^2 - 2x - \frac{4}{x}\Big|_1^4$

$$= (24 - 8 - 1) - (\frac{3}{2} - 2 - 40 = \frac{39}{2}.$$

41. a. $C(300) - C(0) = \int_0^{300} (0.0003x^2 - 0.12x + 20)\, dx = 0.0001x^3 - 0.06x^2 + 20x\big|_0^{300}$

$$= 0.0001(300)^3 - 0.06(300)^2 + 20(300) = 3300.$$

Therefore $C(300) = 3300 + C(0) = 3300 + 800 = 4100$, or \$4100.

b. $\int_{200}^{300} C'(x)\, dx = (0.0001x^3 - 0.06x^2 + 20x)\big|_{200}^{300}$

$$= [0.0001(300)^3 - 0.06(300)^2 + 20(300)]$$
$$-[0.0001(200)^3 - 0.06(200)^2 + 20(200)]$$

$$= 900 \text{ or } \$900.$$

43. a. The profit is $\displaystyle\int_0^{200} (-0.0003x^2 + 0.02x + 20)\,dx + P(0)$

$$= -0.0001x^3 + 0.01x^2 + 20x\big|_0^{200} + P(0)$$

$$= 3600 + P(0) = 3600 - 800, \text{ or } \$2800.$$

  b. $\displaystyle\int_{200}^{220} P'(x)\,dx = P(220) - P(200) = -0.0001x^3 + 0.01x^2 + 20x\big|_{200}^{220}$

$$= 219.20, \text{ or } \$219.20.$$

45. The distance is

$$\int_0^{20} v(t)\,dt = \int_0^{20} (-t^2 + 20t + 440)\,dt = -\tfrac{1}{3}t^3 + 10t^2 + 440t\big|_0^{20} \approx 10{,}133\tfrac{1}{3}\,\text{ft.}$$

47. The number will be

$$\int (0.00933t^3 + 0.019t^2 - 0.10833t + 1.3467)\,dt$$

$$= 0.0023325t^4 + 0.0063333t^3 - 0.054165t^2 + 1.3467\,t\big|_0^{10} = 37.7,$$

or approximately 37.7 million Americans.

49. The average population over the period in question is

$$A = \tfrac{1}{3} \int \frac{85}{1 + 1.859e^{-0.66t}}\,dt$$

Multiplying the integrand by $e^{0.66t}/e^{0.66t}$ gives

$$A = \frac{85}{3} \int_0^3 \frac{e^{0.66t}}{e^{0.66t} + 1.859}\,dt$$

Let $u = 1.859 + e^{0.66t}$, $du = 0.66e^{0.66t}\,dt$, or $e^{0.66t}\,dt = \dfrac{du}{0.66}$

If $t = 0$, then $u = 2.859$. If $t = 3$, then $u = 9.1017$
Substituting

$$A = \frac{85}{3} \int_{2.859}^{9.1017} \frac{du}{(0.66)u} = \frac{85}{3(0.66)} \ln u \Big|_{2.859}^{9.1017}$$

$$= \frac{85}{3(0.66)} (\ln 9.1017 - \ln 2.859) = \frac{85}{3(0.66)} \ln \frac{9.1017}{2.859} \approx 49.712,$$

or approximately 49.7 million people.

51. False. The integrand $f(x) = \dfrac{1}{x^3}$ is discontinuous at $x = 0$.

53. False. $f(x)$ is not nonnegative on $[0, 2]$.

USING TECHNOLOGY EXERCISES 6.4, page 441

1.   6.1787   3.   0.7873   5.   −0.5888   7.   2.7044

9.   3.9973   11. 37.7 million   13. 333,209   15. 903,213

EXERCISES 6.5, page 449

1. Let $u = x^2 - 1$ so that $du = 2x\,dx$ or $x\,dx = \frac{1}{2}\,du$. Also, if $x = 0$,
    then $u = -1$ and if $x = 2$, then $u = 3$. So
    $$\int_0^2 x(x^2 - 1)^3\,dx = \frac{1}{2}\int_{-1}^3 u^3\,du = \frac{1}{8}u^4\Big|_{-1}^3 = \frac{1}{8}(81) - \frac{1}{8}(1) = 10.$$

3. Let $u = 5x^2 + 4$ so that $du = 10x\,dx$ or $x\,dx = \frac{1}{10}\,du$. Also, if
    $x = 0$, then $u = 4$, and if $x = 1$, then $u = 9$. So
    $$\int_0^1 x\sqrt{5x^2 + 4}\,dx = \frac{1}{10}\int_4^9 u^{1/2}\,du = \frac{1}{15}u^{3/2}\Big|_4^9 = \frac{1}{15}(27) - \frac{1}{15}(8) = \frac{19}{15}.$$

5. Let $u = x^3 + 1$ so that $du = 3x^2\,dx$ or $x^2\,dx = \frac{1}{3}\,du$. Also, if $x = 0$,
    then $u = 1$, and if $x = 2$, then $u = 9$. So,
    $$\int_0^2 x^2(x^3 + 1)^{3/2}\,dx = \frac{1}{3}\int_1^9 u^{3/2}\,du = \frac{2}{15}u^{5/2}\Big|_1^9 = \frac{2}{15}(243) - \frac{2}{15}(1) = \frac{484}{15}.$$

7. Let $u = 2x + 1$ so that $du = 2\,dx$ or $dx = \frac{1}{2}\,du$. Also, if $x = 0$,
    then $u = 1$ and if $x = 1$ then $u = 3$. So
    $$\int_0^1 \frac{1}{\sqrt{2x+1}}\,dx = \frac{1}{2}\int_1^3 \frac{1}{\sqrt{u}}\,du = \frac{1}{2}\int_1^3 u^{-1/2}\,du = u^{1/2}\Big|_1^3 = \sqrt{3} - 1.$$

9. $\int_1^2 (2x-1)^4 \, dx$. Put $u = 2x - 1$ so that $du = 2 \, dx$ or $dx = \frac{1}{2} du$.

Then $\int_1^2 (2x-1)^4 \, dx = \frac{1}{2} \int_1^3 u^4 \, du = \frac{1}{10} u^5 \big|_1^3 = \frac{1}{10}(243-1) = \frac{121}{5} = 24\frac{1}{5}$.

11. Let $u = x^3 + 1$ so that $du = 3x^2 \, dx$ or $x^2 \, dx = \frac{1}{3} du$. Also, if $x = -1$,
then $u = 0$ and if $x = 1$, then $u = 2$. So
$$\int_{-1}^1 x^2 (x^3 + 1)^4 \, dx = \frac{1}{3} \int_0^2 u^4 \, du = \frac{1}{15} u^5 \big|_0^2 = \frac{32}{15}.$$

13. Let $u = x - 1$ so that $du = dx$. Then if $x = 1$, $u = 0$, and if $x = 5$, then $u = 4$.
$$\int_1^5 x\sqrt{x-1} \, dx = \int_0^4 (u+1)u^{1/2} \, du = \int_0^4 (u^{3/2} + u^{1/2}) \, du$$
$$= \frac{2}{5} u^{5/2} + \frac{2}{3} u^{3/2} \big|_0^4 = \frac{2}{5}(32) + \frac{2}{3}(8) = 18\frac{2}{15}.$$

15. Let $u = x^2$ so that $du = 2x \, dx$ or $x \, dx = \frac{1}{2} du$. If $x = 0$, $u = 0$ and if
$x = 2$, $u = 4$. So
$$\int_0^2 xe^{x^2} \, dx = \frac{1}{2} \int_0^4 e^u \, du = \frac{1}{2} e^u \big|_0^4 = \frac{1}{2}(e^4 - 1).$$

17. $\int_0^1 (e^{2x} + x^2 + 1) \, dx = \frac{1}{2} e^{2x} + \frac{1}{3} x^3 + x \big|_0^1 = (\frac{1}{2} e^2 + \frac{1}{3} + 1) - \frac{1}{2}$
$$= \frac{1}{2} e^2 + \frac{5}{6}.$$

19. Put $u = x^2 + 1$ so that $du = 2x \, dx$ or $x \, dx = \frac{1}{2} du$. Then
$$\int_{-1}^1 xe^{x^2+1} \, dx = \frac{1}{2} \int_2^2 e^u \, du = \frac{1}{2} e^u \big|_2^2 = 0$$
(Since the upper and lower limits are equal.)

21. Let $u = x - 2$ so that $du = dx$. If $x = 3$, $u = 1$ and if $x = 6$, $u = 4$. So
$$\int_3^6 \frac{2}{x-2} \, dx = 2 \int_1^4 \frac{du}{u} = 2 \ln|u| \big|_1^4 = 2 \ln 4.$$

23. Let $u = x^3 + 3x^2 - 1$ so that $du = (3x^2 + 6x)dx = 3(x^2 + 2x)dx$. If
$x = 1$, $u = 3$, and if $x = 2$, $u = 19$. So

$$\int_1^2 \frac{x^2+2x}{x^3+3x^2-1}\,dx = \frac{1}{3}\int_3^{19}\frac{du}{u} = \frac{1}{3}\ln u \Big|_3^{19} = \frac{1}{3}(\ln 19 - \ln 3).$$

25. $\displaystyle\int_1^2\left(4e^{2u}-\frac{1}{u}\right)du = 2e^{2u}-\ln u\Big|_1^2 = (2e^4-\ln 2)-(2e^2-0) = 2e^4-2e^2-\ln 2.$

27. $\displaystyle\int_1^2 (2e^{-4x}-x^{-2})\,dx = -\frac{1}{2}e^{-4x}+\frac{1}{x}\Big|_1^2 = (-\frac{1}{2}e^{-8}+\frac{1}{2})-(-\frac{1}{2}e^{-4}+1)$

$$= -\frac{1}{2}e^{-8}+\frac{1}{2}e^{-4}-\frac{1}{2}$$

$$= \frac{1}{2}(e^{-4}-e^{-8}-1).$$

29. $\displaystyle \text{AV} = \frac{1}{2}\int_0^2 (2x+3)\,dx = \frac{1}{2}(x^2+3x)\Big|_0^2 = \frac{1}{2}(10) = 5.$

31. $\displaystyle \text{AV} = \frac{1}{2}\int_1^3 (2x^2-3)\,dx = \frac{1}{2}(\frac{2}{3}x^3-3x)\Big|_1^3 = \frac{1}{2}(9+\frac{7}{3}) = \frac{17}{3}.$

33. $\displaystyle \text{AV} = \frac{1}{3}\int_{-1}^2 (x^2+2x-3)\,dx = \frac{1}{3}(\frac{1}{3}x^3+x^2-3x)\Big|_{-1}^2$

$$= \frac{1}{3}[(\frac{8}{3}+4-6)-(-\frac{1}{3}+1+3)] = \frac{1}{3}(\frac{8}{3}-2+\frac{1}{3}-4) = -1.$$

35. $\displaystyle \text{AV} = \frac{1}{4}\int_0^4 (2x+1)^{1/2}\,dx = (\frac{1}{4})(\frac{1}{2})(\frac{2}{3})(2x+1)^{3/2}\Big|_0^4 = \frac{1}{12}(27-1) = \frac{13}{6}$

37. $\displaystyle \text{AV} = \frac{1}{2}\int_0^2 xe^{x^2}\,dx = \frac{1}{4}e^{x^2}\Big|_0^2 = \frac{1}{4}(e^4-1).$

39. The amount produced was

$$\int_0^{20} 3.5e^{0.05t}\,dt = \frac{3.5}{0.05}e^u\Big|_0^{20} \qquad \text{(Use the substitution } u = 0.05t.)$$

$$= 70(e-1) \approx 120.3, \quad \text{or } 120.3 \text{ billion metric tons.}$$

41. The amount is $\int_1^2 t(\frac{1}{2}t^2+1)^{1/2}\,dt$. Let $u = \frac{1}{2}t^2+1$, so that $du = t\,dt$. Therefore,

$$\int_1^2 t(\tfrac{1}{2}t^2+1)^{1/2}\,dt = \int_{3/2}^3 u^{1/2}\,du = \tfrac{2}{3}u^{3/2}\Big|_{3/2}^3 = \tfrac{2}{3}[(3)^{3/2}-(\tfrac{3}{2})^{3/2}]$$
$$\approx 2.24 \text{ million dollars.}$$

43. The tractor will depreciate

$$\int_0^5 13388.61e^{-0.22314t}\,dt = \frac{13388.61}{-0.22314}e^{-0.22314t}\Big|_0^5$$

$$= -60{,}000.94e^{-0.22314t}\Big|_0^5 = -60{,}000.94(-0.672314)$$

$$= 40{,}339.47, \quad \text{or } \$40{,}339.$$

45. $\quad \bar{A} = \tfrac{1}{5}\int (\tfrac{1}{12}t^2+2t+44)dt = \tfrac{1}{5}\left[\tfrac{1}{36}t^3+t^2+44t\Big|_0^5\right]$

$$= \tfrac{1}{5}\left[\tfrac{125}{36}+25+220\right] = \tfrac{125+900+7920}{5(36)} \approx 49.69, \text{ or } 49.7 \text{ ft/sec.}$$

47. The average whale population will be

$$\tfrac{1}{10}\int_0^{10}(3t^3+2t^2-10t+600)\,dt = \tfrac{1}{10}(\tfrac{3}{4}t^4+\tfrac{2}{3}t^3-5t^2+600t\Big|_0^{10}$$

$$\approx \tfrac{1}{10}(7500+666.67-500+6000) \approx 1367 \text{ whales.}$$

49. The average yearly sales of the company over its first 5 years of operation is given

by $\quad \tfrac{1}{5-0}\int_0^5 t(0.2t^2+4)^{1/2}\,dt = \tfrac{1}{5}[(\tfrac{5}{2})(\tfrac{2}{3})(0.2t^2+4)^{3/2}\Big|_0^5 \qquad$ [Let $u = -0.2t^2+4$.]

$$= \tfrac{1}{5}[\tfrac{5}{3}(5+4)^{3/2}-\tfrac{5}{3}(4)^{3/2}] = \tfrac{1}{3}(27-8) = \tfrac{19}{3}, \text{ or } 6\tfrac{1}{3} \text{ million dollars.}$$

51. The average velocity is $\tfrac{1}{4}\int_0^4 3t\sqrt{16-t^2}\,dt = \tfrac{1}{4}(64) = 16$, or 16 ft/sec.

(Using the results of Exercise 44.)

53. $\int_0^5 p\,dt = \tfrac{1}{5}\left[18t+\tfrac{3}{2}e^{-2t}+18e^{-t/3}\right]_0^5$

$$= \tfrac{1}{5}\left[18(5)+\tfrac{3}{2}e^{-10}+18e^{-5/3}-\tfrac{3}{2}-18\right] = 14.78, \text{ or } \$14.78.$$

55. $\displaystyle\int_a^b f(x)\,dx = F(x)\Big|_a^b = F(b) - F(a) = -[F(a) - F(b)]$

$\qquad = -F(x)\big|_b^a = -\displaystyle\int_b^a f(x)\,dx$

57. $\displaystyle\int_a^b cf(x)\,dx = xF(x)\Big|_a^b = c[F(b) - F(a)] = c\displaystyle\int_a^b f(x)\,dx.$

59. $\displaystyle\int_0^1 (1 + x - e^x)\,dx = x + \tfrac{1}{2}x^2 - e^x\Big|_0^1 = (1 + \tfrac{1}{2} - e) + 1 = \tfrac{5}{2} - e.$

$\qquad \displaystyle\int_0^1 dx + \int_0^1 x\,dx - \int_0^1 e^x\,dx = x\big|_0^1 + \tfrac{1}{2}x^2\big|_0^1 - e^x\big|_0^1 = (1 - 0) + (\tfrac{1}{2} - 0) - (e - 1) = \tfrac{5}{2} - e.$

61. $\displaystyle\int_0^3 (1 + x^3)\,dx = x + \tfrac{1}{4}x^4\Big|_0^3 = 3 + \tfrac{81}{4} = \tfrac{93}{4}.$

$\qquad \displaystyle\int_0^1 (1 + x^3)\,dx + \int_1^2 (1 + x^3)\,dx + \int_2^3 (1 + x^3)\,dx$

$\qquad\qquad = (x + \tfrac{1}{4}x^4)\big|_0^1 + (x + \tfrac{1}{4}x^4)\big|_1^2 + (x + \tfrac{1}{4}x^4)\big|_2^3$

$\qquad\qquad = (1 + \tfrac{1}{4}) + (2 + 4) - (1 + \tfrac{1}{4}) + (3 + \tfrac{81}{4}) - (2 + 4) = \tfrac{93}{4}.$

63. $\displaystyle\int_3^0 f(x)\,dx = -\int_0^3 f(x)\,dx = -4.$ (Property 2)

65. a. $\displaystyle\int_{-1}^2 [2f(x) + g(x)]\,dx = 2\int_{-1}^2 f(x)\,dx + \int_{-1}^2 g(x)\,dx = 2(-2) + 3 = -1.$

$\qquad$ b. $\displaystyle\int_{-1}^2 [g(x) - f(x)]\,dx = \int_{-1}^2 g(x)\,dx - \int_{-1}^2 f(x)\,dx = 3 - (-2) = 5.$

$\qquad$ c. $\displaystyle\int_{-1}^2 [2f(x) - 3g(x)]\,dx = 2\int_{-1}^2 f(x)\,dx - 3\int_{-1}^2 g(x)\,dx = 2(-2) - 3(3) = -13.$

67. True. This follows from Property 1 of the definite integral.

69. False. Only a constant can be "moved out" of the integral sign.

71. True. This follows from Properties 3 and 4 of the definite integral.

1. 7.71667   3. 17.5649   5. 10,140   7. 60.5/day

EXERCISES 6.6, page 463

1. $-\int_0^6 (x^3 - 6x^2)\,dx = -\frac{1}{4}x^4 + 2x^3\Big|_0^6 = -\frac{1}{4}(6^4) + 2(6^3) = 108$ sq units.

3. $A = -\int_{-1}^0 x\sqrt{1-x^2}\,dx + \int_0^1 x\sqrt{1-x^2}\,dx = 2\int_0^1 x(1-x^2)^{1/2}\,dx$ (by symmetry). Let
   $u = 1 - x^2$ so that $du = -2x\,dx$ or $x\,dx = -\frac{1}{2}\,du$. Also, if $x = 0$, then $u = 1$ and
   if $x = 1$, $u = 0$. So $A = (2)(-\frac{1}{2})\int_0^1 u^{1/2}\,du = -\frac{2}{3}u^{3/2}\Big|_1^0 = \frac{2}{3}$, or $\frac{2}{3}$ sq units.

5. $A = -\int_0^4 (x - 2\sqrt{x})\,dx = \int_0^4 (-x + 2x^{1/2})\,dx = -\frac{1}{2}x^2 + \frac{4}{3}x^{3/2}\Big|_0^4$
   $= 8 + \frac{32}{3} = \frac{8}{3}$ sq units.

7. The required area is given by
   $\int_{-1}^0 (x^2 - x^{1/3})\,dx + \int_0^1 (x^{1/3} - x^2)\,dx = \frac{1}{3}x^3 - \frac{3}{4}x^{4/3}\Big|_{-1}^0 + \frac{3}{4}x^{4/3} - \frac{1}{3}x^3\Big|_0^1$
   $= -(-\frac{1}{3} - \frac{3}{4}) + (\frac{3}{4} - \frac{1}{3}) = 1\frac{1}{2}$ sq units.

9. The required area is given by
   $-\int_{-1}^2 -x^2\,dx = \frac{1}{3}x^3\Big|_{-1}^2 = \frac{8}{3} + \frac{1}{3} = 3$ sq units.

11. $y = x^2 - 5x + 4 = (x - 4)(x - 1) = 0$
    if $x = 1$ or 4. These give the x-intercepts.
    $A = -\int_1^3 (x^2 - 5x + 4)\,dx = -\frac{1}{3}x^3 + \frac{5}{2}x^2 - 4x\Big|_1^3$
    $= (-9 + \frac{45}{2} - 12) - (-\frac{1}{3} + \frac{5}{2} - 4) = \frac{10}{3} = 3\frac{1}{3}$.

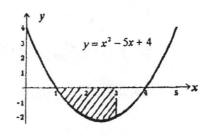

13. The required area is given by

$$-\int_0^9 -(1+\sqrt{x})\,dx = x + \tfrac{2}{3}x^{3/2}\Big|_0^9 = 9 + 18 = 27.$$

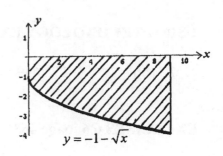

$$y = -1 - \sqrt{x}$$

15. $-\displaystyle\int_{-2}^4 -e^{(1/2)x}\,dx = 2e^{(1/2)x}\Big|_{-2}^4$

$$= 2(e^2 - e^{-1})\,\text{sq units.}$$

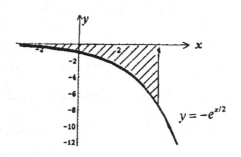

$$y = -e^{x/2}$$

17. $A = \displaystyle\int_1^3 [(x^2+3)-1]\,dx$

$$= \int_1^3 (x^2+2)\,dx = \tfrac{1}{3}x^3 + 2x\Big|_1^3$$

$$= (9+6) - (\tfrac{1}{3}+2) = \tfrac{38}{3}.$$

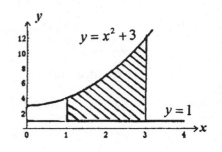

$$y = x^2 + 3$$

$$y = 1$$

19. $A = \displaystyle\int_0^2 (-x^2 + 2x + 3 + x - 3)\,dx$

$$= \int_0^2 (-x^2 + 3x)\,dx$$

$$= -\tfrac{1}{3}x^3 + \tfrac{3}{2}x^2\Big|_0^2 = -\tfrac{1}{3}(8) + \tfrac{3}{2}(4)$$

$$= 6 - \tfrac{8}{3} = \tfrac{10}{3}\ \text{sq units}$$

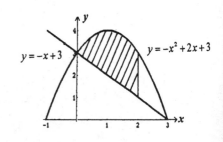

$$y = -x + 3$$

$$y = -x^2 + 2x + 3$$

21. $A = \int_{-1}^{2} [(x^2 + 1) - \frac{1}{3}x^3] dx$

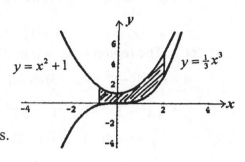

$\quad = \int_{-1}^{2} (-\frac{1}{3}x^3 + x^2 + 1) dx$

$\quad = -\frac{1}{12}x^4 + \frac{1}{3}x^3 + x \Big|_{-1}^{2}$

$\quad = (-\frac{4}{3} + \frac{8}{3} + 2) - (-\frac{1}{12} - \frac{1}{3} - 1) = 4\frac{3}{4}$ sq units.

23. $A = \int_{1}^{4} \left[(2x - 1) - \frac{1}{x} \right] dx = \int_{1}^{4} \left(2x - 1 - \frac{1}{x} \right) dx$

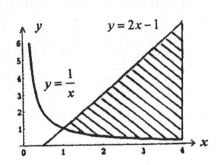

$\quad = (x^2 - x - \ln x) \Big|_{1}^{4}$

$\quad = (16 - 4 - \ln 4) - (1 - 1 - \ln 1)$

$\quad = 12 - \ln 4 \approx 10.6$ sq units.

25. $A = \int_{1}^{2} \left(e^x - \frac{1}{x} \right) dx = e^x - \ln x \Big|_{1}^{2}$

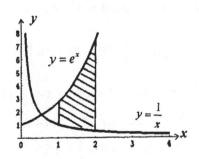

$\quad = (e^2 - \ln 2) - e = (e^2 - e - \ln 2)$ sq units.

27.

$A = -\int_{-1}^{0} x \, dx + \int_{0}^{2} x \, dx$

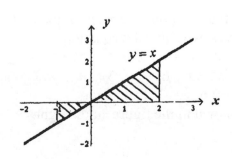

$\quad = -\frac{1}{2}x^2 \Big|_{-1}^{0} + \frac{1}{2}x^2 \Big|_{0}^{2}$

$\quad = \frac{1}{2} + 2 = 2\frac{1}{2}$ sq units.

29. The x–intercepts are found by solving

$x^2 - 4x + 3 = (x - 3)(x - 1) = 0$ giving $x = 1$
or 3. The region is shown in the figure.

$$A = -\int_{-1}^{1}[(-x^2 + 4x - 3)\,dx + \int_{1}^{2}(-x^2 + 4x - 3)\,dx$$

$$= \tfrac{1}{3}x^3 - 2x^2 + 3x\big|_{-1}^{1} + (-\tfrac{1}{3}x^3 + 2x^2 - 3x)\big|_{1}^{2}$$

$$= (\tfrac{1}{3} - 2 + 3) - (-\tfrac{1}{3} - 2 - 3)$$

$$+(-\tfrac{8}{3} + 8 - 6) - (-\tfrac{1}{3} + 2 - 3) = \tfrac{22}{3} \text{ sq units.}$$

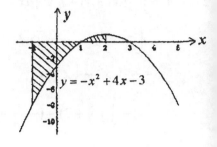

31. The region is shown in the figure at the right.

$$A = \int_{0}^{1}(x^3 - 4x^2 + 3x)\,dx - \int_{1}^{2}(x^3 - 4x^2 + 3x)\,dx$$

$$= (\tfrac{1}{4}x^4 - \tfrac{4}{3}x^3 + \tfrac{3}{2}x^2)\big|_{0}^{1}$$

$$-(\tfrac{1}{4}x^4 - \tfrac{4}{3}x^3 + \tfrac{3}{2}x^2)\big|_{1}^{2} = \tfrac{3}{2}\text{ sq units.}$$

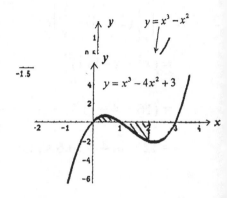

33. The region is shown in the figure at the right.

$$A = -\int_{-1}^{0}(e^x - 1)\,dx + \int_{0}^{3}(e^x - 1)\,dx$$

$$= (-e^x + x)\big|_{-1}^{0} + (e^x - x)\big|_{0}^{3}$$

$$= -1 - (-e^{-1} - 1) + (e^3 - 3) - 1$$

$$= e^3 - 4 + \tfrac{1}{e} \approx 16.5 \quad \text{sq units.}$$

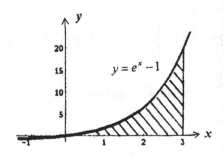

35. To find the points of intersection of the two
curves, we solve the equation
$$x^2 - 4 = x + 2$$
$$x^2 - x - 6 = (x - 3)(x + 2) = 0, \text{ obtaining}$$
$x = -2$ or $x = 3$. The region
is shown in the figure at the right.

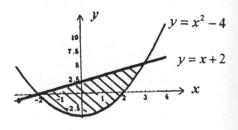

$$A = \int_{-2}^{3}[(x+2)-(x^2-4)]\,dx = \int_{-2}^{3}(-x^2+x+6)\,dx = (-\tfrac{1}{3}x^3+\tfrac{1}{2}x^2+6x)\big|_{-2}^{3}$$
$$= (-9+\tfrac{9}{2}+18)-(\tfrac{8}{3}+2-12) = \tfrac{125}{6}\text{ sq units.}$$

37. To find the points of intersection of the two
curves, we solve the equation $x^3 = x^2$
or $x^3 - x^2 = x^2(x-1) = 0$ giving $x = 0$ or 1.
The region is shown in the figure.

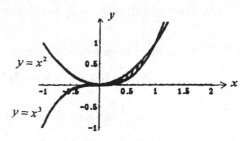

$$A = -\int_{0}^{1}(x^2-x^3)\,dx$$

$$= (\tfrac{1}{3}x^3-\tfrac{1}{4}x^4)\big|_{0}^{1} \ = \tfrac{1}{3}-\tfrac{1}{4} = \tfrac{1}{12}\text{ sq units}$$

39. To find the points of intersection of the two curves,   we solve the equation
$$x^3 - 6x^2 + 9x = x^2 - 3x,$$
or $x^3 - 7x^2 + 12x = x(x-4)(x-3) = 0$
obtaining $x = 0$, 3, or 4.

$$A = \int_{0}^{3}[(x^3-6x^2+9x)-(x^2+3x)]\,dx$$
$$+ \int_{3}^{4}[(x^2-3x)-(x^3-6x^2+9x)]\,dx$$
$$= \int_{0}^{3}(x^3-7x^2+12x)\,dx - \int_{3}^{4}(x^3-7x^2+12x)\,dx$$
$$= (\tfrac{1}{4}x^4-\tfrac{7}{3}x^3+6x^2)\big|_{0}^{3} - (\tfrac{1}{4}x^4-\tfrac{7}{3}x^3+6x^2)\big|_{3}^{4}$$
$$= (\tfrac{81}{4}-63+54)-(64-\tfrac{448}{3}+96)+(\tfrac{81}{4}-63+54) = \tfrac{71}{6}.$$

41. By symmetry, $A = 2\int_{0}^{3} x(9-x^2)^{1/2}\,dx$. We integrate by substitution with

$u = 9-x^2$, $du = -2x\,dx$. If $x = 0$, $u = 9$,   and if $x = 3, u = 0$. So

$$A = 2\int_{9}^{0} -\tfrac{1}{2}u^{1/2}\,du = -\int_{9}^{0}u^{1/2}\,du = -\tfrac{2}{3}u^{3/2}\big|_{9}^{0} = \tfrac{2}{3}(9)^{3/2} = 18 \text{ sq units.}$$

43. S gives the additional revenue that the company would realize if it used a different

advertising agency. $S = \int_{0}^{b}[g(x)-f(x)]\,dx.$

45. Shortfall $= \displaystyle\int_{2010}^{2050} [f(t) - g(t)]\,dt$

47. a. $\displaystyle\int_{T_1}^{T} [g(t) - f(t)]\,dt - \int_{0}^{T_1} [f(t) - g(t)]\,dt = A_2 - A_1.$

b. The number $A_2 - A_1$ gives the distance car 2 is ahead of car 1 after t seconds.

49. The turbo-charged model is moving at

$$A = \int_{0}^{10} [(4 + 1.2t + 0.03t^2) - (4 + 0.8t)]\,dt$$

$$= \int_{0}^{10} (0.4t + 0.03t^2)\,dt = (0.2t^2 + 0.1t^3)\Big|_{0}^{10}$$

$= 20 + 10,$ or 30 ft/sec faster than the standard model.

51. The additional number of cars will be given by

$$\int_{0}^{5} (5e^{0.3t} - 5 - 0.5t^{3/2})\,dt = \frac{5}{0.3}e^{0.3t} - 5t - 0.2t^{5/2}\Big|_{0}^{5}$$

$$= \frac{5}{0.3}e^{1.5} - 25 - 0.2(5)^{5/2} - \frac{5}{0.3} = 74.695 - 25 - 0.2(5)^{5/2} - \frac{50}{3}$$

$\approx 21.85,$ or 21,850 cars. (Remember t is measured in thousands.)

53. True. If $f(x) \geq g(x)$ on $[a, b]$, then the area of the said region is

$$\int_{a}^{b} [f(x) - g(x)]\,dx = \int_{a}^{b} |f(x) - g(x)|\,dx$$

If $f(x) \leq g(x)$ on $[a, b]$, then the area of the region is

$$\int_{a}^{b} [g(x) - f(x)]\,dx = \int_{a}^{b} -[f(x) - g(x)]\,dx = \int_{a}^{b} |f(x) - g(x)|\,dx$$

55. The area of R' is

$$A = \int_{a}^{b} \{[f(x) + C] - [g(x) + C]\}\,dx = \int_{a}^{b} [f(x) + C - g(x) - C]\,dx$$

$$= \int_{a}^{b} [f(x) - g(x)]\,dx$$

USING TECHNOLOGY EXERCISES 6.6, page 464

1. a.

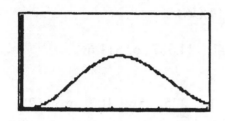

    b. 1074.2857

3. a.

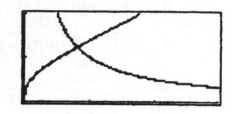

    b. 0.9961 sq units

5. a.

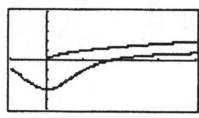

    b. 5.4603 sq units

7. a.

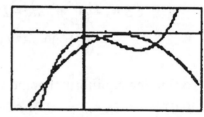

    b. 25.8549 sq units

9. a.

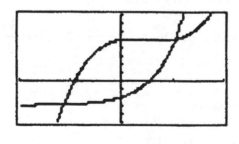

    b. 10.5144 sq units

11. a.

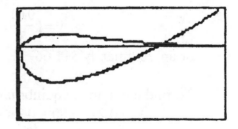

    b. 3.5799 sq units

13. 207.43 sq units

EXERCISES 6.7, page 478

1. When $p = 4$, $-0.01x^2 - 0.1x + 6 = 4$ or $x^2 + 10x - 200 = 0$, $(x - 10)(x + 20) = 0$ and $x = 10$ or -20. We reject the root $x = -20$. The consumers' surplus is

    $$CS = \int_0^{10} (-0.01x^2 - 0.1x + 6)\, dx - (4)(10)$$

    $$= -\frac{0.01}{3} x^3 - 0.05x^2 + 6x \Big|_0^{10} - 40 \approx 11.667 \text{, or } \$11,667.$$

3. Setting $p = 10$, we have $\sqrt{225 - 5x} = 10$, $225 - 5x = 100$, or $x = 25$.

    Then $CS = \int_0^{25} \sqrt{225 - 5x}\, dx - (10)(25) = \int_0^{25} (225 - 5x)^{1/2}\, dx - 250$.

    To evaluate the integral, let $u = 225 - 5x$ so that $du = -5\, dx$ or $dx = -\frac{1}{5} du$. If $x = 0$, $u = 225$ and if $x = 25$, $u = 100$. So

    $$CS = -\frac{1}{5} \int_{225}^{100} u^{1/2}\, du - 250 = -\frac{2}{15} u^{3/2} \Big|_{225}^{100} - 250$$

    $$= -\frac{2}{15}(1000 - 3375) - 250 = 66.667 \text{, or } \$6,667.$$

5. To find the equilibrium point, we solve

    $$0.01x^2 + 0.1x + 3 = -0.01x^2 - 0.2x + 8, \quad \text{or } 0.02x^2 + 0.3x - 5 = 0,$$
    $$2x^2 + 30x - 500 = (2x - 20)(x + 25) = 0$$

    obtaining $x = -25$ or 10. So the equilibrium point is $(10,5)$. Then

    $$PS = (5)(10) - \int_0^{10} (0.01x^2 + 0.1x + 3)\, dx$$

    $$= 50 - (\frac{0.01}{3} x^3 + 0.05x^2 + 3x) \Big|_0^{10} = 50 - \frac{10}{3} - 5 - 30 = \frac{35}{3},$$

    or approximately $\$11,667$.

7. To find the market equilibrium, we solve

    $$-0.2x^2 + 80 = 0.1x^2 + x + 40, \quad 0.3x^2 + x - 40 = 0,$$
    $$3x^2 + 10x - 400 = 0, \ (3x + 40)(x - 10) = 0$$

    giving $x = -\frac{40}{3}$ or $x = 10$. We reject the negative root. The corresponding equilibrium price is $\$60$. The consumers' surplus is

    $$CS = \int_0^{10} (-0.2x^2 + 80)dx - (60)(10) = -\frac{0.2}{3} x^3 + 80x \Big|_0^{10} - 600 = 133\tfrac{1}{3},$$

    or $\$13,333$. The producers' surplus is

$$PS = 600 - \int_0^{10} (0.1x^2 + x + 40)dx = 600 - [\tfrac{0.1}{3}x^3 + \tfrac{1}{2}x^2 + 40x]\Big|_0^{10}$$
$$= 116\tfrac{2}{3}, \text{ or } \$11{,}667.$$

9. Here $P = 200{,}000$, $r = 0.08$, and $T = 5$. So
$$PV = \int_0^5 200{,}000 e^{-0.08t}\,dt = -\frac{200{,}000}{0.08} e^{-0.08t}\Big|_0^5 = -2{,}500{,}000(e^{-0.4} - 1)$$
$$\approx 824{,}199.85, \text{ or } \$824{,}200.$$

11. Here $P = 250$, $m = 12$, $T = 20$, and $r = 0.08$. So
$$A = \frac{mP}{r}(e^{rT} - 1) = \frac{12(250)}{0.08}(e^{1.6} - 1) \approx 148{,}238.70$$
or approximately $\$148{,}239$.

13. Here $P = 150$, $m = 12$, $T = 15$, and $r = 0.08$. So
$$A = \frac{12(150)}{0.08}(e^{1.2} - 1) \approx 52{,}202.60, \text{ or approximately } \$52{,}203.$$

15. Here $P = 2000$, $m = 1$, $T = 15.75$, and $r = 0.1$. So
$$A = \frac{1(2000)}{0.1}(e^{1.575} - 1) \approx 76{,}615, \text{ or approximately } \$76{,}615.$$

17. Here $P = 1200$, $m = 12$, $T = 15$, and $r = 0.1$. So
$$PV = \frac{12(1200)}{0.1}(1 - e^{-1.5}) \approx 111{,}869, \text{ or approximately } \$111{,}869.$$

19. We want the present value of an annuity with $P = 300$, $m = 12$, $T = 10$, and $r = 0.12$. So
$$PV = \frac{12(300)}{0.12}(1 - e^{-1.2}) \approx 20{,}964, \text{ or approximately } \$20{,}964.$$

21. a.

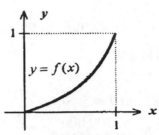

b. $f(0.4) = \frac{15}{16}(0.16) + \frac{1}{16}(0.4) \approx 0.175$; $f(0.9) = \frac{15}{16}(0.81) + \frac{1}{16}(0.9) \approx 0.816$.

So, the lowest 40 percent of the people receive 17.5 percent of the total income and the lowest 90 percent of the people receive 81.6 percent of the income.

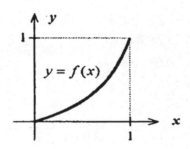

23. a.

b. $f(0.3) = \frac{14}{15}(0.09) + \frac{1}{15}(0.3) = 0.104$

$f(0.7) = \frac{14}{15}(0.49) + \frac{1}{15}(0.7) \approx 0.504$.

USING TECHNOLOGY EXERCISES 6.7, page 481

1. Consumer's surplus: $18,000,000; producer's surplus: $11,700,000.

3. Consumer's surplus: $33,120; producer's surplus: $2,880.

5. Investment A

EXERCISES 6.8, page 526

1. $V = \pi \int_a^b [f(x)]^2 \, dx = \pi \int_0^1 (3x)^2 \, dx = 9\pi \int_0^1 x^2 \, dx = 3\pi x^3 \Big|_0^1 = 3\pi$ cu units.

3. $V = \pi \int_a^b [f(x)]^2 \, dx = \pi \int_1^4 (\sqrt{x})^2 \, dx = \pi \int_1^4 x \, dx = \frac{\pi}{2}(16-1) = \frac{15\pi}{2}$ cu units.

5. $V = \pi \int_a^b [f(x)]^2 dx = \pi \int_0^1 (\sqrt{1+x^2})^2 dx = \pi \int_0^1 (1+x^2) dx = \pi(x + \frac{1}{3}x^3)\big|_0^1$

   $= \pi(1 + \frac{1}{3}) = \frac{4\pi}{3}$ cu units.

7. $V = \pi \int_a^b [f(x)]^2 dx = \pi \int_{-1}^1 (1-x^2)^2 dx = \pi \int_{-1}^1 (1-2x^2+x^4) dx = \pi(x - \frac{2}{3}x^3 + \frac{1}{5})x^5\big|_{-1}^1$

   $= \pi[(1 - \frac{2}{3} + \frac{1}{5}) - (-1 + \frac{2}{3} - \frac{1}{5})] = \frac{16\pi}{15}$ cu units.

9. $V = \pi \int_a^b [f(x)]^2 dx = \pi \int_0^1 (e^x)^2 dx = \pi \int_0^1 e^{2x} dx = \frac{\pi}{2}e^{2x}\big|_0^1 = \frac{\pi}{2}(e^2 - 1)$ cu units.

11. $V = \pi \int_a^b \{[f(x)]^2 - [g(x)]^2\} dx = \pi \int_0^1 [(x)^2 - (x^2)^2] dx = \pi \int_0^1 (x^2 - x^4) dx$

    $= \pi(\frac{1}{3}x^3 - \frac{1}{5}x^5)\big|_0^1 = \pi(\frac{1}{3} - \frac{1}{5}) = \frac{2\pi}{15}$ cu units.

13. $V = \pi \int_a^b \{[f(x)]^2 - [g(x)]^2\} dx = \pi \int_{-1}^1 [(4-x^2)^2 - 3^2] dx$

    $= \pi \int_{-1}^1 (16 - 8x^2 + x^4 - 9) dx = \pi \int_{-1}^1 (7 - 8x^2 + x^4) dx$

    $= \pi(7x - \frac{8}{3}x^3 + \frac{1}{5}x^5)\big|_{-1}^1 = \pi(7 - \frac{8}{3} + \frac{1}{5}) - (-7 + \frac{8}{3} - \frac{1}{5})] = \frac{136\pi}{15}$ cu units.

15. $V = \pi \int_a^b \{[f(x)]^2 - [g(x)]^2\} dx = \pi \int_0^{2\sqrt{2}} [(\sqrt{16-x^2})^2 - (x)^2] dx$

    $= \pi \int_0^{2\sqrt{2}} (16 - x^2 - x^2) dx = \pi \int_0^{2\sqrt{2}} (16 - 2x^2) dx = \pi(16x - \frac{2}{3}x^3)\big|_0^{2\sqrt{2}}$

    $= \pi[32\sqrt{2} - \frac{2}{3}(2\sqrt{2})^3] = \pi[32\sqrt{2} - \frac{32\sqrt{2}}{3}] = \frac{64\sqrt{2}\pi}{3}$ cu units.

17. $V = \pi \int_a^b \{[f(x)]^2 - [g(x)]^2\} dx = \pi \int_0^1 [(e^x)^2 - (e^{-x})^2] dx = \pi \int_0^1 (e^{3x} - e^{-2x}) dx$

    $= \pi \int_0^1 (e^{3x} - e^{-2x}) dx = \pi(\frac{1}{2}e^{2x} + \frac{1}{2}e^{-2x})\big|_0^1 = \pi[(\frac{1}{2}e^2 + \frac{1}{2}e^{-2}) - (\frac{1}{2} + \frac{1}{2})]$

    $= \frac{\pi}{2}(e^2 - 2 + e^{-2})$ cu units.

19. The region is shown in the figure at the right. The points of intersection of the curves are found by solving the simultaneous system of equations $y = x$ and $y = \sqrt{x}$, giving

$$x = \sqrt{x}$$
$$x^2 = x$$
$$x^2 - x = 0$$
$$x(x-1) = 0, \text{ or } x = 0 \text{ and } x = 1.$$

$$V = \pi \int_a^b \left\{[f(x)]^2 - [g(x)]^2\, dx\right\} = \pi \int_0^1 \left[(\sqrt{x})^2 - x^2\right]dx$$

$$\pi \int_0^1 (x - x^2)\,dx = \pi\left(\tfrac{1}{2}x^2 - \tfrac{1}{3}x^3\right)\Big|_0^1 = \pi\left(\tfrac{1}{2} - \tfrac{1}{3}\right) = \tfrac{\pi}{6} \text{ cu units.}$$

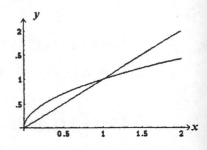

21. The region is shown in the figure. The points of intersection of the curves are found by solving the simultaneous system of equations

$$y = x^2 \text{ and } y = \tfrac{1}{2}x + 3$$
$$2x^2 - x - 6 = 0$$
$$(2x + 3)(x - 2) = 0,$$

or $x = -\tfrac{3}{2}$ and $x = 2$.

$$V = \pi \int_a^b \left\{[f(x)]^2 - [g(x)]^2\, dx\right\} = \pi \int_{-3/2}^2 \left[(\tfrac{1}{2}x + 3)^2 - (x)^2\right]dx$$

$$= \pi \int_{-3/2}^2 (\tfrac{1}{4}x^2 + 3x + 9 - x^4)\,dx = \pi\left(\tfrac{1}{12}x^3 + \tfrac{3}{2}x^2 + 9x - \tfrac{1}{5}x^5\right)\Big|_{-3/2}^2$$

$$= \left[(\tfrac{2}{3} + 6 + 18 - \tfrac{32}{5}) - (-\tfrac{27}{96} + \tfrac{27}{8} - \tfrac{27}{2} + \tfrac{243}{160})\right] = \tfrac{6517\pi}{240}, \text{ or } \tfrac{6517\pi}{240} \text{ cu units.}$$

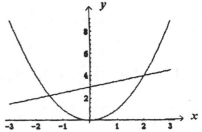

23. The region is shown in the figure. To find where the graphs intersect, we solve $y = x^2$ and $y = 4 - x^2$, simultaneously, obtaining

$$x^2 = 4 - x^2, \ 2x^2 = 4, \ x^2 = 2$$

giving $x = \pm\sqrt{2}$. The required volume is

$$V = \pi \int_a^b \left\{ [f(x)]^2 - [g(x)]^2 \right\} dx$$

$$= \pi \int_{-\sqrt{2}}^{\sqrt{2}} \left[(4 - x^2)^2 - (x^2)^2 \right] dx$$

$$= 2\pi \int_0^{\sqrt{2}} \left[16 - 8x^2 + x^4 - x^4 \right] dx$$

$$= \pi \int_{-\sqrt{2}}^{\sqrt{2}} \left[(4 - x^2)^2 - (x^2)^2 \right] dx$$

$$= 16\pi \left[2\sqrt{2} - \tfrac{1}{3}(2\sqrt{2}) \right] = \frac{64\sqrt{2}\pi}{3} \text{ cu units.}$$

25. The region is shown in the figure. To find the
points of intersection of $y = 2x$ and $y = \dfrac{1}{x}$,
we solve

$$2x = \frac{1}{x}$$

$$2x^2 = 1$$

or $\; x = \pm \dfrac{1}{\sqrt{2}} = \pm \dfrac{\sqrt{2}}{2}.$

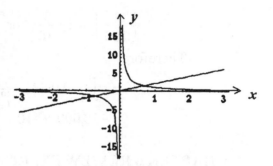

To find the points of intersection of $y = x$
and

$y = \dfrac{1}{x}$, we solve

$$x = \frac{1}{x}$$

giving $x = \pm 1$. By symmetry

$$V = 2\pi \int_0^{\sqrt{2}/2} \left[(2x)^2 - (x)^2 \right] dx + 2\pi \int_{\sqrt{2}/2}^1 \left[\left(\frac{1}{x} \right)^2 - (x)^2 \right] dx$$

$$= 2\pi \int_0^{\sqrt{2}/2} \left[4x^2 - x^2 \right] dx + 2\pi \int_{\sqrt{2}/2}^1 \left[\frac{1}{x^2} - x^2 \right] dx$$

$$= 2\pi \int_0^{\sqrt{2}/2} 3x^2 \, dx + 2\pi \int_{\sqrt{2}/2}^1 (x^{-2} - x^2) \, dx = 2\pi x^3 \Big|_0^{\sqrt{2}/2} + 2\pi \left(-\frac{1}{x} - \frac{1}{3} x^3 \right) \Big|_{\sqrt{2}/2}^1$$

$$= 2\pi\left(\frac{2\sqrt{2}}{8}\right) + 2\pi\left[(-1-\tfrac{1}{3}) - \left(-\frac{2}{\sqrt{2}} - \frac{\sqrt{2}}{12}\right)\right] = \frac{8(\sqrt{2}-1)\pi}{3} \text{ cu units.}$$

27. $\displaystyle V = \pi\int_a^b [f(x)]^2\,dx = \pi\int_{-r}^r \left[\sqrt{r^2 - x^2}\right]^2 dx = \pi\int_{-r}^r (r^2 - x^2)\,dx$

$\displaystyle \qquad = 2\pi\int_0^r (r^2 - x^2)\,dx = 2\pi(r^2 x - \tfrac{1}{3}x^3)\big|_0^r = 2\pi(r^3 - \tfrac{1}{3}x^3) = \tfrac{4}{3}\pi r^3$ cu units.

29. $\displaystyle V = \pi\int_{-10}^0 x^2\,dy = \pi\int_{-10}^0 [f(y)]^2\,dy.$

Solving the given equation for x in terms of y, we have

$$\frac{y}{10} = \left(\frac{x}{100}\right)^2 - 1$$

$$\left(\frac{x}{100}\right)^2 = 1 + \frac{y}{10}, \quad \text{or} \quad x = 100\sqrt{1 + \frac{y}{10}}.$$

Therefore,

$$V = \pi\int_{-10}^0 10{,}000\left(1 + \tfrac{y}{10}\right) dy = 10{,}000\pi(y + \tfrac{1}{20}y^2)\big|_{-10}^0$$

$$= -10{,}000\pi(-10 + \tfrac{100}{20}) = 50{,}000\pi \text{ cu ft.}$$

CHAPTER 6 REVIEW EXERCISES, page 490

1. $\displaystyle \int (x^3 + 2x^2 - x)\,dx = \tfrac{1}{4}x^4 + \tfrac{2}{3}x^3 - \tfrac{1}{2}x^2 + C.$

3. $\displaystyle \int\left(x^4 - 2x^3 + \frac{1}{x^2}\right) dx = \frac{x^5}{5} - \frac{1}{2}x^4 - \frac{1}{x} + C$

5. $\displaystyle \int x(2x^2 + x^{1/2})\,dx = \int (2x^3 + x^{3/2})\,dx = \tfrac{1}{2}x^4 + \tfrac{2}{5}x^{5/2} + C.$

7. $\displaystyle \int \left(x^2 - x + \tfrac{2}{x} + 5\right) dx = \int x^2\,dx - \int x\,dx + 2\int \frac{dx}{x} + 5\int dx$

$\displaystyle \qquad = \tfrac{1}{3}x^3 - \tfrac{1}{2}x^2 + 2\ln|x| + 5x + C.$

9. Let $u = 3x^2 - 2x + 1$ so that $du = (6x - 2)\,dx = 2(3x - 1)\,dx$ or $(3x - 1)\,dx = \tfrac{1}{2}\,du.$

So $\int (3x-1)(3x^2-2x+1)^{1/3}\,dx = \frac{1}{2}\int u^{1/3}\,du = \frac{3}{8}u^{4/3}+C = \frac{3}{8}(3x^2-2x+1)^{4/3}+C.$

11. Let $u = x^2 - 2x + 5$ so that $du = 2(x-1)\,dx$ or $(x-1)\,dx = \frac{1}{2}\,du$.

$$\int \frac{x-1}{x^2-2x+5}\,dx = \frac{1}{2}\int \frac{du}{u} = \frac{1}{2}\ln|u| + C = \frac{1}{2}\ln(x^2-2x+5) + C.$$

13. Put $u = x^2 + x + 1$ so that $du = (2x+1)\,dx = 2(x+\frac{1}{2})\,dx$ and $(x+\frac{1}{2})dx = \frac{1}{2}\,du$.

$$\int (x+\tfrac{1}{2})e^{x^2+x+1}dx = \frac{1}{2}\int e^u\,du = \frac{1}{2}e^u + C = \frac{1}{2}e^{x^2+x+1} + C.$$

15. Let $u = \ln x$ so that $du = \frac{1}{x}\,dx$. Then

$$\int \frac{(\ln x)^5}{x}\,dx = \int u^5\,du = \frac{1}{6}u^6 + C = \frac{1}{6}(\ln x)^6 + C.$$

17. Let $u = x^2 + 1$ so that $du = 2x\,dx$ or $x\,dx = \frac{1}{2}\,du$. Then

$$\int x^3(x^2+1)^{10}\,dx = \frac{1}{2}\int (u-1)u^{10}\,du \qquad (x^2 = u - 1)$$

$$= \frac{1}{2}\int (u^{11} - u^{10})\,du = \frac{1}{2}(\tfrac{1}{12}u^{12} - \tfrac{1}{11}u^{11}) + C$$

$$= \frac{1}{264}u^{11}(11u - 12) + C = \frac{1}{264}(x^2+1)^{11}(11x^2 - 1) + C.$$

19. Put $u = x - 2$ so that $du = dx$. Then $x = u + 2$ and

$$\int \frac{x}{\sqrt{x-2}}\,dx = \int \frac{u+2}{\sqrt{u}}\,du = \int (u^{1/2} + 2u^{-1/2})\,du = \int u^{1/2}\,du + 2\int u^{-1/2}\,du$$

$$= \frac{2}{3}u^{3/2} + 4u^{1/2} + C = \frac{2}{3}u^{1/2}(u+6) + C$$

$$= \frac{2}{3}\sqrt{x-2}(x-2+6) + C$$

$$= \frac{2}{3}(x+4)\sqrt{x-2} + C.$$

21. $\int_0^1 (2x^3 - 3x^2 + 1)\,dx = \frac{1}{2}x^4 - x^3 + x\Big|_0^1 = \frac{1}{2} - 1 + 1 = \frac{1}{2}.$

23. $\int_1^4 (x^{1/2} + x^{-3/2}) \, dx = \frac{2}{3}x^{3/2} - 2x^{-1/2}\Big|_1^4 = \frac{2}{3}x^{3/2} - \frac{2}{\sqrt{x}}\Big|_1^4 = (\frac{16}{3} - 1) - (\frac{2}{3} - 2) = \frac{17}{3}.$

25. Put $u = x^3 - 3x^2 + 1$ so that $du = (3x^2 - 6x) \, dx = 3(x^2 - 2x) \, dx$ or $(x^2 - 2x) \, dx = \frac{1}{3} \, du$. Then if $x = -1$, $u = -3$, and if $x = 0$, $u = 1$,

$$\int_{-1}^0 12(x^2 - 2x)(x^3 - 3x^2 + 1)^3 \, dx = (12)(\tfrac{1}{3})\int_{-3}^1 u^3 \, du = 4(\tfrac{1}{4})u^4\Big|_{-3}^1$$
$$= 1 - 81 = -80.$$

27. Let $u = x^2 + 1$ so that $du = 2x \, dx$ or $x \, dx = \frac{1}{2} \, du$. Then, if $x = 0$, $u = 1$, and if $x = 2$, $u = 5$, so

$$\int_0^2 \frac{x}{x^2 + 1} \, dx = \frac{1}{2}\int_1^5 \frac{du}{u} = \frac{1}{2}\ln u\Big|_1^5 = \frac{1}{2}\ln 5.$$

29. Let $u = 1 + 2x^2$ so that $du = 4x \, dx$ or $x \, dx = \frac{1}{4} \, du$. If $x = 0$, then $u = 1$ and if $x = 2$, then $u = 9$.

$$\int_0^2 \frac{4x}{\sqrt{1 + 2x^2}} \, dx = \int_1^9 \frac{du}{u^{1/2}} = 2u^{1/2}\Big|_1^9 = 2(3 - 1) = 4.$$

31. Let $u = 1 + e^{-x}$ so that $du = -e^{-x} \, dx$ and $e^{-x} \, dx = -\, du$. Then

$$\int_{-1}^0 \frac{e^{-x}}{(1 + e^{-x})^2} \, dx = -\int_{1+e}^2 \frac{du}{u^2} = \frac{1}{u}\Big|_{1+e}^2 = \frac{1}{2} - \frac{1}{1+e} = \frac{e-1}{2(1+e)}.$$

33. $f(x) = \int f'(x) \, dx = \int (3x^2 - 4x + 1) \, dx = 3\int x^2 \, dx - 4\int x \, dx + \int dx$
$$= x^3 - 2x^2 + x + C.$$
The given condition implies that $f(1) = 1$ or $1 - 2 + 1 + C = 1$, and $C = 1$. Therefore, the required function is $f(x) = x^3 - 2x^2 + x + 1$.

35. $f(x) = \int f'(x) \, dx = \int (1 - e^{-x}) \, dx = x + e^{-x} + C$, $f(0) = 2$ implies $0 + 1 + C = 2$ c $C = 1$. So $f(x) = x + e^{-x} + 1$.

37. $\Delta x = \frac{2-1}{5} = \frac{1}{5}$; $x_1 = \frac{6}{5}$, $x_2 = \frac{7}{5}$, $x_3 = \frac{8}{5}$, $x_4 = \frac{9}{5}$, $x_5 = \frac{10}{5}$. The Riemann sum is

$$f(x_1)\Delta x + \cdots + f(x_5)\Delta x = \left\{\left[-2(\tfrac{6}{5})^2 + 1\right] + \left[-2(\tfrac{7}{5})^2 + 1\right] + \cdots + \left[-2(\tfrac{10}{5})^2 + 1\right]\right\}(\tfrac{1}{5})$$
$$= \tfrac{1}{5}(-1.88 - 2.92 - 4.12 - 5.48 - 7) = -4.28.$$

39. a. $R(x) = \int R'(x)\,dx = \int (-0.03x + 60)\,dx = -0.015x^2 + 60x + C$.

$R(0) = 0$ implies that $C = 0$. So, $R(x) = -0.015x^2 + 60x$.

b. From $R(x) = px$, we have $-0.015x^2 + 60x = px$ or $p = -0.015x + 60$.

41. The total number of systems that Vista may expect to sell t months from the time they are put on the market is given by $f(t) = 3000t - 50,000(1 - e^{-0.04t})$.

The number is $\displaystyle\int_0^{12} (3000 - 2000e^{-0.04t})\,dt = \left(3000t - \frac{2000}{-0.04}e^{-0.04t}\right)\Bigg|_0^{12}$

$$= 3000(12) + 50,000e^{-0.48} - 50,000 = 16,939.$$

43. $C(x) = \displaystyle\int C'(x)\,dx = \int (0.00003x^2 - 0.03x + 10)\,dx$

$$= 0.00001x^3 - 0.015x^2 + 10x + k.$$

But $C(0) = 600$ and this implies that $k = 600$. Therefore,
$$C(x) = 0.00001x^3 - 0.015x^2 + 10x + 600.$$
The total cost incurred in producing the first 500 corn poppers is
$$C(500) = 0.00001(500)^3 - 0.015(500)^2 + 10(500) + 600$$
$$= 3,100, \text{ or } \$3,100.$$

45. $A = \displaystyle\int_{-1}^{2} (3x^2 + 2x + 1)\,dx = x^3 + x^2 + x\big|_{-1}^{2} = [2^3 + 2^2 + 2] - [(-1)^3 + 1 - 1]$

$$= 14 - (-1) = 15.$$

47. $A = \displaystyle\int_1^3 \frac{1}{x^2}\,dx = \int_1^3 x^{-2}\,dx = -\frac{1}{x}\bigg|_1^3 = -\frac{1}{3} + 1 = \frac{2}{3}.$

49.

$$A = \int_a^b [f(x) - g(x)]\,dx$$

$$= \int_0^2 (e^x - x)\,dx$$

$$= \left(e^x - \frac{1}{2}x^2 \right)\Big|_0^2$$

$$= (e^2 - 2) - (1 - 0) = e^2 - 3.$$

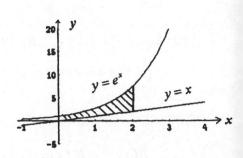

51.

$$A = \int_0^1 (x^3 - 3x^2 + 2x)\,dx - \int_1^2 (x^3 - 3x^2 + 2x)\,dx$$

$$= \frac{x^4}{4} - x^3 + x^2 \Big|_0^1 - \left(\frac{x^4}{4} - x^3 + x^2 \right)\Big|_1^2$$

$$= \tfrac{1}{4} - 1 + 1 - [(4 - 8 + 4) - (\tfrac{1}{4} - 1 + 1)]$$

$$= \tfrac{1}{4} + \tfrac{1}{4} = \tfrac{1}{2}.$$

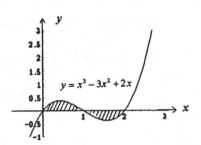

53.

$$A = \frac{1}{3}\int_0^3 \frac{x}{\sqrt{x^2 + 16}}\,dx = \frac{1}{3}\cdot\frac{1}{2}\cdot 2(x^2 + 16)^{1/2}\Big|_0^3$$

$$= \frac{1}{3}(x^2 + 16)^{1/2}\Big|_0^3 = \frac{1}{3}(5 - 4) = \frac{1}{3} \text{ sq units.}$$

55. To find the equilibrium point, we solve $0.1x^2 + 2x + 20 = -0.1x^2 - x + 40$
$0.2x^2 + 3x - 20 = 0$, $x^2 + 15x - 100 = 0$, $(x + 20)(x - 5) = 0$, or $x = 5$.
Therefore, $p = -0.1(25) - 5 + 40 = 32.5$.

$$CS = \int_0^5 (-0.1x^2 - x + 40)\,dx - (5)(32.5) = -\frac{0.1}{3}x^3 - \frac{1}{2}x^2 + 40x\Big|_0^5 - 162.5$$

$$= 20.833, \text{ or } \$2083.$$

$$PS = (5)(32.5) - \int_0^5 (0.1x^2 + 2x + 20)\,dx = 162.5 - \frac{0.1}{3}x^3 + x^2 + 20x\Big)\Big|_0^5$$

$$= 33.333, \text{ or } \$3,333.$$

57. Use Equation (18) with $P = 925$, $m = 12$, $T = 30$, and $r = 0.12$, obtaining

$$PV = \frac{mP}{r}(1-e^{-rT}) = \frac{(12)(925)}{(0.12)}(1-e^{-0.12(30)}) = 89972.56,$$

and we conclude that the present value of the purchase price of the house is $89,972.56 + $9000 , or $98,972.56.

59. a.

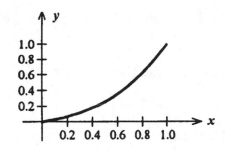

b. $f(0.3) = \frac{17}{18}(0.3)^2 + \frac{1}{18}(0.3) \approx 0.1$ so that 30 percent of the people receive 10

percent of the total income. $f(0.6) = \frac{17}{18}(0.6)^2 + \frac{1}{18}(0.6) \approx 0.37$ so that 60 percent of

the people receive 37 percent of the total revenue.

c. The coefficient of inequality for this curve is

$$L = 2\int_0^1 [x - \frac{17}{18}x^2 - \frac{1}{18}x]dx = \frac{17}{9}\int_0^1 (x - x^2)\,dx = \frac{17}{9}\left(\frac{1}{2}x^2 - \frac{1}{3}x^3\right)\Big|_0^1$$

$$= \frac{17}{54} \approx 0.315.$$

61. $V = \pi\int_1^3 \frac{dx}{x^2} = -\frac{\pi}{x}\Big|_1^3 = \pi(-\frac{1}{3}+1) = \frac{2\pi}{3}$ cu units.

CHAPTER 7

EXERCISES 7.1, page 499

1. $I = \int xe^{2x}\,dx$. Let $u = x$ and $dv = e^{2x}\,dx$. Then $du = dx$ and $v = \frac{1}{2}e^{2x}$. Therefore,

$$I = uv - \int v\,du = \frac{1}{2}xe^{2x} - \int \frac{1}{2}e^{2x}\,dx = \frac{1}{2}xe^{2x} - \frac{1}{4}e^{2x} = \frac{1}{4}e^{2x}(2x-1) + C.$$

3. $I = \int xe^{x/4}dx.$ Let $u = x$ and $dv = e^{x/4}\,dx$. Then $du = dx$ and $v = 4e^{x/4}$.

$$\int xe^{x/4}\,dx = uv - \int v\,du = 4xe^{x/4} - 4\int e^{x/4}\,dx = 4xe^{x/4} - 16e^{x/4} + C$$
$$= 4(x-4)e^{x/4} + C.$$

5. $\int (e^x - x)^2\,dx = \int (e^{2x} - 2xe^x + x^2)\,dx = \int e^{2x}\,dx - 2\int xe^x\,dx + \int x^2\,dx.$

Using the result $\int xe^x\,dx = (x-1)e^x + k,$ from Example 1, we see that

$$\int (e^x - x)^2\,dx = \frac{1}{2}e^{2x} - 2(x-1)e^x + \frac{1}{3}x^3 + C.$$

7. $I = \int (x+1)e^x\,dx.$ Let $u = x + 1$, $dv = e^x\,dx$. Then $du = dx$ and $v = e^x$. Therefore,

$$I = (x+1)e^x - \int e^x\,dx = (x+1)e^x - e^x + C = xe^x + C.$$

9. Let $u = x$ and $dv = (x + 1)^{-3/2}\,dx$. Then $du = dx$ and $v = -2(x+1)^{-1/2}$.

$$\int x(x+1)^{-3/2}\,dx = uv - \int v\,du = -2x(x+1)^{-1/2} + 2\int (x+1)^{-1/2}\,dx$$
$$= -2x(x+1)^{-1/2} + 4(x+1)^{1/2} + C$$
$$= 2(x+1)^{-1/2}[-x + 2(x+1)] + C = \frac{2(x+2)}{\sqrt{x+1}} + C.$$

11. $I = \int x(x-5)^{1/2}\,dx.$ Let $u = x$ and $dv = (x - 5)^{1/2}\,dx$. Then $du = dx$ and

$v = \frac{2}{3}(x - 5)^{3/2}$. Therefore,

$$I = \tfrac{2}{3}x(x-5)^{3/2} - \int \tfrac{2}{3}(x-5)^{3/2}\,dx = \tfrac{2}{3}x(x-5)^{3/2} - \tfrac{2}{3}\cdot\tfrac{2}{5}(x-5)^{5/2} + C$$

$$= \tfrac{2}{3}(x-5)^{3/2}[x - \tfrac{2}{5}(x-5)] + C = \tfrac{2}{15}(x-5)^{3/2}(5x - 2x + 10) + C$$

$$= \tfrac{2}{15}(x-5)^{3/2}(3x+10) + C.$$

13. $I = \displaystyle\int x \ln 2x\,dx.$ Let $u = \ln 2x$ and $dv = x\,dx$. Then $du = \tfrac{1}{x}\,dx$ and $v = \tfrac{1}{2}x^2$.

Therefore, $I = \tfrac{1}{2}x^2 \ln 2x - \displaystyle\int \tfrac{1}{2}x\,dx = \tfrac{1}{2}x^2 \ln 2x - \tfrac{1}{4}x^2 + C = \tfrac{1}{4}x^2(2\ln 2x - 1) + C.$

15. Let $u = \ln x$ and $dv = x^3\,dx$, then $du = \tfrac{1}{x}\,dx$, and $v = \tfrac{1}{4}x^4$.

$$\int x^3 \ln x\,dx = \tfrac{1}{4}x^4 \ln x - \tfrac{1}{4}\int x^3\,dx = \tfrac{1}{4}x^4 \ln x - \tfrac{1}{16}x^4 + C$$

$$= \tfrac{1}{16}x^4(4\ln x - 1) + C.$$

17. Let $u = \ln x^{1/2}$ and $dv = x^{1/2}\,dx$. Then $du = \tfrac{1}{2x}\,dx$ and $v = \tfrac{2}{3}x^{3/2}$,

and $\displaystyle\int \sqrt{x}\ln\sqrt{x}\,dx = uv - \int v\,du = \tfrac{2}{3}x^{3/2}\ln x^{1/2} - \tfrac{1}{3}\int x^{1/2}\,dx$

$$= \tfrac{2}{3}x^{3/2}\ln x^{1/2} - \tfrac{2}{9}x^{3/2} + C = \tfrac{2}{9}x\sqrt{x}(3\ln\sqrt{x} - 1) + C.$$

9. Let $u = \ln x$ and $dv = x^{-2}\,dx$. Then $du = \tfrac{1}{x}\,dx$ and $v = -x^{-1}$,

$$\int \frac{\ln x}{x^2}\,dx = uv - \int v\,du = -\frac{\ln x}{x} + \int x^{-2}\,dx = -\frac{\ln x}{x} - \frac{1}{x} + C$$

$$= -\frac{1}{x}(\ln x + 1) + C.$$

1. Let $u = \ln x$ and $dv = dx$. Then $du = \tfrac{1}{x}\,dx$ and $v = x$ and

$$\int \ln x\,dx = uv - \int v\,du = x\ln x - \int dx = x\ln x - x + C = x(\ln x - 1) + C.$$

3. Let $u = x^2$ and $dv = e^{-x}\,dx$. Then $du = 2x\,dx$ and $v = -e^{-x}$, and

$$\int x^2 e^{-x}\,dx = uv - \int v\,du = -x^2 e^{-x} + 2\int xe^{-x}\,dx.$$

We can integrate by parts again, or, using the result of Problem 2, we find

$$\int x^2 e^{-x}\,dx = -x^2 e^{-x} + 2[-(x+1)e^{-x}] + C = -x^2 e^{-x} - 2(x+1)e^{-x} + C$$

$$= -(x^2 + 2x + 2)e^{-x} + C.$$

25. $I = \int x(\ln x)^2\,dx$. Let $u = (\ln x)^2$ and $dv = x\,dx$, so that

$du = 2(\ln x)\left(\dfrac{1}{x}\right) = \dfrac{2\ln x}{x}$ and $v = \frac{1}{2}x^2$. Then $I = \frac{1}{2}x^2(\ln x)^2 - \int x\ln x\,dx$.

Next, we evaluate $\int x\ln x\,dx$, by letting $u = \ln x$ and $dv = x\,dx$, so that $du = \frac{1}{x}\,dx$

and $v = \frac{1}{2}x^2$. Then $\int x\ln x\,dx = \frac{1}{2}x^2(\ln x) - \frac{1}{2}\int x\,dx = \frac{1}{2}x^2\ln x - \frac{1}{4}x^2 + C.$

Therefore, $\int x(\ln x)^2\,dx = \frac{1}{2}x^2(\ln x)^2 - \frac{1}{2}x^2\ln x + \frac{1}{4}x^2 + C$

$$= \frac{1}{4}x^2[2(\ln x)^2 - 2\ln x + 1] + C.$$

27. $\displaystyle\int_0^{\ln 2} xe^x\,dx = (x-1)e^x\Big|_0^{\ln 2}$ \qquad (Using the results of Example 1.)

$$= (\ln 2 - 1)e^{\ln 2} - (-e^0) = 2(\ln 2 - 1) + 1\ \text{(Recall } e^{\ln 2} = 2.) = 2\ln 2 - 1.$$

29. We first integrate $I = \int \ln x\,dx$. Integrating by parts with $u = \ln x$ and $dv = dx$ so

that $du = \frac{1}{x}\,dx$ and $v = x$, we find

$$I = x\ln x - \int dx = x\ln x - x + C = x(\ln x - 1) + C.$$

Therefore, $\displaystyle\int_1^4 \ln x\,dx = x(\ln x - 1)\Big|_1^4 = 4(\ln 4 - 1) - 1(\ln 1 - 1) = 4\ln 4 - 3.$

31. Let $u = x$ and $dv = e^{2x}\,dx$. Then $u = dx$ and $v = \frac{1}{2}e^{2x}$ and

$$\int_0^2 xe^{2x}\,dx = \frac{1}{2}xe^{2x}\Big|_0^2 - \frac{1}{2}\int_0^2 e^{2x}\,dx = e^4 - \frac{1}{4}e^{2x}\Big|_0^2$$

$$= e^4 - \frac{1}{4}e^4 + \frac{1}{4} = \frac{1}{4}(3e^4 + 1).$$

33. Let $u = x$ and $dv = e^{-2x}\,dx$, so that $du = dx$ and $v = -\frac{1}{2}e^{-2x}$.

$f(x) = \int xe^{-2x}\,dx = -\frac{1}{2}xe^{-2x} - \frac{1}{4}e^{-2x} + C$; $f(0) = -\frac{1}{4} + C = 3$ and $C = \frac{13}{4}$.

Therefore, $y = -\frac{1}{2}xe^{-2x} - \frac{1}{4}e^{-2x} + \frac{13}{4}$.

35. The required area is given by $\int_1^5 \ln x\,dx$. We first find $\int \ln x\,dx$. Using the technique of integration by parts with $u = \ln x$ and $dv = dx$ so that $du = \frac{1}{x}\,dx$ and $v = x$, we have

$$\int \ln x\,dx = x\ln x - \int dx = x\ln x - x = x(\ln x - 1) + C.$$

Therefore, $\int_1^5 \ln x\,dx = x(\ln x - 1)\Big|_1^5 = 5(\ln 5 - 1) - 1(\ln 1 - 1) = 5\,\ln 5 - 4$

and the required area is $(5 \ln 5 - 4)$ sq units.

37. The distance covered is given by $\int_0^{10} 100te^{-0.2t}\,dt = 100\int_0^{10} te^{-0.2t}\,dt.$
We integrate by parts, letting $u = t$ and $dv = e^{-0.2t}\,dt$ so that $du = dt$ and
$v = -\dfrac{1}{0.2}e^{-0.2t} = -5e^{-0.2t}$. Therefore,

$$100\int_0^{10} te^{-0.2t}\,dt = 100\left[-5te^{-0.2t}\Big|_0^{10}\right] + 5\int_0^{10} e^{-0.2t}\,dt$$

$$= 100[-5te^{-0.2t} - 25e^{-0.2t}]\Big|_0^{10} = -500e^{-0.2t}(t+5)\Big|_0^{10}$$

$$= -500e^{-2}(15) + 500(5) = 1485,\ \text{or } 1485 \text{ feet}.$$

39. The average concentration is $C = \dfrac{1}{12}\int_0^{12} 3te^{-t/3}\,dt = \dfrac{1}{4}\int_0^{12} te^{-t/3}\,dt.$
Let $u = t$ and $dv = e^{-t/3}\,dt$. So $du = dt$ and $v = -3e^{-t/3}$. Then

$$C = \frac{1}{4}\left[-3te^{-t/3}\Big|_0^{12} + 3\int_0^{12} e^{-t/3}\,dt\right] = \frac{1}{4}\left\{-36e^{-4} - \left[9e^{-t/3}\Big|_0^{12}\right]\right\}$$

$$= \frac{1}{4}(-36e^{-4} - 9e^{-4} + 9) \approx 2.04\ \text{mg/ml}.$$

1. $N = 2\int te^{-0.1t}\,dt$. Let $u = t$ and $dv = e^{-0.1t}$, so that $du = dt$ and $v = -10e^{-0.1t}$. Then
$v = -10e^{-0.1t}$. Then

$$N(t) = 2[-10te^{-0.1t} + 10\int e - 0.1t \, dt] = 2(-10te^{-0.1t} - 100e^{-0.1t}) + C$$

$$= -20e^{-0.1t}(t + 10) + 200. \qquad\qquad [N(0) = 0]$$

43. $$PV = \int_0^5 (30{,}000 + 800t)e^{-0.08t} \, dt = 30{,}000\int_0^5 e^{-0.08t} \, dt + 800\int_0^5 te^{-0.08t} \, dt \,.$$

Let $I = \int te^{-0.08t} \, dt.$ To evaluate I by parts, let $u = t, \, dv = e^{-0.08t} \, dt$

and $du = dt, \, v = -\dfrac{1}{0.08}e^{-0.08t} = -12.5e^{-0.08t}.$

Therefore, $I = -12.5te^{-0.08t} + 12.5\int e^{-0.08t} \, dt = -12.5te^{-0.08t} - 156.25e^{-0.08t} + C.$

$$PV = \left[-\frac{30{,}000}{0.08}e^{-0.08t} - 800(12.5)te^{-0.08t} - 800(156.25)e^{-0.08t} \right]_0^5$$

$$= -375{,}000\,e^{-0.4} + 375{,}000 - 50{,}000e^{-0.4} - 125{,}000e^{-0.4} + 125{,}000$$

$$= 500{,}000 - 550{,}000e^{-0.4} = 131{,}323.97, \text{ or approximately } \$131{,}324.$$

45. The membership will be
$$N(5) = N(0) + \int_0^5 9\sqrt{t+1} \ln\sqrt{t+1} \, dt = 50 + 9\int_0^5 \sqrt{t+1} \ \ln\sqrt{t+1} \, dt$$
To evaluate the integral, let $u = t + 1$ so that $du = dt$. Also, if $t = 0$, then $u = 1$ and
if $t = 5$, then $u = 6$. So $9\int_0^5 \sqrt{t+1} \ \ln\sqrt{t+1} \, dt = 9\int_1^6 \sqrt{u} \ln\sqrt{u} \, du.$

Using the results of Problem 17, we find $9\int_1^6 \sqrt{u} \ln\sqrt{u} \, du = 2u\sqrt{u}(3\ln\sqrt{u} - 1)\Big|_1^6.$
Therefore, $N = 50 + 51.606 \approx 101.606$ or $101{,}606$ people.

47. True. This is just the integration by parts formula.

EXERCISES 7.2, page 507

1. First we note that
$$\int \frac{2x}{2+3x} \, dx = 2\int \frac{x}{2+3x} \, dx.$$
Next, we use Formula 1 with $a = 2$, $b = 3$, and $u = x$. Then

$$\int \frac{2x}{2+3x}\,dx = \frac{2}{9}[2+3x-2\ln|2+3x|]+C.$$

$$\int \frac{3x^2}{2+4x}\,dx = \frac{3}{2}\int \frac{x^2}{1+2x}\,dx.$$

Use Formula 2 with $a = 1$ and $b = 2$ obtaining

$$\int \frac{3x^2}{2+4x}\,dx = \frac{3}{32}[(1+2x)^2 - 4(1+2x)+2\ln|1+2x|]+C.$$

$$\int x^2\sqrt{9+4x^2}\,dx = \int x^2\sqrt{4(\tfrac{9}{4})+x^2)}\,dx = 2\int x^2\sqrt{(\tfrac{3}{2})^2 + x^2}\,dx.$$

Use Formula 8 with $a = 3/2$, we find that

$$\int x^2\sqrt{9+4x^2}\,dx == 2[\tfrac{x}{8}(\tfrac{9}{4}+2x^2)\sqrt{\tfrac{9}{4}+x^2} - \tfrac{81}{128}\ln\left|x+\sqrt{\tfrac{9}{4}+x^2}\right|+C.$$

Use Formula 6 with $a = 1$, $b = 4$, and $u = x$, then

$$\int \frac{dx}{x\sqrt{1+4x}} = \ln\left|\frac{\sqrt{1+4x}-1}{\sqrt{1+4x}+1}\right|+C.$$

Use Formula 9 with $a = 3$ and $u = 2x$. Then $du = 2\,dx$ and

$$\int_0^2 \frac{dx}{\sqrt{9+4x^2}} = \frac{1}{2}\int_0^4 \frac{du}{\sqrt{3^2+u^2}} = \frac{1}{2}\ln\left|u+\sqrt{9+u^2}\right|\Big|_0^4$$

$$= \frac{1}{2}(\ln 9 - \ln 3) = \frac{1}{2}\ln 3.$$

Note that the limits of integration have been changed from $x = 0$ to $x = 2$ and from $u = 0$ to $u = 4$.

1. Using Formula 22 with $a = 3$, we see that $\displaystyle\int \frac{dx}{(9-x^2)^{3/2}} = \frac{x}{9\sqrt{9-x^2}}+C.$

3. $\displaystyle\int x^2\sqrt{x^2-4}\,dx.$

Use Formula 14 with $a = 2$ and $u = x$, obtaining

$$\int x^2\sqrt{x^2-4}\,dx = \tfrac{x}{8}(2x^2-4)\sqrt{x^2-4} - 2\ln\left|x+\sqrt{x^2-4}\right|+C.$$

15. Using Formula 19 with $a = 2$ and $u = x$, we have

$$\int \frac{\sqrt{4-x^2}}{x} dx = \sqrt{4-x^2} - 2\ln\left|\frac{2+\sqrt{4-x^2}}{x}\right| + C.$$

17. $\int xe^{2x}\, dx.$

   Use Formula 23 with $u = x$ and $a = 2$, obtaining

$$\int xe^{2x}\, dx = \frac{1}{4}(2x-1)e^{2x} + C.$$

19. $\displaystyle\int \frac{dx}{(x+1)\ln(x+1)}.$

   Let $u = x + 1$ so that $du = dx$. Then $\displaystyle\int \frac{dx}{(x+1)\ln(x+1)} = \int \frac{du}{u\ln u}.$

   Use Formula 28 with $u = x$, obtaining $\displaystyle\int \frac{du}{u\ln u} = \ln|\ln u| + C.$

   Therefore, $\displaystyle\int \frac{dx}{(x+1)\ln(x+1)} = \ln|\ln(x+1)| + C$

21. $\displaystyle\int \frac{e^{2x}}{(1+3e^x)^2} dx.$

   Put $u = e^x$ then $du = e^x dx.$ Then we use Formula 3 with $a = 1, b = 3.$ Then

$$I = \int \frac{u}{(1+3u)^2}\, du = \frac{1}{9}\left[\frac{1}{1+3u} + \ln|1+3u|\right] + C = \frac{1}{9}\left[\frac{1}{1+3e^x} + \ln(1+3e^x)\right] + C$$

23. $\displaystyle\int \frac{3e^x}{1+e^{x/2}} dx = 3\int \frac{e^{x/2}}{e^{-x/2}+1}\, dx .$

   Let $v = e^{x/2}$ so that $dv = \frac{1}{2}e^{x/2}dx$ or $e^{x/2}\, dx = 2\, dv.$ Then

$$\int \frac{3e^x}{1+e^{x/2}} dx = 6\int \frac{dv}{\frac{1}{v}+1} = 6\int \frac{v}{v+1}\, dv.$$

   Use Formula 1 with $a = 1$, $b = 1$, and $u = v$, obtaining

$6\int \dfrac{v}{v+1}dv = 6[1+v-\ln|1+v|]+C.$ So $\int \dfrac{3e^x}{1+e^{x/2}}dx = 6[1+e^{x/2}-\ln(1+e^{x/2})]+C.$

This answer may be written in the form $6[e^{x/2}-\ln(1+e^{x/2})]+C$ since C is an arbitrary constant.

25. $\int \dfrac{\ln x}{x(2+3\ln x)}dx.$ Let $v = \ln x$ so that $dv = \dfrac{1}{x}dx.$ Then

$\int \dfrac{\ln x}{x(2+3\ln x)}dx = \int \dfrac{v}{2+3v}dv.$

Use Formula 1 with $a = 2$, $b = 3$, and $u = v$ to obtain

$\int \dfrac{v}{2+3v}dv = \tfrac{1}{9}[2+3\ln x - 2\ln|2+3\ln x|]+C.$ So

$\int \dfrac{\ln x}{x(2+3\ln x)}dx = \tfrac{1}{9}[2+3\ln x - 2\ln|2+3\ln x|]+C.$

27. Using Formula 24 with $a = 1$, $n = 2$, and $u = x.$ Then

$\int_0^1 x^2 e^x \, dx = x^2 e^x \Big|_0^1 - 2\int_0^1 xe^x \, dx = x^2 e^x - 2(xe^x - e^x)\Big|_0^1$

$= x^2 e^x - 2xe^x + 2e^x \Big|_0^1 = e - 2e + 2e - 2 = e - 2.$

29. $\int x^2 \ln x \, dx.$ Use Formula 27 with $n = 2$ and $u = x$, obtaining

$\int x^2 \ln x \, dx = \dfrac{x^3}{9}(3\ln x - 1)+C.$

31. $\int (\ln x)^3 dx.$ Use Formula 29 with $n = 3$ to write

$\int (\ln x)^3 \, dx = x(\ln x)^3 - 3\int (\ln x)^2 \, dx.$ Using Formula 29 again with $n = 2$, we obtain

$\int (\ln x)^3 dx = x(\ln x)^3 - 3[x(\ln x)^2 - 2\int \ln x \, dx].$

Using Formula 29 one more time with $n = 1$ gives

$$\int (\ln x)^3 dx = x(\ln x)^3 - 3x(\ln x)^2 + 6(x \ln x - x) + C$$

$$= x(\ln x)^3 - 3x(\ln x)^2 + 6x \ln x - 6x + C.$$

33. Letting $p = 50$ gives $50 = \dfrac{250}{\sqrt{16+x^2}}$

from which we deduce that $\sqrt{16+x^2} = 5$, $16 + x^2 = 25$, and $x = 3$.
Using Formula 9 with $u = 3$, we see that

$$CS = \int_0^3 \frac{250}{\sqrt{16+x^2}} dx = 50(3) = 250 \int_0^3 \frac{1}{\sqrt{16+x^2}} dx - 150$$

$$= 250 \ln\left| x + \sqrt{16+x^2} \right|_0^3 - 150 = 250[\ln 8 - \ln 4] - 150$$

$$= 23.286795, \text{ or approximately } \$2,329.$$

35. The number of visitors admitted to the amusement park by noon is found by evaluating the integral

$$\int_0^3 \frac{60}{(2+t^2)^{3/2}} dt = 60 \int_0^3 \frac{dt}{(2+t)^{3/2}}.$$

Using Formula 12 with $a = \sqrt{2}$ and $u = t$, we find

$$60 \int_0^3 \frac{dt}{(2+t^2)^{3/2}} = 60 \left[\frac{t}{2\sqrt{2+t^3}} \right]_0^3 = 60 \left[\frac{3}{2\sqrt{11}-0} \right] = \frac{90}{\sqrt{11}} = 27.136, \text{ or } 27,136.$$

37. In the first 10 days

$$\frac{1}{10} \int_0^{10} \frac{1000}{1+24e^{-0.02t}} dt = 100 \int_0^{10} \frac{1}{1+24e^{-0.02t}} dt = 100 \left[t + \frac{1}{0.02} \ln(1 + 24e^{-0.02t}) \right]_0^{10}$$

(Use Formula 25 with $a = 0.02$ and $b = 24$.)

$$= 100[10 + 50 \ln 20.64953807 - 50 \ln 25] = 44.0856,$$

or approximately 44 fruitflies. In the first 20 days

$$\frac{1}{20} \int_0^{20} \frac{1000}{1+24e^{-0.02t}} dt = 50 \int_0^{10} \frac{1}{1+24e^{-0.02t}} dt$$

$$= 500[t + \ln(1 + 24e^{-0.02t})]_0^{20}$$

$$= 50[20 + 50 \ln 17.0876822 - 50 \ln 25] = 48.71$$

or approximately 49 fruitflies.

39. $\dfrac{1}{5}\displaystyle\int_0^5 \dfrac{100,000}{2(1+1.5e^{-0.2t})}\,dt = 10,000\int_0^5 \dfrac{1}{1+1.5e^{-0.2t}}\,dt$

$$= 10,000[t + 5\ln(1+1.5e^{-0.2t})]\Big|_0^5$$

(Use Formula 25 with a = -0.2 and b = 1.5.)

$$= 10,000[5 + 5\ln 1.551819162 - 5\ln 2.5] \approx 26157,$$

or approximately 26,157 people.

41. $\displaystyle\int_0^5 20,000te^{0.15t}\,dt = 20,000\int_0^5 te^{0.15t}\,dt = 20,000\left[\dfrac{1}{(0.15)^2}(0.15t-1)e^{0.15t}\right]\Big|_0^5$

(Use Formula 23 with a = 0.15.)

$$= 888,888.8889[-0.25e^{0.75} + 1] = \$418,444.$$

EXERCISES 7.3, page 520

1. $\Delta x = \frac{2}{6} = \frac{1}{3}, x_0 = 0, x_1 = \frac{1}{3}, x_2 = \frac{2}{3}, x_3 = 1, x_4 = \frac{4}{3}, x_5 = \frac{5}{3}, x_6 = 2.$

Trapezoidal Rule:

$\displaystyle\int_0^2 x^2\,dx \approx \frac{1}{6}\left[0 + 2(\frac{1}{3})^2 + 2(\frac{2}{3})^2 + 2(1)^2 + 2(\frac{4}{3})^2 + 2(\frac{5}{3})^2 + 2^2\right]$

$\approx \frac{1}{6}\,(0.22222 + 0.88889 + 2 + 3.55556 + 5.55556 + 4) \approx 2.7037.$

Simpson's Rule:

$\displaystyle\int x^2\,dx = \frac{1}{9}[0 + 4(\frac{1}{3})^2 + 2(\frac{2}{3})^2 + 4(1)^2 + 2(\frac{4}{3})^2 + 4(\frac{5}{3})^2 + 2^2]$

$\approx \frac{1}{9}\,(0.44444 + 0.88889 + 4 + 3.55556 + 11.11111 + 4) \approx 2.6667.$

Exact Value: $\displaystyle\int_0^2 x^2\,dx = \frac{1}{3}x^3\Big|_0^2 = \frac{8}{3} = 2\frac{2}{3}.$

3. $\Delta x = \dfrac{b-a}{n} = \dfrac{1-0}{4} = \frac{1}{4}; x_0 = 0, x_1 = \frac{1}{4}, x_2 = \frac{1}{2}, x_3 = \frac{3}{4}, x_4 = 1.$

Trapezoidal Rule:

$\displaystyle\int_0^1 x^3\,dx \approx \frac{1}{8}\left[0 + 2(\frac{1}{4})^3 + 2(\frac{1}{2})^3 + 2(\frac{3}{4})^3 + 1^3\right] \approx \frac{1}{8}(0 + 0.3125 + 0.25 + 0.8)$

$\approx 0.265625.$

Simpson's Rule:

$$\int_0^1 x^3\,dx \approx \tfrac{1}{12}\left[0+4(\tfrac{1}{4})^3+2(\tfrac{1}{2})^3+4(\tfrac{3}{4})^3+1\right] \approx \tfrac{1}{12}[0+0.625+0.25+1.6875+1]$$
$$\approx 0.25.$$

Exact Value: $\int_0^1 x^3\,dx = \tfrac{1}{4}x^4\Big|_0^1 = \tfrac{1}{4}-0 = \tfrac{1}{4}$.

5.   a. Here $a=1$, $b=2$, and $n=4$; so $\Delta x = \tfrac{2-1}{4} = \tfrac{1}{4} = 0.25$, and $x_0=1$, $x_1=1.25$, $x_2=1.5$, $x_3=1.75$, $x_4=2$.

Trapezoidal Rule:
$$\int_1^2 \frac{1}{x}\,dx \approx \frac{0.25}{2}\left[1+2\left(\frac{1}{1.25}\right)+2\left(\frac{1}{1.5}\right)+2\left(\frac{1}{1.75}\right)+\frac{1}{2}\right] \approx 0.697.$$

Simpson's Rule:
$$\int_1^2 \frac{1}{x}\,dx \approx \frac{0.25}{3}\left[1+4\left(\frac{1}{1.25}\right)+2\left(\frac{1}{1.5}\right)+4\left(\frac{1}{1.75}\right)+\frac{1}{2}\right] \approx 0.6933.$$
$$\int_1^2 \frac{1}{x}\,dx = \ln x\Big|_1^2 = \ln 2 - \ln 1 \approx 0.6931.$$

7. $\Delta x = \tfrac{1}{4}$, $x_0=1$, $x_1=\tfrac{5}{4}$, $x_2=\tfrac{3}{2}$, $x_3=\tfrac{7}{4}$, $x_4=2$.

Trapezoidal Rule:
$$\int_1^2 \frac{1}{x^2}\,dx \approx \tfrac{1}{8}\left[1+2(\tfrac{4}{5})^2+2(\tfrac{2}{3})^2+2(\tfrac{4}{7})^2+(\tfrac{1}{2})^2\right] \approx 0.5090.$$

Simpson's Rule:
$$\int_1^2 \frac{1}{x^2}\,dx \approx \tfrac{1}{12}\left[1+4(\tfrac{4}{5})^2+2(\tfrac{2}{3})^2+4(\tfrac{4}{7})^2+(\tfrac{1}{2})^2\right] \approx 0.5004.$$

Exact Value: $\int_1^2 \frac{1}{x^2}\,dx = -\frac{1}{x}\Big|_1^2 = -\frac{1}{2}+1 = \frac{1}{2}$.

9. $\Delta x = \frac{b-a}{n} = \frac{4-0}{8} = \tfrac{1}{2}$; $x_0=0, x_1=\tfrac{1}{2}, x_2=\tfrac{2}{2}, x_3=\tfrac{3}{2},\ldots, x_8=\tfrac{8}{2}$.

Trapezoidal Rule:
$$\int_0^4 \sqrt{x}\,dx \approx \frac{\frac{1}{2}}{2}\left[0+2\sqrt{0.5}+2\sqrt{1}+2\sqrt{1.5}+\cdots+2\sqrt{3.5}+\sqrt{4}\right] \approx 5.26504.$$

Simpson's Rule:
$$\int_0^4 \sqrt{x}\,dx \approx \frac{\frac{1}{2}}{3}\left[0+4\sqrt{0.5}+2\sqrt{1}+4\sqrt{1.5}+\cdots+4\sqrt{3.5}+\sqrt{4}\right] \approx 5.30463.$$

The actual value is $\displaystyle\int_0^4 \sqrt{x}\,dx \approx \frac{2}{3}x^{3/2}\Big|_0^4 = \frac{2}{3}(8) = \frac{16}{3} \approx 5.333333.$

11. $\Delta x = \frac{1-0}{6} = \frac{1}{6}; x_0 = 0, x_1 = \frac{1}{6}, x_2 = \frac{2}{6}, \ldots, x_6 = \frac{6}{6}.$

Trapezoidal Rule:

$\displaystyle\int_0^1 e^{-x}\,dx \approx \frac{\frac{1}{6}}{2}[1 + 2e^{-1/6} + 2e^{-2/6} + \cdots + 2e^{-5/6} + e^{-1}] \approx 0.633583.$

Simpson's Rule:

$\displaystyle\int_0^1 e^{-x}\,dx \approx \frac{\frac{1}{6}}{3}[1 + 4e^{-1/6} + 2e^{-2/6} + \cdots + 4e^{-5/6} + e^{-1}] \approx 0.632123.$

The actual value is $\displaystyle\int_0^1 e^{-x}\,dx = -e^{-x}\Big|_0^1 = -e^{-1} + 1 \approx 0.632121.$

13. $\Delta x = \frac{1}{4}; x_0 = 0, x_1 = \frac{5}{4}, x_2 = \frac{3}{2}, x_3 = \frac{7}{4}, x_4 = 2.$

Trapezoidal Rule:

$\displaystyle\int_1^2 \ln x\,dx \approx \frac{1}{8}[\ln 1 + 2\ln\frac{5}{4} + 2\ln\frac{3}{2} + 2\ln\frac{7}{4} + \ln 2] \approx 0.38370.$

Simpson's Rule:

$\displaystyle\int_1^2 \ln x\,dx \approx \frac{1}{12}[\ln 1 + 4\ln\frac{5}{4} + 2\ln\frac{3}{2} + 4\ln\frac{7}{4} + \ln 2] \approx 0.38626.$

Exact Value: $\displaystyle\int_1^2 \ln x\,dx \approx x(\ln x - 1)\Big|_1^2 = 2(\ln 2 - 1) + 1 = 2\ln 2 - 1.$

15. $\Delta x = \frac{1-0}{4} = \frac{1}{4}; x_0 = 0, x_1 = \frac{1}{4}, x_2 = \frac{2}{4}, x_3 = \frac{3}{4}, x_4 = \frac{4}{4}.$

Trapezoidal Rule:

$\displaystyle\int_0^1 \sqrt{1+x^3}\,dx \approx \frac{\frac{1}{4}}{2}\left[\sqrt{1} + 2\sqrt{1+(\frac{1}{4})^3} + \cdots + 2\sqrt{1+(\frac{3}{4})^3} + \sqrt{2}\right] \approx 1.1170.$

Simpson's Rule:

$\displaystyle\int_0^1 \sqrt{1+x^3}\,dx \approx \frac{\frac{1}{4}}{3}\left[\sqrt{1} + 4\sqrt{1+(\frac{1}{4})^3} + 2\sqrt{1+(\frac{2}{4})^3} \cdots + 4\sqrt{1+(\frac{3}{4})^3} + \sqrt{2}\right] \approx 1.1114.$

17. $\Delta x = \frac{2-0}{4} = \frac{1}{2}; x_0 = 0, x_1 = \frac{1}{2}, x_2 = \frac{2}{2}, x_3 = \frac{3}{2}, x_4 = \frac{4}{2}.$

Trapezoidal Rule:

$\displaystyle\int_0^2 \frac{1}{\sqrt{x^3+1}}\,dx = \frac{\frac{1}{2}}{2}\left[1 + \frac{2}{\sqrt{(\frac{1}{2})^3+1}} + \frac{2}{\sqrt{(1)^3+1}} + \frac{2}{\sqrt{(\frac{3}{2})^3+1}} + \frac{1}{\sqrt{(2)^3+1}}\right]$

$$\approx 1.3973$$

Simpson's Rule:

$$\int_0^2 \frac{1}{\sqrt{x^3+1}}\, dx = \frac{\frac{1}{2}}{3}\left[1 + \frac{4}{\sqrt{(\frac{1}{2})^3+1}} + \frac{2}{\sqrt{(1)^3+1}} + \frac{4}{\sqrt{(\frac{3}{2})^3+1}} + \frac{1}{\sqrt{(2)^3+1}}\right]$$

$$\approx 1.4052$$

19. $\Delta x = \frac{2}{4} = \frac{1}{2}; x_0 = 0, x_1 = \frac{1}{2}, x_2 = 1, x_3 = \frac{3}{2}, x_4 = 2.$

Trapezoidal Rule:

$$\int_0^2 e^{-x^2}\, dx = \frac{1}{4}[e^{-0} + 2e^{-(1/2)^2} + 2e^{-1} + 2e^{-(3/2)^2} + e^{-4}] \approx 0.8806.$$

Simpson's Rule:

$$\int_0^2 e^{-x^2}\, dx = \frac{1}{6}[e^{-0} + 4e^{-(1/2)^2} + 2e^{-1} + 4e^{-(3/2)^2} + e^{-4}] \approx 0.8818.$$

21. $\Delta x = \frac{2-1}{4} = \frac{1}{4}; x_0 = 1, x_1 = \frac{5}{4}, x_2 = \frac{6}{4}, x_3 = \frac{7}{4}, x_4 = \frac{8}{4}.$

Trapezoidal Rule:

$$\int_1^2 x^{-1/2} e^x\, dx = \frac{\frac{1}{4}}{2}\left[e + \frac{2e^{5/4}}{\sqrt{\frac{5}{4}}} + \cdots + \frac{2e^{7/4}}{\sqrt{\frac{7}{4}}} + \frac{e^2}{\sqrt{2}}\right] \approx 3.7757.$$

Simpson's Rule:

$$\int_1^2 x^{-1/2} e^x\, dx = \frac{\frac{1}{4}}{3}\left[e + \frac{4e^{5/4}}{\sqrt{\frac{5}{4}}} + \cdots + \frac{4e^{7/4}}{\sqrt{\frac{7}{4}}} + \frac{e^2}{\sqrt{2}}\right] \approx 3.7625.$$

23. a. Here $a = -1$, $b = 2$, $n = 10$, and $f(x) = x^5$. $f'(x) = 5x^4$ and $f''(x) = 20x^3$.
Because $f'''(x) = 60x^2 > 0$ on $(-1,0) \cup (0,2)$, we see that $f''(x)$ is increasing on
$(-1,0) \cup (0,2)$. So, we take $M = f''(2) = 20(2^3) = 160$.
Using (7), we see that the maximum error incurred is

$$\frac{M(b-a)^3}{12n^2} = \frac{160[2-(-1)]^3}{12(100)} = 3.6.$$

b. We compute $f''' = 60x^2$ and $f^{(iv)}(x) = 120x$. $f^{(iv)}(x)$ is clearly increasing on

(-1,2), so we can take $M = f^{(iv)}(2) = 240$. Therefore, using (8), we see that an

error bound is $\dfrac{M(b-a)^3}{180n^4} = \dfrac{240(3)^5}{180(10^4)} \approx 0.0324$.

25. a. Here $a = 1$, $b = 3$, $n = 10$, and $f(x) = \dfrac{1}{x}$. We find $f'(x) = -\dfrac{1}{x^2}$, $f''(x) = \dfrac{2}{x^3}$.

Since $f'''(x) = -\dfrac{6}{x^4} < 0$ on $(1,3)$, we see that $f''(x)$ is decreasing there. We

may take $M = f''(1) = 2$. Using (7), we find an error bound is

$\dfrac{M(b-a)^3}{12n^2} = \dfrac{2(3-1)^3}{12(100)} \approx 0.013$.

b. $f'''(x) = -\dfrac{6}{x^4}$ and $f^{(iv)}(x) = \dfrac{24}{x^5}$. $f^{(iv)}(x)$ is decreasing on $(1,3)$, so we can

take $M = f^{(iv)}(1) = 24$. Using (8), we find an error bound is $\dfrac{24(3-1)^5}{180(10^4)} \approx 0.00043$.

27. a. Here $a = 0$, $b = 2$, $n = 8$, and $f(x) = (1+x)^{-1/2}$. We find

$\qquad f'(x) = -\tfrac{1}{2}(1+x)^{-3/2}$, $f''(x) = \tfrac{3}{4}(1+x)^{-5/2}$.

Since f'' is positive and decreasing on $(0,2)$, we see that $|f''(x)| \le \tfrac{3}{4}$.

So the maximum error is $\dfrac{\tfrac{3}{4}(2-0)^3}{12(8)^2} = 0.0078125$.

b. $f''' = -\tfrac{15x}{8}(1+x)^{-7/2}$ and $f^{(4)}(x) = \dfrac{105}{16}(1+x)^{-9/2}$. Since $f^{(4)}$ is positive

and decreasing on $(0,2)$, we find $|f^{(4)}(x)| \le \tfrac{105}{16}$.

Therefore, the maximum error is $\dfrac{\tfrac{105}{16}(2-0)^5}{180(8)^4} = 0.000285$.

29. The distance covered is given by

$d = \displaystyle\int_0^2 V(t)\,dt = \tfrac{\tfrac{1}{4}}{2}\big[V(0) + 2V(\tfrac{1}{4}) + \cdots + 2V(\tfrac{7}{4}) + V(2)\big]$

$\qquad = \tfrac{1}{8}[19.5 + 2(24.3) + 2(34.2) + 2(40.5) + 2(38.4) + 2(26.2)$

$\qquad\qquad + 2(18) + 2(16) + 8] \approx 52.84$, or 52.84 miles.

31. $\dfrac{1}{13}\displaystyle\int_0^{13} f(t)\,dt = (\tfrac{1}{13})(\tfrac{1}{2})\{13.2 + 2[14.8 + 16.6 + 17.2 + 18.7 + 19.3 + 22.6 + 24.2 + 25$

$$+24.6 + 25.6 + 26.4 + 26.6] + 26.6\} \approx 21.65.$$

33. Think of the "upper" curve as the graph of f and the lower curve as the graph of g. Then the required area is given by

$$A = \int_0^{150} [f(x) - g(x)]dx = \int_0^{150} h(x)dx$$

where $h = f - g$. Using Simpson's rule,

$A \approx \tfrac{15}{3}[h(0) + 4h(1) + 2f(2) + 4f(3) + 2f(4) + \cdots 4f(9) + f(10)]$

$\quad = 5[0 + 4(25) + 2(40) + 4(70) + 2(80) + 4(90) +$

$\quad\quad 2(65) + 4(50) + 2(60) + 4(35) + 0]$

$\quad = 7850 \ \text{ or } \ 7850 \text{ sq ft.}$

35. We solve the equation $8 = \sqrt{0.01x^2 + 0.11x + 38}$.

$64 = 0.01x^2 + 0.11x + 38,\ 0.01x^2 + 0.11x - 26 = 0,\ x^2 + 11x - 2600 = 0,$

and $x = \dfrac{-11 \pm \sqrt{121 + 10{,}400}}{2} \approx 45.786$. Therefore

$PS = (8)(45.786) - \displaystyle\int_0^{45.786} \sqrt{0.01x^2 + 0.11x + 38}\,dx.$

a. $\Delta x = \dfrac{45.786}{8} = 5.72;\ x_0 = 0,\ x_1 = 5.72,\ x_2 = 11.44,\ ...,\ x_8 = 45.79$

$PS = 366.288 - \dfrac{5.72}{2}\Big[\sqrt{38} + 2\sqrt{0.01(5.72)^2 + 0.11(5.72) + 38} + \cdots$

$\quad\quad\quad + \sqrt{0.01(45.79)^2 + 0.11(45.79) + 38}\Big] \quad \approx 51{,}558, \text{ or } \$51{,}558.$

$PS = 366.288 - \dfrac{5.72}{2}\Big[\sqrt{38} + 4\sqrt{0.01(5.72)^2 + 0.11(5.72) + 38} + \cdots$

$\quad\quad\quad + \sqrt{0.01(45.79)^2 + 0.11(45.79) + 38}\Big] \quad \approx 51{,}708, \text{ or } \$51{,}708.$

37. The percentage of the nonfarm work force in a certain country, will continue to grow at the rate of $A = 30 + \displaystyle\int_0^1 5e^{1/(t+1)}\,dt$ percent, t decades from now.

$\Delta t = \tfrac{1}{10} = 0.1,\ t_0 = 0,\ t_1 = 0.1,\ ...,\ t_{10} = 1.$

Using Simpson's Rule we have

$$A = 30 + \tfrac{1}{3}(5e^1 + 4 \cdot 5e^{1/1.1} + 2 \cdot 5e^{1/1.2} + 4 \cdot 5e^{1/1.3} + \cdots + 4 \cdot 5e^{1/1.9} + 5e^{1/2})$$

$$= 40.1004, \quad \text{or approximately 40.1 percent.}$$

39. $\Delta x = \dfrac{40,000-30,000}{10} = 1000; \; x_0 = 30,000, \; x_1 = 31,000, \; x_2, \ldots, x_{10} = 40,000.$

$$P = \frac{100}{2000\sqrt{2\pi}} \int_{30,000}^{40,000} e^{-0.5[x-40,000)/[2000]^2} \, dx$$

$$P = \frac{100(1000)}{2000\sqrt{2\pi}} \left[e^{-0.5[30,000-40,000)/[2000]^2} + 4e^{-0.5[(31,000-40,000)/2000]^2} + \cdots + 1] \right]$$

$$\approx 0.50, \text{ or 50 percent.}$$

41. False. The number n can be odd or even.

43. True.

45. Taking the limit and recalling the definition of the Riemann sum, we find

$$\lim_{\Delta t \to 0}[c(t_1)R\Delta t + c(t_2)R\Delta t + \cdots + c(t_n)R\Delta t]/60 = D$$

$$\frac{R}{60}\lim_{\Delta t \to 0}[c(t_1)\Delta t + c(t_2)\Delta t + \cdots + c(t_n)\Delta t] = D$$

$$\frac{R}{60}\int_0^T c(t)\,dt = D, \text{ or } R = \frac{60D}{\int_0^T c(t)\,dt}.$$

EXERCISES 7.4, page 531

1. The required area is given by

$$\int_3^\infty \frac{2}{x^2}\,dx = \lim_{b \to \infty}\int_3^b \frac{2}{x^2}\,dx = \lim_{b \to \infty}\left(-\frac{2}{x}\right)\Big|_3^b = \lim_{b \to \infty}\left(-\frac{2}{b} + \frac{2}{3}\right) = \frac{2}{3} \text{ or } 2/3 \text{ sq units.}$$

3. $$A = \int_3^\infty \frac{1}{(x-2)^2}\,dx = \lim_{b \to \infty}\int_3^b (x-2)^{-2}\,dx = \lim_{b \to \infty}-\frac{1}{x-2}\Big|_3^b = \lim_{b \to \infty}\left(-\frac{1}{b-2} + 1\right) = 1.$$

5. $A = \displaystyle\int_1^\infty \frac{1}{x^{3/2}}\,dx = \lim_{b\to\infty}\int_1^b x^{-3/2}\,dx = \lim_{b\to\infty} -\frac{2}{\sqrt{x}}\Big|_1^b = \lim_{b\to\infty}\left(-\frac{2}{\sqrt{b}}+2\right) = 2.$

7. $A = \displaystyle\int_0^\infty \frac{1}{(x+1)^{5/2}}\,dx = \lim_{b\to\infty}\int_1^b (x+1)^{-5/2}\,dx = \lim_{b\to\infty} -\frac{2}{3}(x+1)^{-3/2}\Big|_0^b$

$\qquad = \displaystyle\lim_{b\to\infty}\left[-\frac{2}{3(b+1)^{3/2}}+\frac{2}{3}\right] = \frac{2}{3}.$

9. $A = \displaystyle\int_{-\infty}^2 e^{2x}\,dx = \lim_{a\to-\infty}\int_a^2 e^{2x}\,dx = \lim_{a\to-\infty}\tfrac{1}{2}e^{2x}\Big|_a^2 = \lim_{a\to-\infty}\left(\tfrac{1}{2}e^4-\tfrac{1}{2}e^{2a}\right) = \tfrac{1}{2}e^4.$

11. Using symmetry, the required area is given by

$$2\int_0^\infty \frac{x}{(1+x^2)^2}\,dx = 2\lim_{b\to\infty}\int_0^\infty \frac{x}{(1+x^2)^2}\,dx.$$

To evaluate the indefinite integral $\displaystyle\int \frac{x}{(1+x^2)^2}\,dx$, put $u = 1+x^2$ so that

$du = 2x\,dx$ or $x\,dx = \tfrac{1}{2}\,du.$

Then $\displaystyle\int \frac{x}{(1+x^2)^2}\,dx = \frac{1}{2}\int \frac{du}{u^2} = -\frac{1}{2u}+C = -\frac{1}{2(1+x^2)}+C.$

Therefore, $2\displaystyle\lim_{b\to\infty}\int_0^b \frac{x}{(1+x^2)}\,dx = \lim_{b\to\infty}-\frac{1}{(1+x^2)^2}\Big|_0^b = \lim_{b\to\infty}\left[-\frac{1}{(1+b^2)}+1\right] = 1,$

or 1 sq unit.

13. a. $I(b) = \displaystyle\int_0^b \sqrt{x}\,dx = \tfrac{2}{3}x^{3/2}\Big|_0^b = \tfrac{2}{3}b^{3/2}.$ b. $\displaystyle\lim_{b\to\infty}I(b) = \lim_{b\to\infty}\tfrac{2}{3}b^{3/2} = \infty.$

15. $\displaystyle\int_1^\infty \frac{3}{x^4}\,dx = \lim_{b\to\infty}\int_1^b 3x^{-4}\,dx = \lim_{b\to\infty}\left(-\frac{1}{x^3}\right)\Big|_1^b = \lim_{b\to\infty}\left(-\frac{1}{b^3}+1\right) = 1.$

17. $A = \displaystyle\int_4^\infty \frac{2}{x^{3/2}}\,dx = \lim_{b\to\infty}\int_4^b 2x^{-3/2}\,dx = \lim_{b\to\infty}-4x^{-1/2}\Big|_4^b = \lim_{b\to\infty}\left(-\frac{4}{\sqrt{b}}+2\right) = 2.$

19. $\displaystyle\int_1^\infty \frac{4}{x}\,dx = \lim_{b\to\infty}\int_1^b \frac{4}{x}\,dx = \lim_{b\to\infty}4\ln x\Big|_1^b = \lim_{b\to\infty}(4\ln b) = \infty.$

21. $\displaystyle\int_{-\infty}^{0}(x-2)^{-3}\,dx=\lim_{a\to-\infty}\int_{a}^{0}(x-2)^{-3}\,dx=\lim_{a\to-\infty}-\frac{1}{2(x-2)^{2}}\Big|_{a}^{0}=-\frac{1}{8}.$

23. $\displaystyle\int_{1}^{\infty}\frac{1}{(2x-1)^{3/2}}\,dx=\lim_{b\to\infty}\int_{1}^{b}(2x-1)^{-3/2}\,dx=\lim_{b\to\infty}-\frac{1}{(2x-1)^{1/2}}\Big|_{1}^{b}$

$$=\lim_{b\to\infty}\left(-\frac{1}{\sqrt{2b-1}}+1\right)=1.$$

25. $\displaystyle\int_{0}^{\infty}e^{-x}\,dx=\lim_{b\to\infty}\int_{0}^{b}e^{-x}\,dx=\lim_{b\to\infty}-e^{-x}\Big|_{0}^{b}=\lim_{b\to\infty}(-e^{-b}+1)=1.$

27. $\displaystyle\int_{-\infty}^{0}e^{2x}\,dx=\lim_{a\to-\infty}\tfrac{1}{2}e^{2x}\Big|_{a}^{0}=\lim_{a\to-\infty}\left(\tfrac{1}{2}-\tfrac{1}{2}e^{2a}\right)=\tfrac{1}{2}.$

29. $\displaystyle\int_{1}^{\infty}\frac{e^{\sqrt{x}}}{\sqrt{x}}\,dx=\lim_{b\to\infty}\int_{1}^{b}\frac{e^{\sqrt{x}}}{\sqrt{x}}\,dx=\lim_{b\to\infty}-2e^{\sqrt{x}}\Big|_{1}^{b}$ (Integrate by substitution: $u=\sqrt{x}$.)

$$=\lim_{b\to\infty}(2e^{\sqrt{b}}-2e)=\infty,\text{ and so it diverges.}$$

31. $\displaystyle\int_{-\infty}^{0}xe^{x}\,dx=\lim_{a\to-\infty}\int_{a}^{0}xe^{x}\,dx=\lim_{a\to-\infty}(x-1)e^{x}\Big|_{a}^{0}=\lim_{a\to-\infty}[-1+(a-1)e^{a}]=-1.$
Note: We have used integration by parts to evaluate the integral.

33. $\displaystyle\int_{-\infty}^{\infty}x\,dx=\lim_{a\to-\infty}\tfrac{1}{2}x^{2}\Big|_{a}^{0}+\lim_{b\to\infty}\tfrac{1}{2}x^{2}\Big|_{0}^{b}$ both of which diverge and so the integral diverges.

35. $\displaystyle\int_{-\infty}^{\infty}x^{3}(1+x^{4})^{-2}\,dx=\int_{-\infty}^{0}x^{3}(1+x^{4})^{-2}\,dx+\int_{0}^{\infty}x^{3}(1+x^{4})^{-2}\,dx$

$$=\lim_{a\to-\infty}\int_{a}^{0}x^{3}(1+x^{4})^{-2}\,dx+\lim_{b\to\infty}\int_{0}^{b}x^{3}(1+x^{4})^{-2}\,dx$$

$$=\lim_{a\to-\infty}\left[-\frac{1}{4}(1+x^{4})^{-1}\Big|_{a}^{0}\right]+\lim_{b\to\infty}\left[-\frac{1}{4}(1+x^{4})^{-1}\Big|_{0}^{b}\right]$$

$$=\lim_{a\to-\infty}\left[-\frac{1}{4}+\frac{1}{4(1+a^{4})}\right]+\lim_{b\to\infty}\left[-\frac{1}{4(1+b^{4})}+\frac{1}{4}\right]$$

$$=-\tfrac{1}{4}+\tfrac{1}{4}=0.$$

37. $\int_{-\infty}^{\infty} xe^{1-x^2}\,dx = \lim_{a\to-\infty}\int_a^0 xe^{1-x^2}\,dx + \lim_{b\to\infty}\int_0^b xe^{1-x^2}\,dx$

$$= \lim_{a\to-\infty} -\tfrac{1}{2}e^{1-x^2}\Big|_a^0 + \lim_{b\to\infty}-\tfrac{1}{2}e^{1-x^2}\Big|_0^b$$

$$= \lim_{a\to-\infty}\left(-\tfrac{1}{2}e + \tfrac{1}{2}e^{1-a^2}\right) + \lim_{b\to\infty}\left(-\tfrac{1}{2}e^{1-b^2} + \tfrac{1}{2}e\right) = 0.$$

39. $\int_{-\infty}^{\infty}\dfrac{e^{-x}}{1+e^{-x}}\,dx = \lim_{a\to-\infty} -\ln(1+e^{-x})\Big|_a^0 + \lim_{b\to\infty} -\ln(1+e^{-x})\Big|_0^b = \infty$, and it is divergent.

41. First, we find the indefinite integral $I = \displaystyle\int \dfrac{1}{x\ln^3 x}\,dx$. Let $u = \ln x$ so that

$du = \dfrac{1}{x}\,dx$. Therefore, $I = \displaystyle\int \dfrac{du}{u^3} = -\dfrac{1}{2u^2} + C = -\dfrac{1}{2\ln^2 x} + C.$ So

$$\int_e^{\infty}\dfrac{1}{x\ln^3 x}\,dx = \lim_{b\to\infty}\int_e^b \dfrac{1}{x\ln^3 x}\,dx$$

$$= \lim_{b\to\infty}\left[-\dfrac{1}{2\ln^2 x}\Big|_e^b\right] = \lim_{b\to\infty}\left(-\dfrac{1}{2(\ln b)^2} + \dfrac{1}{2}\right) = \dfrac{1}{2}$$

and so the given integral is convergent.

43. We want the present value PV of a perpetuity with $m = 1$, $P = 1500$, and $r = 0.08$.

We find $PV = \dfrac{(1)(1500)}{0.08} = 18{,}750$, or $\$18{,}750$.

45. $PV = \displaystyle\int_0^{\infty}(10{,}000 + 4000t)e^{-rt}\,dt = 10{,}000\int_0^{\infty}e^{-rt}\,dt + 4000\int_0^{\infty}te^{-rt}\,dt$

$$= \lim_{b\to\infty}\left(-\dfrac{10{,}000}{r}e^{-rt}\Big|_0^b\right) + 4000\left(\dfrac{1}{r^2}\right)(-rt-1)e^{-rt}\Big|_0^b$$

(Integrating by parts.)

$$= \dfrac{10{,}000}{r} + \dfrac{4000}{r^2} = \dfrac{10{,}000r + 4000}{r^2} \text{ dollars.}$$

47. True. $\displaystyle\int_a^{\infty} f(x)\,dx = \int_a^b f(x)\,dx + \int_b^{\infty} f(x)\,dx$. So if $\displaystyle\int_a^{\infty} f(x)\,dx$ exists then

$$\int_b^\infty f(x)\,dx = \int_a^\infty f(x)\,dx - \int_a^b f(x)\,dx.$$

49. False. Let $f(x) = \begin{cases} e^{2x} & \text{if } -\infty < x \le 0 \\ e^{-x} & \text{if } 0 < x < \infty \end{cases}$. Then

$$\int_{-\infty}^\infty f(x)\,dx = \int_{-\infty}^0 e^{2x}\,dx + \int_0^\infty e^{-x}\,dx = \frac{1}{2} + 1 = \frac{3}{2}. \text{ But } 2\int_0^\infty f(x)\,dx = 2\int_0^\infty e^{-x}\,dx = 2.$$

51. $\displaystyle\int_0^\infty e^{-px}\,dx = \lim_{b\to\infty}\int_a^b e^{-px}\,dx = \lim_{b\to\infty}\left[-\frac{1}{p}e^{-px}\Big|_a^b\right] = \lim_{b\to\infty}\left(-\frac{1}{p}e^{-pb} + \frac{1}{p}e^{-pa}\right)$

$$= \frac{1}{pe^{pa}} \quad \text{if } p > 0 \text{ and is divergent if } p < 0.$$

CHAPTER 7 REVIEW EXERCISES, page 534

1. Let $u = 2x$ and $dv = e^{-x}\,dx$ so that $du = 2\,dx$ and $v = -e^{-x}$. Then
$$\int 2xe^{-x}\,dx = uv - \int v\,du = -2xe^{-x} + 2\int e^{-x}\,dx$$
$$= -2xe^{-x} - 2e^{-x} + C = -2(1+x)e^{-x} + C.$$

3. Let $u = \ln 5x$ and $dv = dx$, so that $du = \frac{1}{x}\,dx$ and $v = x$. Then
$$\int \ln 5x\,dx = x\ln 5x\,dx - \int dx = x\ln 5x - x + C = x(\ln 5x - 1) + C.$$

5. Let $u = x$ and $dv = e^{-2x}\,dx$ so that $du = dx$ and $v = -\frac{1}{2}e^{-2x}$. Then
$$\int_0^1 xe^{-2x}\,dx = -\frac{1}{2}xe^{-2x}\Big|_0^1 + \frac{1}{2}\int_0^1 e^{-2x}\,dx = -\frac{1}{2}e^{-2} - \frac{1}{4}e^{-2x}\Big|_0^1$$
$$= -\frac{1}{2}e^{-2} - \frac{1}{4}e^{-2} + \frac{1}{4} = \frac{1}{4}(1 - 3e^{-2}).$$

7. $f(x) = \displaystyle\int f'(x)\,dx = \int \frac{\ln x}{\sqrt{x}}\,dx$. To evaluate the integral, we integrate by parts

with $u = \ln x$, $dv = x^{-1/2}\,dx$, $du = \frac{1}{x}\,dx$ and $v = 2x^{1/2}\,dx$. Then

$$\int \frac{\ln x}{x^{1/2}}\,dx = 2x^{1/2}\ln x - \int 2x^{-1/2}\,dx = 2x^{1/2}\ln x - 4x^{1/2} + C$$

$$= 2x^{1/2}(\ln x - 2) + C = 2\sqrt{x}(\ln x - 2) + C.$$

But $f(1) = -2$ and this gives $2\sqrt{1}(\ln 1 - 2) + C = -2$, or $C = 2$. Therefore, $f(x) = 2\sqrt{x}(\ln x - 2) + 2$.

9. Using Formula 4 with $a = 3$ and $b = 2$, we obtain

$$\int \frac{x^2}{(3+2x)^2}\,dx = \frac{1}{8}\left[3 + 2x - \frac{9}{3+2x} - 6\ln|3+2x|\right] + C.$$

11. Use Formula 24 with $a = 4$ and $n = 2$, obtaining $\int x^2 e^{4x}\,dx = \frac{1}{4}x^2 e^{4x} - \frac{1}{2}\int xe^{4x}\,dx.$

Use Formula 23 to obtain

$$\int x^2 e^{4x}\,dx = \frac{1}{4}x^2 e^{4x} - \frac{1}{2}\left[\frac{1}{16}(4x-1)e^{4x}\right] + C$$

$$= \frac{1}{32}(8x^2 - 4x + 1)e^{4x} + C.$$

13. Use Formula 17 with $a = 2$ obtaining $\int \frac{dx}{x^2\sqrt{x^2-4}} = \frac{\sqrt{x^2-4}}{4x} + C.$

15. $\int_0^\infty e^{-2x}\,dx = \lim_{b\to\infty}\int_0^b e^{-2x}\,dx = \lim_{b\to\infty}(-\frac{1}{2}e^{-2x})\Big|_0^b = \lim_{b\to\infty}(-\frac{1}{2}e^{-2b} + \frac{1}{2}) = \frac{1}{2}.$

17. $\int_3^\infty \frac{2}{x}\,dx = \lim_{b\to\infty}\int_3^b \frac{2}{x}\,dx = \lim_{b\to\infty} 2\ln x\Big|_3^b = \lim_{b\to\infty}(2\ln b - 2\ln 3) = \infty.$

19. $\int_2^\infty \frac{dx}{(1+2x)^2} = \lim_{b\to\infty}\int_2^b (1+2x)^{-2}\,dx = \lim_{b\to\infty}(\frac{1}{2})(-1)(1+2x)^{-1}\Big|_2^b$

$$= \lim_{b\to\infty}\left(-\frac{1}{2(1+2b)} + \frac{1}{2(5)}\right) = \frac{1}{10}.$$

21. $\Delta x = \frac{b-a}{n} = \frac{3-1}{4} = \frac{1}{2}; x_0 = 1, x_1 = \frac{3}{2}, x_2 = 2, x_3 = \frac{5}{2}, x_4 = 3.$

Trapezoidal Rule:

$$\int_1^3 \frac{dx}{1+\sqrt{x}} \approx \frac{\frac{1}{2}}{2}\left[\frac{1}{2} + \frac{2}{1+\sqrt{1.5}} + \frac{2}{1+\sqrt{2}} + \frac{2}{1+\sqrt{2.5}} + \frac{1}{1+\sqrt{3}}\right] \approx 0.8421.$$

Simpson's Rule

$$\int_1^3 \frac{dx}{1+\sqrt{x}} \approx \frac{\frac{1}{2}}{3}\left[\frac{1}{2} + \frac{4}{1+\sqrt{1.5}} + \frac{2}{1+\sqrt{2}} + \frac{4}{1+\sqrt{2.5}} + \frac{1}{1+\sqrt{3}}\right] \approx 0.8404.$$

23. $\Delta x = \frac{1-(-1)}{4} = \frac{1}{2}$; $x_0 = -1$, $x_1 = -\frac{1}{2}$, $x_2 = 0$, $x_3 = \frac{1}{2}$, $x_4 = 1$.

Trapezoidal Rule:

$$\int_{-1}^1 \sqrt{1+x^4}\, dx \approx \frac{0.5}{2}\left[\sqrt{2} + 2\sqrt{1+(-0.5)^4} + 2 + 2\sqrt{1+(0.5)^4} + \sqrt{2}\right] \approx 2.2379.$$

Simpson's Rule:

$$\int_{-1}^1 \sqrt{1+x^4}\, dx \approx \frac{0.5}{3}\left[\sqrt{2} + 4\sqrt{1+(-0.5)^4} + 2 + 4\sqrt{1+(0.5)^4} + \sqrt{2}\right]$$
$$\approx 2.1791.$$

25. The producer's surplus is given by $PS = \bar{p}\bar{x} - \int_0^{\bar{x}} s(x)\, dx$, where $\bar{x}$ is found by
    solving the equation $2\sqrt{25+x^2} = 13$. Then $\sqrt{25+x^2} = 13$, $25+x^2 = 169$, and
    $x = \pm 12$. So $\bar{x} = 12$. Therefore, $PS = (26)(12) - 2\int_0^{12}(25+x^2)^{1/2}\, dx$.
    Using Formula 7 with $a = 5$, we obtain
    $$PS = (26)(12) - 2\int_0^{12}(25+x^2)^{1/2}\, dx$$
    $$= 312 - 2\left(\frac{x}{2}(25+x^2)^{1/2} + \frac{25}{2}\ln\left|x+(25+x^2)^{1/2}\right|\right)\Big|_0^{12}$$
    $$= 312 - 2[6(13) + \frac{25}{2}\ln(12+13) - \frac{25}{2}\ln 5] \approx 115.76405,$$
    or $\$1,157,641$.

27. If $p = 30$, we have $2\sqrt{325-x^2} = 30$, $\sqrt{325-x^2} = 15$, or $325 - x^2 = 225$, $x^2 = 100$,
    or $x = \pm 10$. So the equilibrium point is $(10, 30)$.
    $$CS = \int_0^{10} 2\sqrt{325-x^2}\, dx - (30)(10).$$
    To evaluate the integral using Simpson's Rule with $n = 10$, we have

    $\Delta x = \frac{10-0}{10} = 1$; $x_0 = 0$, $x_1 = 1$, $x_2 = 2$, ..., $x_{10} = 10$.

    $$2\int_0^{10} \sqrt{325-x^2}\, dx$$

$$\approx \tfrac{2}{3}\left[\sqrt{325} + 4\sqrt{325-1} + 2\sqrt{325-4} + \cdots + 4\sqrt{325-81} + \sqrt{325-100}\right]$$

Therefore, $CS \approx 341.0 - 300 \approx 41.1$, or \$41,100.

29. We want the present value of a perpetuity with $m = 1$, $P = 10,000$, and $r = 0.09$.
    We find $PV = \dfrac{(1)(10,000)}{0.09} \approx 111,111$ or approximately \$111,111.

CHAPTER 8

EXERCISES 8.1, page 544

1. $f(0, 0) = 2(0) + 3(0) - 4 = -4.$ $f(1, 0) = 2(1) + 3(0) - 4 = -2.$
   $f(0, 1) = 2(0) + 3(1) - 4 = -1.$ $f(1, 2) = 2(1) + 3(2) - 4 = 4.$
   $f(2,-1) = 2(2) + 3(-1) - 4 = -3.$

3. $f(1, 2) = 1^2 + 2(1)(2) - 1 + 3 = 7; \ f(2, 1) = 2^2 + 2(2)(1) - 2 + 3 = 9$
   $f(-1, 2) = (-1)^2 + 2(-1)(2) - (-1) + 3 = 1; f(2, -1) = 2^2 + 2(2)(-1) - 2 + 3 = 1.$

5. $g(s,t) = 3s\sqrt{t} + t\sqrt{s} + 2; \ g(1,2) = 3(1)\sqrt{2} + 2\sqrt{1} + 2 = 4 + 3\sqrt{2}$
   $g(2,1) = 3(2)\sqrt{1} + \sqrt{2} + 2 = 8 + \sqrt{2};$
   $g(0, 4) = 0 + 0 + 2 = 2, \ g(4,9) = 3(4)\sqrt{9} + 9\sqrt{4} + 2 = 56.$

7. $h(1,e) = \ln e - e \ln 1 = \ln e = 1; \ h(e,1) = e \ln 1 - \ln e = -1;$
   $h(e,e) = e \ln e - e \ln e = 0.$

9. $g(r,s,t) = re^{s/t}; \ g(1,1,1) = e, \ g(1,0,1) = 1, \ g(-1,-1,-1) = -e^{-1/(-1)} = -e.$

11. The domain of f is the set of all ordered pairs (x, y) where x and y are real numbers.

13. All real values of u and v except those satisfying the equation $u = v$.

15. The domain of g is the set of all ordered pairs (r,s) satisfying $rs \geq 0$, that is the set of all ordered pairs where both $r \geq 0$ and $s \geq 0$, or in which both $r \leq 0$ and $s \leq 0$.

17. The domain of h is the set of all ordered pairs (x, y) such that $x + y > 5$.

19. The level curves of $z = f(x, y) = 2x + 3y$ for $z = -2, -1, 0, 1, 2$, follow.

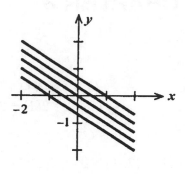

21. The level curves of $z = -4, -2, 2, 4$ are shown below.

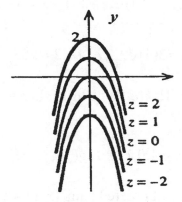

23. The level curves of $f(x,y) = \sqrt{16 - x^2 - y^2}$
    for $z = 0, 1, 2, 3, 4$

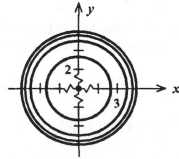

25. $V = f(1.5, 4) = \pi(1.5)^2(4) = 9\pi$, or 9π cu ft

27. a. $M = \dfrac{80}{(1.8)^2} = 24.69$.

 b. $\dfrac{w}{(1.8)^2} < 25$; that is, $w < 2.5(1.8)^2 = 81$, that is less than 81 kg.

29. a. $R(x,y) = xp + yq = x(200 - \frac{1}{5}x - \frac{1}{10}y) + y(160 - \frac{1}{10}x - \frac{1}{4}y)$

$= -\frac{1}{5}x^2 - \frac{1}{4}y^2 - \frac{1}{5}xy + 200x + 160y$.

 b. The domain of R is the set of all points (x,y) satisfying
$200 - \frac{1}{5}x - \frac{1}{10}y \geq 0$, $160 - \frac{1}{10}x - \frac{1}{4}y \geq 0$

31. a. $R(x,y) = xp + yq = 20x - 0.005x^2 - 0.001xy + 15y - 0.001xy - 0.003y^2$

$= -0.005x^2 - 0.003y^2 - 0.002xy + 20x + 15y$.

 b. Since p and q must both be nonnegative, the domain of R is the set of all ordered pairs (x, y) for which
$$20 - 0.005x - 0.001y \geq 0$$
and $\qquad\qquad 15 - 0.001x - 0.003y \geq 0$.

33. a. The domain of V is the set of all ordered pairs (P,T) where P and T are positive real numbers.

 b. $V = \dfrac{30.9(273)}{760} = 11.10$ liters.

35. The number of suspicious fires is
$$N(100,20) = \dfrac{100[1000 + 0.03(100^2)(20)]^{1/2}}{[5 + 0.2(20)]^2} = 103.29, \text{ or approximately } 103.$$

37. a. $P = f(100,000, 0.08, 30) = \dfrac{100,000(0.08)}{12\left[1 - \left(1 + \dfrac{0.08}{12}\right)^{-360}\right]} \approx 733.76,$ or $\$733.76$.

$P = f(100,000, 0.1, 30) = \dfrac{100,000(0.1)}{12\left[1 - \left(1 + \dfrac{0.1}{12}\right)^{-360}\right]} \approx 877.57,$ or $\$877.57$.

b. $P = f(100{,}000, 0.08, 20) = \dfrac{100{,}000(0{,}08)}{12\left[1-\left(1+\dfrac{0.08}{12}\right)^{-240}\right]} \approx 836.44, \text{ or } \$836.44.$

39. $f(M, 600, 10) = \dfrac{\pi^2(360{,}000)\,M(10)}{900} \approx 39{,}478.42\,M$

   or $\dfrac{39{,}478.42}{980} \approx 40.28$ times gravity.

41. False. Let $h(x, y) = xy$. Then there are no functions f and g such that $h(x, y) = f(x) + g(y)$.

43. False. Since $x^2 - y^2 = (x+y)(x-y)$, we see that $x^2 - y^2 = 0$ if $y = \pm x$. Therefore, the domain of f is $\{(x, y)\,|\,y \neq \pm x\}$.

EXERCISES 8.2, page 557

1. $f_x = 2,\ f_y = 3$ 3. $g_x = 4x,\ g_y = 4$ 5. $f_x = -\dfrac{4y}{x^3};\ f_y = \dfrac{2}{x^2}.$

7. $g(u, v) = \dfrac{u-v}{u+v};\ \dfrac{\partial g}{\partial u} = \dfrac{(u+v)(1)-(u-v)(1)}{(u+v)^2} = \dfrac{2v}{(u+v)^2}.$

   $\dfrac{\partial g}{\partial v} = \dfrac{(u+v)(-1)-(u-v)(1)}{(u+v)^2} = -\dfrac{2u}{(u+v)^2}.$

9. $f(s, t) = (s^2 - st + t^2)^3;\ f_s = 3(s^2 - st + t^2)^2(2s - t)$ and $f_t = 3(s^2 - st + t^2)^2(2t - s)$

11. $f(x, y) = (x^2 + y^2)^{2/3};\ f_x = \tfrac{2}{3}(x^2 + y^2)^{-1/3}(2x) = \tfrac{4}{3}x(x^2 + y^2)^{-1/3}.$ Similarly, $f_y = \tfrac{4}{3}y(x^2 + y^2)^{-1/3}.$

13. $f(x, y) = e^{xy+1};\ f_x = ye^{xy+1},\ f_y = xe^{xy+1}.$

15. $f(x, y) = x\ln y + y\ln x;\ f_x = \ln y + \dfrac{y}{x},\ f_y = \dfrac{x}{y} + \ln x.$

17. $g(u,v) = e^u \ln v.\ g_u = e^u \ln v,\ g_v = \dfrac{e^u}{v}$.

19. $f(x,y,z) = xyz + xy^2 + yz^2 + zx^2;\ f_x = yz + y^2 + 2xz,$
$f_y = xz + 2xy + z^2,\ f_z = xy + 2yz + x^2$.

21. $h(r,s,t) = e^{rst};\ h_r = ste^{rst},\ h_s = rte^{rst},\ h_t = rse^{rst}$.

23. $f(x,y) = x^2 y + xy^2;\ f_x(1,2) = 2xy + y^2\big|_{(1,2)} = 8;\ f_y(1,2) = x^2 + 2xy\big|_{(1,2)} = 5.$

25. $f(x,y) = x\sqrt{y} + y^2 = xy^{1/2} + y^2;\ f_x(2,1) = \sqrt{y}\ \big|_{(2,1)} = 1,$

$f_y(2,1) = \dfrac{x}{2\sqrt{y}} + 2y\ \big|_{(2,1)} = 3.$

27. $f(x,y) = \dfrac{x}{y};\ f_x(1,2) = \dfrac{1}{y}\bigg|_{(1,2)} = \dfrac{1}{2},\ f_y(1,2) = -\dfrac{x}{y^2}\bigg|_{(1,2)} = -\dfrac{1}{4}.$

29. $f(x,y) = e^{xy}.\ f_x(1,1) = ye^{xy}\big|_{(1,1)} = e,\ f_y(1,1) = xe^{xy}\big|_{(1,1)} = e.$

31. $f(x,y,z) = x^2 yz^3;\ f_x(1,0,2) = 2xyz^3\big|_{(1,0,2)} = 0;\ f_y(1,0,2) = x^2 z^3\big|_{(1,0,2)} = 8.$
$f_z(1,0,2) = 3x^2 yz^2\big|_{(1,0,2)} = 0.$

33. $f(x,y) = x^2 y + xy^3;\ f_x = 2xy + y^3,\ f_y = x^2 + 3xy^2.$
Therefore, $f_{xx} = 2y,\ f_{xy} = 2x + 3y^2 = f_{yx},\ f_{yy} = 6xy.$

35. $f(x,y) = x^2 - 2xy + 2y^2 + x - 2y;\ f_x = 2x - 2y + 1,\ f_y = -2x + 4y - 2;\ f_{xx} = 2,$
$f_{xy} = -2,\ f_{yx} = -2,\ f_{yy} = 4.$

37. $f(x,y) = (x^2 + y^2)^{1/2};\ f_x = \tfrac{1}{2}(x^2 + y^2)^{-1/2}(2x) = x(x^2 + y^2)^{-1/2};$

$$f_y = y(x^2 + y^2)^{-1/2}.$$

$$f_{xx} = (x^2 + y^2)^{-1/2} + x(-\tfrac{1}{2})(x^2 + y^2)^{-3/2}(2x) = (x^2 + y^2)^{-1/2} - x^2(x^2 + y^2)^{-3/2}$$

$$= (x^2 + y^2)^{-3/2}(x^2 + y^2 - x^2) = \frac{y^2}{(x^2 + y^2)^{3/2}}.$$

$$f_{xy} = x(-\tfrac{1}{2})(x^2 + y^2)^{-3/2}(2y) = -\frac{xy}{(x^2 + y^2)^{3/2}} = f_{yx}.$$

$$f_{yy} = (x^2 + y^2)^{-1/2} + y(-\tfrac{1}{2})(x^2 + y^2)^{-3/2}(2y) = (x^2 + y^2)^{-1/2} - y^2(x^2 + y^2)^{-3/2}$$

$$= (x^2 + y^2)^{-3/2}(x^2 + y^2 - y^2) = \frac{x^2}{(x^2 + y^2)^{3/2}}.$$

39. $f(x,y) = e^{-x/y};\ f_x = -\dfrac{1}{y}e^{-x/y};\ f_y = \dfrac{x}{y^2}e^{-x/y};\ f_{xx} = \dfrac{1}{y^2}e^{-x/y};$

$$f_{xy} = -\frac{x}{y^3}e^{-x/y} + \frac{1}{y^2}e^{-x/y} = \left(\frac{-x+y}{y^3}\right)e^{-x/y} = f_{yx}.$$

$$f_{yy} = -\frac{2x}{y^3}e^{-x/y} + \frac{x^2}{y^4}e^{-x/y} = \frac{x}{y^3}\left(\frac{x}{y} - 2\right)e^{-x/y}.$$

41. a. $f(x,y) = 20x^{3/4}y^{1/4}.\ f_x(256,16) = 15\left(\dfrac{y}{x}\right)^{1/4}\Bigg|_{(256,16)}$

$$= 15\left(\frac{16}{256}\right)^{1/4} = 15\left(\frac{2}{4}\right) = 7.5.$$

$$f_y(256,16) = 5\left(\frac{x}{y}\right)^{3/4}\Bigg|_{(256,16)} = 5\left(\frac{256}{16}\right)^{3/4} = 5(80) = 40.$$

  b. Yes.

43. $p(x,y) = 200 - 10(x - \tfrac{1}{2})^2 - 15(y-1)^2.\ \dfrac{\partial p}{\partial x}(0,1) = -20(x - \tfrac{1}{2})\big|_{(0,1)} = 10;$

At the location $(0,1)$ in the figure, the price of land is changing at the rate of $10 per sq ft per mile change to the right.

$$\frac{\partial p}{\partial y}(0,1) = -30(y-1)\big|_{(0,1)} = 0;$$

At the location $(0,1)$ in the figure, the price of land is constant per mile change upwards.

45. $f(p,q) = 10,000 - 10p - e^{0.5q}$; $g(p,q) = 50,000 - 4000q - 10p$.

$$\frac{\partial f}{\partial q} = -0.5e^{0.5q} < 0 \text{ and } \frac{\partial g}{\partial p} = -10 < 0$$

and so the two commodities are complementary commodities.

47. $R(x,y) = -0.2x^2 - 0.25y^2 - 0.2xy + 200x + 160y$.

$$\frac{\partial R}{\partial x}(300,250) = -0.4x - 0.2y + 200\big|_{(300,250)}$$

$$= -0.4(300) - 0.2(250) + 200 = 30$$

and this says that at a sales level of 300 finished and 250 unfinished units the revenue is increasing at the rate of $30 per week per unit increase in the finished units.

$$\frac{\partial R}{\partial y}(300,250) = -0.5y - 0.2x + 160\big|_{(300,250)}$$

$$= -0.5(250) - 0.2(300) + 160 = -25$$

and this says that at a level of 300 finished and 250 unfinished units the revenue is decreasing at the rate of $25 per week per increase in the unfinished units.

49.   a. $T = f(32,20) = 35.74 + 0.6215(32) - 35.75(20^{0.16}) + 0.4275(32)(20^{0.16})$
    ≈ 19.99,   or approximately $20°F$.

    b. $\dfrac{\partial T}{\partial s} = -35.75(0.16S^{-0.84}) + 0.4275t(0.16S^{-0.84})$

    $= 0.16(-35.75 + 0.4275t)s^{-0.84}$

    $\dfrac{\partial T}{\partial s}\bigg|_{(32,20)} = 0.16[-35.75 + 0.4275(32)]20^{-0.84} \approx -0.285$;

    that is, the wind chill will drop by 0.3 degrees for each 1 mph increase in wind speed.

51. $V = \dfrac{30.9T}{P}$. $\dfrac{\partial V}{\partial T} = \dfrac{30.9}{P}$ and $\dfrac{\partial V}{\partial P} = -\dfrac{30.9T}{P^2}$.

    Therefore, $\dfrac{\partial V}{\partial T}\bigg|_{T=300,P=800} = \dfrac{30.9}{800} = 0.039$, or 0.039 liters/degree.

$$\left.\frac{\partial V}{\partial P}\right|_{T=300, P=800} = -\frac{(30.9)(300)}{800^2} = -0.015$$

or -0.015 liters/mm of mercury.

53. $V = \dfrac{kT}{P}$ and $\dfrac{\partial V}{\partial T} = \dfrac{k}{P}$; $T = \dfrac{VP}{k}$ and $\dfrac{\partial T}{\partial P} = \dfrac{V}{k} = \dfrac{T}{P}$; and

$P = \dfrac{kT}{V}$ and $\dfrac{\partial P}{\partial V} = -\dfrac{kT}{V^2} = -kT \cdot \dfrac{P^2}{(kT)^2} = -\dfrac{P^2}{kT}$

Therefore $\dfrac{\partial V}{\partial T} \cdot \dfrac{\partial T}{\partial P} \cdot \dfrac{\partial P}{\partial V} = \dfrac{k}{P} \cdot \dfrac{T}{P} \cdot -\dfrac{P^2}{kT} = -1.$

55.   False. Let $f(x, y) = xy^{1/2}$. Then $f_x = y^{1/2}$ is defined at $(0, 0)$. But

$f_y = \dfrac{1}{2} xy^{-1/2} = \dfrac{x}{2y^{1/2}}$ is not defined at $(0, 0)$.

57.   True. See Section 8.2.

USING TECHNOLOGY EXERCISES 8.2, page 560

1.   1.3124; 0.4038   3. -1.8889; 0.7778   5. -0.3863; -0.8497

EXERCISES 8.3, page 570

1. $f(x, y) = 1 - 2x^2 - 3y^2$. To find the critical point(s) of f, we solve the system
$$\begin{cases} f_x = -4x = 0 \\ f_y = -6y = 0 \end{cases}$$
obtaining $(0,0)$ as the only critical point of f. Next,
$$f_{xx} = -4, f_{xy} = 0, \text{ and } f_{yy} = -6.$$
In particular, $f_{xx}(0,0) = -4, f_{xy}(0,0) = 0$, and $f_{yy}(0,0) = -6$, giving
$$D(0,0) = (-4)(-6) - 0^2 = 24 > 0.$$
Since $f_{xx}(0,0) < 0$, the Second Derivative Test implies that $(0,0)$ gives rise to a relative maximum of f. Finally, the relative maximum of f is $f(0,0) = 1$.

3.   To find the critical points of f, we solve the system

$$\begin{cases} f_x = \quad 2x-2=0 \\ f_y = -2y+4=0 \end{cases}$$

obtaining $x=1$ and $y=2$ so that $(1,2)$ is the only critical point.
$$f_{xx}=2, f_{xy}=0, \text{ and } f_{yy}=-2.$$
So $D(x,y)=f_{xx}f_{yy}-f_{xy}^2=-4$. In particular, $D(1,2)=-4<0$ and so
$(1,2)$ affords a saddle point of f and $f(1,2)=4$.

5. $f(x,y)=x^2+2xy+2y^2-4x+8y-1$. To find the critical point(s) of f, we solve
    the system $\quad \begin{cases} f_x = 2x+2y-4=0 \\ f_y = 2x+4y+8=0 \end{cases}$

    obtaining $(8,-6)$ as the critical point of f. Next, $f_{xx}=2, f_{xy}=2, f_{yy}=4$. In
    particular, $f_{xx}(8,-6)=2, f_{xy}(8,-6)=2, f_{yy}(8,-6)=4$, giving $D=2(4)-4=4>0$.
    Since $f_{xx}(8,-6)>0$, $(8,-6)$ gives rise to a relative minimum of f. Finally, the
    relative minimum value of f is $f(8,-6)=-41$.

7. $f(x,y)=2x^3+y^2-9x^2-4y+12x-2.$. To find the critical points of f, we solve
    the system
    $$\begin{cases} f_x = 6x^2-18x+12=0 \\ f_y = 2y-4=0 \end{cases}$$

    The first equation is equivalent to $x^2-3x+2=0$, or $(x-2)(x-1)=0$ which
    gives $x=1$ or 2. The second equation of the system gives $y=2$. Therefore, there
    are two critical points, $(1,2)$ and $(2,2)$. Next, we compute
    $\quad f_{xx}=12x-18=6(2x-3), f_{xy}=0, f_{yy}=2.$
    At the point $(1,2)$:
    $\quad f_{xx}(1,2)=6(2-3)=-6, f_{xy}(1,2)=0, \text{ and } f_{yy}(1,2)=2.$
    Therefore, $D=(-6)(2)-0=-12<0$ and we conclude that $(1,2)$ gives rise to a
    saddle point of f. At the point $(2,2)$:
    $\quad f_{xx}(2,2)=6(4-3)=6, f_{xy}(2,2)=0, \text{ and } f_{yy}(2,2)=2.$
    Therefore, $D=(6)(2)-0=12>0$. Since $f_{xx}(2,2)>0$, we see that $(2,2)$ gives rise
    to a relative minimum with value $f(2,2)=-2.$

9. To find the critical points of f, we solve the system
    $$\begin{cases} f_x = 3x^2-2y+7=0 \\ f_y = 2y-2x-8=0 \end{cases}$$

Adding the two equations gives $3x^2 - 2x - 1 = 0$, or $(3x+1)(x-1) = 0$. Therefore, $x = -1/3$ or 1. Substituting each of these values of x into the second equation gives $y = 8/3$ and $y = 5$, respectively. Therefore, $(-\frac{1}{3}, \frac{11}{3})$ and $(1,5)$ are critical points of f.

Next, $f_{xx} = 6x, f_{xy} = -2$, and $f_{yy} = 2$. So $D(x,y) = 12x - 4 = 4(3x-1)$. Then
$$D(-\tfrac{1}{3}, \tfrac{11}{3}) = 4(-1-1) = -8 < 0$$
and so $(-\frac{1}{3}, \frac{11}{3})$ gives a saddle point. Next, $D(1,5) = 4(3-1) = 8 > 0$ and since $f_{xx}(1,5) = 6 > 0$, we see that $(1,5)$ gives rise to a relative minimum.

11. To find the critical points of f, we solve the system
$$\begin{cases} f_x = 3x^2 - 3y = 0 \\ f_y = -3x + 3y^2 = 0 \end{cases}$$
The first equation gives $y = x^2$ which when substituted into the second equation gives $-3x + 3x^4 = 3x(x^3 - 1) = 0$. Therefore, $x = 0$ or 1. Substituting these values of x into the first equation gives $y = 0$ and $y = 1$, respectively. Therefore, $(0,0)$ and $(1,1)$ are critical points of f. Next, we find $f_{xx} = 6x, f_{xy} = -3$, and $f_{yy} = 6y$. So $D = f_{xx}f_{yy} - f_{xy}^2 = 36xy - 9$. Since $D(0,0) = -9 < 0$, we see that $(0,0)$ gives a saddle point of f. Next, $D(1,1) = 36 - 9 = 27 > 0$ and since $f_{xx}(1,1) = 6 > 0$, we see that $f(1,1) = -3$ is a relative minimum value of f.

13. Solving the system of equations
$$\begin{cases} f_x = y - \frac{4}{x^2} = 0 \\ f_y = x - \frac{2}{y^2} = 0 \end{cases}$$

we obtain $y = \frac{4}{x^2}$. Therefore, $x - 2\left(\frac{x^4}{16}\right) = 0$ and $8x - x^4 = x(8 - x^3) = 0$, and $x = 0$, or $x = 2$. Since $x = 0$ is not in the domain of f, $(2,1)$ is the only critical point of f. Next, $f_{xx} = \frac{8}{y^3}, f_{xy} = 1$, and $f_{yy} = \frac{4}{y^3}$. Therefore,

$D(2,1) = \frac{32}{x^3 y^3} - 1 \Big|_{(2,1)} = 4 - 1 = 3 > 0$ and $f_{xx}(2,1) = 1 > 0$. Therefore, the relative minimum value of f is $f(2,1) = 2 + 4/2 + 2/1 = 6$.

15. Solving the system of equations $f_x = 2x = 0$ and $f_y = -2ye^{y^2} = 0$, we obtain $x = 0$ and $y = 0$. Therefore, $(0,0)$ is the only critical point of f. Next,

$f_{xx} = 2, f_{xy} = 0, f_{yy} = -2e^{y^2} - 4y^2 e^{y^2}$.

Therefore, $D(0,0) = -4e^{y^2}(1+2y^2)\Big|_{(0,0)} = -4(1) < 0$, and we conclude that $(0,0)$ is a saddle point.

17. $f(x,y) = e^{x^2+y^2}$
Solving the system
$$\begin{cases} f_x = 2xe^{x^2+y^2} = 0 \\ f_y = 2ye^{x^2+y^2} = 0 \end{cases}$$
we see that $x = 0$ and $y = 0$ (recall that $e^{x^2+y^2} \neq 0$). Therefore, $(0,0)$ is the only critical point of f. Next, we compute
$$f_{xx} = 2e^{x^2+y^2} + 2x(2x)e^{x^2+y^2} = 2(1+2x^2)e^{x^2+y^2}$$
$$f_{xy} = 2x(2y)e^{x^2+y^2} = 4xye^{x^2+y^2}$$
$$f_{yy} = 2(1+2y^2)e^{x^2+y^2}.$$
In particular, at the point $(0,0)$, $f_{xx}(0,0) = 2$, $f_{xy}(0,0) = 0$, and $f_{yy}(0,0) = 2$.
Therefore, $D = (2)(2) - 0 = 4 > 0$. Furthermore, since $f_{xx}(0,0) > 0$, we conclude that $(0,0)$ gives rise to a relative minimum of f. The relative minimum value of f is $f(0,0) = 1$.

19. $f(x,y) = \ln(1+x^2+y^2)$. We solve the system of equations
$$f_x = \frac{2x}{1+x^2+y^2} = 0 \text{ and } f_y = \frac{2y}{1+x^2+y^2} = 0,$$
obtaining $x = 0$ and $y = 0$. Therefore, $(0,0)$ is the only critical point of f. Next,
$$f_{xx} = \frac{(1+x^2+y^2)2-(2x)(2x)}{(1+x^2+y^2)^2} = \frac{2+2y^2-2x^2}{(1+x^2+y^2)^2}$$
$$f_{yy} = \frac{(1+x^2+y^2)2-(2y)(2y)}{(1+x^2+y^2)^2} = \frac{2+2x^2-2y^2}{(1+x^2+y^2)^2}$$
$$f_{xy} = -2x(1+x^2+y^2)^{-2}(2y) = -\frac{4xy}{(1+x^2+y^2)^2}.$$
Therefore, $D(x,y) = \frac{(2+2y^2-2x^2)(2+2x^2-2y^2)}{(1+x^2+y^2)^4} - \frac{16x^2y^2}{(1+x^2+y^2)^4}$.
Since $D(0,0) = \frac{4}{1} > 0$ and $f_{xx}(0,0) = 2 > 0$, $f(0,0) = 0$ is a relative minimum value.

21. $P(x) = -0.2x^2 - 0.25y^2 - 0.2xy + 200x + 160y - 100x - 70y - 4000$
    $\qquad = -0.2x^2 - 0.25y^2 - 0.2xy + 100x + 90y - 4000.$

Then $\begin{cases} P_x = -0.4x - 0.2y + 100 = 0 \\ P_y = -0.5y - 0.2x + 90 = 0 \end{cases}$

implies that $\begin{cases} 4x + 2y = 1000 \\ 2x + 5y = 900 \end{cases}$. Solving, we find $x = 200$ and $y = 100$.

Next, $P_{xx} = -0.4, P_{yy} = -0.5, P_{xy} = -0.2$, and

$D(200,100) = (-0.4)(-0.5) - (-0.2)^2 > 0.$ Since $P_{xx}(200, 100) < 0$, we conclude
that $(200,100)$ is a relative maximum of P. Thus, the company should
manufacture 200 finished and 100 unfinished units per week. The maximum
profit is

$\qquad P(200,100) = -0.2(200)^2 - 0.25(100)^2 - 0.2(100)(200) + 100(200) + 90(100) - 4000$
$\qquad\qquad = 10,500$, or \$10,500 dollars.

23. $p(x,y) = 200 - 10(x - \frac{1}{2})^2 - 15(y - 1)^2$. Solving the system of equations
    $\begin{cases} p_x = -20(x - \frac{1}{2}) = 0 \\ p_y = -30(y - 1) = 0 \end{cases}$

we obtain $x = 1/2$, $y = 1$. We conclude that the only critical point of f is $(\frac{1}{2},1)$.

Next, $\qquad p_{xx} = -20, p_{xy} = 0, p_{yy} = -30$

so $\qquad D(\frac{1}{2},1) = (-20)(-30) = 600 > 0.$

Since $p_{xx} = -20 < 0$, we conclude that $f(\frac{1}{2},1)$ gives a relative maximum. So we
conclude that the price of land is highest at $(\frac{1}{2},1)$.

25. We want to minimize
    $f(x,y) = D^2 = (x - 5)^2 + (y - 2)^2 + (x + 4)^2 + (y - 4)^2 + (x + 1)^2 + (y + 3)^2.$

Next, $\begin{cases} f_x = 2(x - 5) + 2(x + 4) + 2(x + 1) = 6x = 0, \\ f_y = 2(y - 2) + 2(y - 4) + 2(y + 3) = 6y - 6 = 0 \end{cases}$

and we conclude that $x = 0$ and $y = 1$. Also,
    $\qquad f_{xx} = 6, f_{xy} = 0, f_{yy} = 6$ and $D(x,y) = (6)(6) = 36 > 0.$
Since $f_{xx} > 0$, we conclude that the function is minimized at $(0,1)$ and so $(0,1)$
gives the desired location.

27. Refer to the figure in the text.

$$xy + 2xz + 2yz = 300; \quad z(2x + 2y) = 300 - xy; \quad \text{and } z = \frac{300 - xy}{2(x+y)}.$$

Then the volume is given by

$$V = xyz = xy\left[\frac{300 - xy}{2(x+y)}\right] = \frac{300xy - x^2y^2}{2(x+y)}.$$

We find

$$\frac{\partial V}{\partial x} = \frac{1}{2}\frac{(x+y)(300y - 2xy^2) - (300xy - x^2y^2)}{(x+y)^2}$$

$$= \frac{300xy - 2x^2y^2 + 300y^2 - 2xy^3 - 300xy + x^2y^2}{2(x+y)^2}$$

$$= \frac{300y^2 - 2xy^3 - x^2y^2}{2(x+y)^2} = \frac{y^2(300 - 2xy - x^2)}{2(x+y)^2}.$$

Similarly, $\dfrac{\partial V}{\partial y} = \dfrac{x^2(300 - 2xy - y^2)}{2(x+y)^2}$. Setting both $\partial V / \partial x$ and $\partial V / \partial y$ equal t

zero and observing that both $x > 0$ and $y > 0$, we have the system

$\begin{cases} 2yx + x^2 = 300 \\ 2yx + y^2 = 300 \end{cases}$. Subtracting, we find $y^2 - x^2 = 0$; that is $(y - x)(y + x) = 0$. So

$y = x$ or $y = -x$. The latter is not possible since $x, y > 0$. Therefore, $y = x$.
Substituting this value into the first equation in the system gives

$$2x^2 + x^2 = 300; \quad x^2 = 100; \quad \text{and } x = 10.$$

Therefore, $y = 10$. Substituting this value into the expression for z

gives $z = \dfrac{300 - 10^2}{2(10 + 10)} = 5$. So the dimensions are $10'' \times 10'' \times 5''$. The volume is

500 cu in.

29. The heating cost is $C = 2xy + 8xz + 6yz$. But $xyz = 12{,}000$ or $z = 12{,}000/xy$.
Therefore,

$$C = f(x, y) = 2xy + 8x\left(\frac{12{,}000}{xy}\right) + 6y\left(\frac{12{,}000}{xy}\right) = 2xy + \frac{96{,}000}{y} + \frac{72{,}000}{x}.$$

To find the minimum of f, we find the critical point of f by solving the system

$$\begin{cases} f_x = 2y - \dfrac{72,000}{x^2} = 0 \\ f_y = 2x - \dfrac{96,000}{y^2} = 0 \end{cases}.$$

The first equation gives $y = 36000/x^2$, which when substituted into the second equation yields

$$2x - 96,000\left(\frac{x^2}{36,000}\right)^2 = 0, \ \ (36,000)^2 x - 48,000 x^4 = 0$$

$$x(27,000 - x^3) = 0.$$

Solving this equation, we have $x = 0$, or $x = 30$. We reject the first root because $x \neq 0$. With $x = 30$, we find $y = 40$ and

$$f_{xx} = \frac{144,000}{x^3}, \ f_{xy} = 2, \text{ and } f_{yy} = \frac{192,000}{y^3}.$$

In particular, $f_{xx}(30,40) = 5.33$, $f_{xy} = (30,40) = 2$, and $f_{yy}(30,40) = 3$. So
$$D(30,40) = (5.33)(3) - 4 = 11.99 > 0$$

and since $f_{xx}(30,40) > 0$, we see that $(30,40)$ gives a relative minimum. Physical considerations tell us that this is an absolute minimum. The minimal annual heating cost is

$$f(30,40) = 2(30)(40) + \frac{96,000}{40} + \frac{72,000}{30} = 7200, \text{ or } \$7,200.$$

31. False. Let $f(x,y) = xy$. Then $f_x(0,0) = 0$ and $f_y(0,0) = 0$. But $(0,0)$ does not afford a relative extremum of $(0,0)$. In fact, $f_{xx} = 0$, $f_{yy} = 0$, and $f_{xy} = 1$. Therefore, $D(x,y) = f_{xx}f_{yy} - f_{xy}^2 = -1$ and so $D(0,0) = -1$ which shows that $(0, 0, 0)$ is a saddle point.

EXERCISES 8.4, page 578

1. a. We first summarize the data:

x	y	x^2	xy
1	4	1	4
2	6	4	12
3	8	9	24
4	11	16	44
10	29	30	84

The normal equations are $4b + 10m = 29$
$$10b + 30m = 84.$$
Solving this system of equations, we obtain $m = 2.3$ and $b = 1.5$. So an equation is $y = 2.3x + 1.5$.

b. The scatter diagram and the least squares line for this data follow:

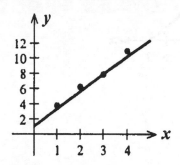

a. We first summarize the data:

x	y	x^2	xy
1	4.5	1	4.5
2	5	4	10
3	3	9	9
4	2	16	8
4	3.5	16	14
6	1	36	6
20	19	82	51.5

The normal equations are $6b + 20m = 19$

$$20b + 82m = 51.5.$$

The solutions are $m \approx -0.7717$ and $b \approx 5.7391$ and so a required equation is $y = -0.772x + 5.739$.

b. The scatter diagram and the least-squares line for these data follow.

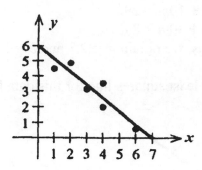

5. a. We first summarize the data:

x	y	x^2	xy
1	3	1	3
2	5	4	10
3	5	9	15
4	7	16	28
5	8	25	40
15	28	55	96

The normal equations are $55m + 15b = 96$
$$15m + 5b = 28.$$
Solving, we find $m = 1.2$ and $b = 2$, so that the required equation is $y = 1.2x + 2$.

b. The scatter diagram and the least-squares line for the given data follow.

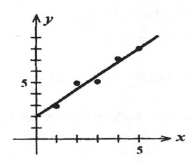

7.   a. We first summarize the data:

x	y	x^2	xy
4	0.5	16	2
4.5	0.6	20.25	2.7
5	0.8	25	4
5.5	0.9	30.25	4.95
6	1.2	36	7.2
25	4	127.5	20.85

The normal equations are
$$5b + 25m = 4$$
$$25b + 127.5m = 20.85.$$
The solutions are $m = 0.34$ and $b = -0.9$, and so a required equation is
$y = 0.34x - 0.9$.

b. The scatter diagram and the least-squares line for these data follow.

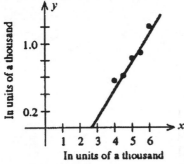

c. If $x = 6.4$, then $y = 0.34(6.4) - 0.9 = 1.276$ and so 1276 completed applications might be expected.

9.   a. We first summarize the data:

x	y	x^2	xy
1	436	1	436
2	438	4	876
3	428	9	1284
4	430	16	1720
5	426	25	2130
15	2158	55	6446

The normal equations are

$$5b + 15m = 2158$$
$$15b + 55m = 6446.$$

Solving this system, we find $m = -2.8$ and $b = 440$.
Thus, the equation of the least-squares line is $y = -2.8x + 440$.

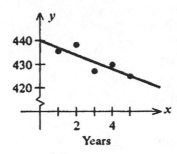

b. The scatter diagram and the least-squares line for this data are shown in the figure that follows.
c. Two years from now, the average SAT verbal score in that area will be
$y = -2.8(7) + 440 = 420.4$.

11. a. We first summarize the data:

x	y	x^2	xy
0	168	0	0
10	213	100	2130
20	297	400	5940
30	374	900	11220
40	427	1600	17080
57	471	3249	26847
157	1950	6249	63217

The normal equations are $\begin{cases} 6b + 157m = 1950 \\ 157b + 6249m = 63217 \end{cases}$.

The solutions are $m = 5.69$ and $b = 176$ and so a required equation is
$y = 5.69x + 176$.
b. In 1985, $x = 45$, $y = 6.226(45) + 176 \approx 432$. Hence, the expected size of the average farm will be 432 acres.

13. a. We first summarize the data:

x	y	x^2	xy
1	20	1	20
2	24	4	48
3	26	9	78
4	28	16	112
5	32	25	160
15	130	55	418

The normal equations are $5b + 15m = 130$
$$15b + 55m = 418.$$
The solutions are $m = 2.8$ and $b = 17.6$, and so an equation of the line is
$$y = 2.8x + 17.6.$$
b. When $x = 8$, $y = 2.8(8) + 17.6 = 40$. Hence, the state subsidy is expected to be $40 million for the eighth year.

15. a. We first summarize the data:

x	y	x^2	xy
1	16.7	1	16.7
3	26	9	78
5	33.3	25	166.5
7	48.3	49	338.1
9	57	81	513
11	65.8	121	723.8
13	74.2	169	964.6
15	83.3	225	1249.5
64	404.6	680	4050.2

The normal equations are $64b + 680m = 4050.2.$
$$8b + 64m = 404.6$$

The solutions are $m = 4.8417$ and $b = 11.8417$ and so a required equation is
$y = 4.842x + 11.842$.
b. In 1993, $x = 19$, and so $y = 4.842(19) + 11.842 = 103.84$. Hence the estimated number of cans produced in 1993 is 103.8 billion.

17. a. We first summarize the given data:

x	y	x^2	xy
0	16.3	0	0
1	21.0	1	21.0
2	25.0	4	50
3	28.8	9	86.4
4	32.7	16	130.8
10	123.8	30	288.2

The normal equations are
$$5b + 10m = 123.8$$
$$10b + 30m = 288.2$$
The solutions are $b = 16.6$ and $m = 4.1$. Therefore, $y = 4.1t + 16.6$
b. In 2004 $t = 3$, and $y = 4.1(3) + 16.6 = 28.9$, or \$28.9 billion.

19. a. We first summarize the given data:

x	y	x^2	xy
0	15.9	0	0
10	16.8	100	168
20	17.6	400	352
30	18.5	900	555
40	19.3	1600	772
50	20.3	2500	1015
150	108.4	5500	2862

The normal equations are
$$6b + 150m = 108.4$$
$$150b + 5500m = 2862$$
The solutions are $b = 15.90$ and $m = 0.09$. Therefore, $y = 0.09x + 15.9$
b. The life expectancy at 65 of a male in 2040 is
$$y = 0.9(40) + 15.9 = 19.50 \quad \text{or} \quad 19.50 \text{ years}$$
The datum gives a life expectancy of 19.5 years.
c. The life expectancy at 65 of a male in 2030 is
$$y = 0.09(30) + 15.9 = 18.6 \quad \text{or} \quad 18.6 \text{ years}$$

21. a. We first summarize the data:

x	y	x^2	xy
0	21.7	0	0
1	32.1	1	32.1
2	45.0	4	90
3	58.3	9	174.9
4	69.6	16	278.4
10	226.7	30	575.4

The normal equations are $\begin{cases} 5b + 10m = 226.7 \\ 10b + 30m = 575.4 \end{cases}$.

The solutions are $m = 12.2$ and $b = 20.9$ and so a required equation is $y = 12.2x + 20.9$.

b. In 2003, $x = 5$ so $y = 12.2(5) + 20.9 = 81.9$, or 81.9 million computers are expected to be connected to the internet in Europe in that year.

23. a. We first summarize the given data:

x	y	x^2	xy
0	90.4	0	0
1	100.0	1	100
2	110.4	4	220.8
3	120.4	9	361.2
4	130.8	16	523.2
5	140.4	25	702
6	150	36	900
21	842.4	91	2807.2

The normal equations are $\begin{cases} 7b + 21m = 842.4 \\ 21b + 91m = 2807.2 \end{cases}$.

The solutions are $m = 10$ and $b = 90.34$. Therefore, the required equation is $y = 10x + 90.34$.

b. If $x = 6$, then $y = 10(6) + 90.34 = 150.34$, or 150,340,000. This compares well with the actual data for that year-- 150,000,000 subscribers.

25. False. See Example 1, page 575.

27. True

USING TECHNOLOGY EXERCISES 8.4, page 581

1. $y = 2.3596x + 3.8639$ 3. $y = -1.1948x + 3.5525$

5. a. $y = 0.45x + 1.28$ b. \$3.08 billion

7. a. $y = 13.321x + 72.57$ b. 192 million tons

9. a. $1.95x + 12.19$; \$23.89 billion   11. a. $y = 3.76x + 87.46$ b. \$140,100

EXERCISES 8.5, page 594

1.   We form the Lagrangian function $F(x,y,\lambda) = x^2 + 3y^2 + \lambda(x + y - 1)$. We solve the
   system
   $$\begin{cases} F_x = 2x + \lambda = 0 \\ F_y = 6y + \lambda = 0 \\ F_\lambda = x + y - 1 = 0. \end{cases}$$
   Solving the first and the second equations for x and y in terms of λ we obtain
   $x = -\frac{\lambda}{2}$ and $y = -\frac{\lambda}{6}$ which, upon substitution into the third equation, yields
   $-\frac{\lambda}{2} - \frac{\lambda}{6} - 1 = 0$ or $\lambda = -\frac{3}{2}$. Therefore, $x = \frac{3}{4}$ and $y = \frac{1}{4}$ which gives the point
   $(\frac{3}{4}, \frac{1}{4})$ as the sole critical point of F. Therefore, $(\frac{3}{4}, \frac{1}{4}) = \frac{3}{4}$ is a minimum of F.

3.   We form the Lagrangian function $F(x,y,\lambda) = 2x + 3y - x^2 - y^2 + \lambda(x + 2y - 9)$. We
   then solve the system
   $$\begin{cases} F_x = 2 - 2x + \lambda = 0 \\ F_y = 3 - 2y + 2\lambda = 0. \\ F_\lambda = x + 2y - 9 = 0 \end{cases}$$
   Solving the first equation λ, we obtain $\lambda = 2x - 2$. Substituting into the second
   equation, we have $3 - 2y + 4x - 4 = 0$, or $4x - 2y - 1 = 0$.   Adding this
   equation to the third equation in the system, we have $5x - 10 = 0$, or $x = 2$.
   Therefore, $y = 7/2$ and $f(2, \frac{7}{2}) = -\frac{7}{4}$ is the maximum value of f.

5. Form the Lagrangian function $F(x,y,\lambda) = x^2 + 4y^2 + \lambda(xy - 1)$. We then solve the

system $\begin{cases} F_x = 2x + \lambda y = 0 \\ F_y = 8y + \lambda x = 0. \\ F_\lambda = xy - 1 = 0 \end{cases}$

Multiplying the first and second equations by x and y, respectively, and subtracting the resulting equations, we obtain $2x^2 - 8y^2 = 0$, or $x = \pm 2y$. Substituting this into the third equation gives $2y^2 - 1 = 0$ or $y = \pm\frac{\sqrt{2}}{2}$. We conclude that $f(-\sqrt{2}, -\frac{\sqrt{2}}{2}) = f(\sqrt{2}, \frac{\sqrt{2}}{2}) = 4$ is the minimum value of f.

7. We form the Lagrangian function
$$F(x,y,\lambda) = x + 5y - 2xy - x^2 - 2y^2 + \lambda(2x + y - 4).$$
Next, we solve the system
$$\begin{cases} F_x = 1 - 2y - 2x + 2\lambda = 0 \\ F_y = 5 - 2x - 4y + \lambda = 0 \\ F_\lambda = 2x + y - 4 = 0 \end{cases}$$
Solving the last two equations for x and y in terms of λ, we obtain
$$y = \tfrac{1}{3}(1 + \lambda) \text{ and } x = \tfrac{1}{6}(11 - \lambda)$$
which, upon substitution into the first equation, yields
$$1 - \tfrac{2}{3}(1 + \lambda) - \tfrac{1}{3}(11 - \lambda) + 2\lambda = 0$$
or $1 - \tfrac{2}{3} - \tfrac{2}{3}\lambda - \tfrac{11}{3} + \tfrac{\lambda}{3} + 2\lambda = 0$
or $\lambda = 2$. Therefore, $x = 3/2$ and $y = 1$. The maximum of f is
$$f(\tfrac{3}{2}, 1) = \tfrac{3}{2} + 5 - 2(\tfrac{3}{2}) - (\tfrac{3}{2})^2 - 2 = -\tfrac{3}{4}.$$

9. Form the Lagrangian $F(x,y,\lambda) = xy^2 + \lambda(9x^2 + y^2 - 9)$. We then solve
$$\begin{cases} F_x = y^2 + 18\lambda x = 0 \\ F_y = 2xy + 2\lambda y = 0 \\ F_\lambda = 9x^2 + y^2 - 9 = 0. \end{cases}$$

The first equation gives $\lambda = -\dfrac{y^2}{18x}$. Substituting into the second gives
$$2xy + 2y\left(-\dfrac{y^2}{18x}\right) = 0, \text{ or } 18x^2y - y^3 = y(18x^2 - y^2) = 0,$$

giving $y = 0$ or $y = \pm 3\sqrt{2}x$. If $y = 0$, then the third equation gives $9x^2 - 9 = 0$ or $x = \pm 1$. If $y = \pm 3\sqrt{3}/3$. Therefore, the points $(-1,0), (-\sqrt{3}/3, -\sqrt{6})$, $(-\sqrt{3}/3, \sqrt{6}), (\sqrt{3}/3, -\sqrt{6})$ and $(\sqrt{3}/3, \sqrt{6})$ give rise to extreme values of f subject to the given constraint. Evaluating $f(x,y)$ at each of these points, we see that $f(\sqrt{3}/3, -\sqrt{6}) = (\sqrt{3}/3, \sqrt{6}) = 2\sqrt{3}$ is the maximum value of f.

11. We form the Lagrangian function $F(x,y,\lambda) = xy + \lambda(x^2 + y^2 - 16)$. To find the critical points of F, we solve the system

$$\begin{cases} F_x = y + 2\lambda x = 0 \\ F_y = x + 2\lambda y = 0 \\ F_\lambda = x^2 + y^2 - 16 = 0 \end{cases}$$

Solving the first equation for λ and substituting this value into the second

equation yields $x - 2\left(\dfrac{y}{2x}\right)y = 0$, or $x^2 = y^2$. Substituting the last equation into

the third equation in the system, yields $x^2 + x^2 - 16 = 0$, or $x^2 = 8$, that is, $x = \pm 2\sqrt{2}$. The corresponding values of y are $y = \pm 2\sqrt{2}$. Therefore the critical points of F are $(-2\sqrt{2}, -2\sqrt{2}), (-2\sqrt{2}, 2\sqrt{2}), (2\sqrt{2}, -2\sqrt{2})(2\sqrt{2}, 2\sqrt{2})$. Evaluating f at each of these values, we find that $f(-2\sqrt{2}, 2\sqrt{2}) = -8$ and $f(2\sqrt{2}, -2\sqrt{2}) = -8$ are relative minimum values and $f(-2\sqrt{2}, -2\sqrt{2}) = 8$ and $f(2\sqrt{2}, 2\sqrt{2}) = 8$, are relative maximum values.

13. We form the Lagrangian function $F(x,y,\lambda) = xy^2 + \lambda(x^2 + y^2 - 1)$. Next, we solve the system

$$\begin{cases} F_x = y^2 + 2x\lambda = 0 \\ F_y = 2xy + 2y\lambda = 0 \\ F_\lambda = x^2 + y^2 - 1 = 0 \end{cases}.$$

We find that $x = \pm\sqrt{3}/3$ and $y = \pm\sqrt{6}/3$ and $x = \pm 1$, $y = 0$. Evaluating f at each of the critical points $(-\frac{\sqrt{3}}{3}, -\frac{\sqrt{6}}{3}), (-\frac{\sqrt{3}}{3}, \frac{\sqrt{6}}{3})(\frac{\sqrt{3}}{3}, -\frac{\sqrt{6}}{3})(\frac{\sqrt{3}}{3}, \frac{\sqrt{6}}{3}), (-1,0)$, and $(1,0)$, we find that $f(-\frac{\sqrt{3}}{3}, -\frac{\sqrt{6}}{3}) = -\frac{2\sqrt{3}}{9}$ and $f(-\frac{\sqrt{3}}{3}, \frac{\sqrt{6}}{3}) = -\frac{2\sqrt{3}}{9}$ are relative minimum values and $f(\frac{\sqrt{3}}{3}, -\frac{\sqrt{6}}{3}) = \frac{2\sqrt{3}}{9}$ and $f(\frac{\sqrt{3}}{3}, \frac{\sqrt{6}}{3}) = \frac{2\sqrt{3}}{9}$ are relative maximum values.

15. Form the Lagrangian function $F(x,y,z,\lambda) = x^2 + y^2 + z^2 + \lambda(3x + 2y + z - 6)$. We solve the system

$$\begin{cases} F_x = 2x + 3\lambda = 0 \\ F_y = 2y + 2\lambda = 0 \\ F_z = 2x + \lambda = 0 \\ F_\lambda = 3x + 2y + z - 6 = 0. \end{cases}$$

The third equation give $\lambda = -2z$. Substituting into the first two equations gives

$$\begin{cases} 2x - 6z = 0 \\ 2y - 4z = 0. \end{cases}$$

So $x = 3z$ and $y = 2z$. Substituting into the third equation yields
$9z + 4z + z - 6 = 0$, or $z = 3/7$. Therefore, $x = 9/7$ and $y = 6/7$. Therefore,
$f(\frac{9}{7}, \frac{6}{7}, \frac{3}{7}) = \frac{18}{7}$ is the minimum value of F.

17. We want to maximize P subject to the constraint $x + y = 200$. The Lagrangian function is

$$F(x, y, \lambda) = -0.2x^2 - 0.25y^2 - 0.2xy + 100x + 90y - 4000 + \lambda(x + y - 200).$$

Next, we solve

$$\begin{cases} F_x = -0.4x - 0.2y + 100 + \lambda = 0 \\ F_y = -0.5y - 0.2x + 90 + \lambda = 0 \\ F_\lambda = x + y - 200 = 0. \end{cases}$$

Subtracting the first equation from the second yields
$0.2x - 0.3y - 10 = 0$, or $2x - 3y - 100 = 0$.
Multiplying the third equation in the system by 2 and subtracting the resulting equation from the last equation, we find $-5y + 300 = 0$ or $y = 60$. So $x = 140$ and the company should make 140 finished and 60 unfinished units.

19. Suppose each of the sides made of pine board is x feet long and those of steel are y feet long. Then $xy = 800$. The cost is $C = 12x + 3y$ and is to be minimized subject to the condition $xy = 800$. We form the Lagrangian function
$$F(x,y,\lambda) = 12x + 3y + \lambda(xy - 800).$$
We solve the system

$$\begin{cases} F_x = 12 + \lambda y = 0 \\ F_y = 3 + \lambda x = 0 \\ F_\lambda = xy - 800 = 0. \end{cases}$$

Multiplying the first equation by x and the second equation by y and subtracting the resulting equations, we obtain $12x - 3y = 0$, or $y = 4x$. Substituting this into the third equation of the system, we obtain

$$4x^2 - 800 = 0, \text{ or } x = \pm\sqrt{200} = \pm10\sqrt{2}.$$

Since x must be positive, we take $x = 10\sqrt{2}$. So $y = 40\sqrt{2}$. So the dimensions are approximately 14.14ft by 56.56 ft.

21. We want to minimize the function $C(r,h)$ subject to the constraint $\pi r^2 h - 64 = 0$. We form the Lagrangian function $F(r,h,\lambda) = 8\pi rh + 6\pi r^2 - \lambda(\pi r^2 h - 64)$. Then we solve the system

$$\begin{cases} F_r = 8\pi h + 12\pi r - 2\lambda \pi rh = 0 \\ F_h = 8\pi r - \lambda r^2 = 0 \\ F_\lambda = \pi r^2 h - 64 = 0 \end{cases}$$

Solving the second equation for λ yields $\lambda = 8/r$, which when substituted into the first equation yields

$$8\pi h + 12\pi r - 2\pi rh(\tfrac{8}{r}) = 0$$
$$12\pi r = 8\pi h$$
$$h = \tfrac{3r}{2}.$$

Substituting this value of h into the third equation of the system, we find

$$3r^2(\tfrac{3r}{2}) = 64, \quad r^3 = \frac{128}{3\pi}, \text{ or } r = \frac{4}{3}\sqrt[3]{\frac{18}{\pi}} \text{ and } h = 2\sqrt[3]{\frac{18}{\pi}}.$$

23. Let the box have dimensions x' by y' by z'. Then $xyz = 4$. We want to minimize
$$C = 2xz + 2yz + \tfrac{3}{2}(2xy) = 2xz + 2yz + 3xy.$$
Form the Lagrangian function
$$F(x,y,z,\lambda) = 2xz + 2yz + 3xy + \lambda(xyz - 4).$$
Next, we solve the system

$$\begin{cases} F_x = 2z + 3y + \lambda yz = 0 \\ F_y = 2z + 3x + \lambda xz = 0 \\ F_z = 2x + 2y + \lambda xy = 0 \\ F_\lambda = xyz - 4 = 0. \end{cases}$$

Multiplying the first, second, and third equations by x, y, and z, respectively, we have

$$\begin{cases} 2xz + 3xy + \lambda xyz = 0 \\ 2yz + 3xy + \lambda xyz = 0 \\ 2xz + 2yz + \lambda xyz = 0. \end{cases}$$

The first two equations imply that $2z(x - y) = 0$. Since $z \neq 0$, we see that $x = y$. The second and third equations imply that $x(3y - 2z) = 0$ or $x = (3/2)y$. Substituting these values into the fourth equation in the system, we find

$$y^2 \left(\tfrac{3}{2}y\right) = 4 \quad \text{or} \quad y^3 = \tfrac{8}{3}.$$

Therefore, $y = \dfrac{2}{3^{1/3}} = \dfrac{2}{3}\sqrt[3]{9}$ and $x = \dfrac{2}{3}\sqrt[3]{9}$, and $z = \sqrt[3]{9}$.

So the dimensions are $\dfrac{2}{3}\sqrt[3]{9} \times \dfrac{2}{3}\sqrt[3]{9} \times \sqrt[3]{9}$.

25. We want to maximize $f(x,y) = 100x^{3/4}y^{1/4}$ subject to $100x + 200y = 200,000$.
Form the Lagrangian function
$$F(x,y,\lambda) = 100x^{3/4}y^{1/4} + \lambda(100x + 200y - 200,000).$$
We solve the system

$$\begin{cases} F_x = 75x^{-1/4}y^{1/4} + 100\lambda = 0 \\ F_y = 25x^{3/4}y^{-3/4} + 200\lambda = 0 \\ F_\lambda = 100x + 200y - 200,000 = 0. \end{cases}$$

The first two equations imply that $150x^{-1/4}y^{1/4} - 25x^{3/4}y^{-3/4} = 0$ or upon multiplying by $x^{1/4}y^{3/4}$, $150y - 25x = 0$, which implies that $x = 6y$. Substituting this value of x into the third equation of the system, we have
$$600y + 200y - 200,000 = 0$$
giving $y = 250$ and therefore $x = 1500$. So to maximize production, he should spend 1500 units on labor and 250 units of capital.

27. False. See Example 1, Section 8.5.

EXERCISES 8.6, page 600

1. $f(x, y) = x^2 + 2y$; $df = 2x\,dx + 2\,dy$

3. $f(x, y) = 2x^2 - 3xy + 4x$; $df = (4x - 3y + 4)dx - 3x\,dy$

5. $f(x,y)=\sqrt{x^2+y^2}$; $df=\frac{1}{2}(x^2+y^2)^{-1/2}(2x)\,dx+\frac{1}{2}(x^2+y^2)^{-1/2}(2y)\,dy$

$$=\frac{x}{\sqrt{x^2+y^2}}\,dx+\frac{y}{\sqrt{x^2+y^2}}\,dy.$$

7. $f(x,y)=\dfrac{5y}{x-y}$;

$$df=\frac{\partial}{\partial x}[5y(x-y)^{-1}]dx+\frac{\partial}{\partial x}\left(\frac{5y}{x-y}\right)dy$$

$$=5y(-1)(x-y)^{-2}\,dx+\frac{(x-y)(5)-5y(-1)}{(x-y)^2}\,dy$$

$$=-\frac{5y}{(x-y)^2}\,dx+\frac{5x}{(x-y)^2}\,dy.$$

9. $f(x,y)=2x^5-ye^{-3x}$; $df=(10x^4+3ye^{-3x})dx-e^{-3x}dy.$

11. $f(x,y)=x^2e^y+y\ln x$; $df=(2xe^y+\dfrac{y}{x})dx+(x^2e^y+\ln x)dy.$

13. $f(x,y,z)=xy^2z^3$; $df=y^2z^3\,dx+2xyz^3\,dy+3xy^2z^2\,dz.$

15. $f(x,y,z)=\dfrac{x}{y+z}$; $df=\dfrac{1}{y+z}\,dx+x(-1)(y+z)^{-2}\,dy+x(-1)(y+z)^{-2}\,dz$

$$=\frac{1}{y+z}\,dx-\frac{x}{(y+z)^2}\,dy-\frac{x}{(y+z)^2}\,dz.$$

17. $f(x,y,z)=xyz+xe^{yz}$; $df=(yz+e^{yz})dx+(xz+xze^{yz})dy+(xy+xye^{yz})dz$

$$=(yz+e^{yz})dx+xz(1+e^{yz})dy+xy(1+e^{yz})dz.$$

19. $\Delta z\approx dz=(8x-y)dx-x\,dy\big|_{\substack{x=1,\,y=2\\ dx=0.01,\,dy=0.02}}\qquad =(8-2)(0.01)-0.02=0.04.$

21. $\Delta z\approx dz=\frac{2}{3}x^{-1/3}y^{1/2}dx+\frac{1}{2}x^{2/3}y^{-1/2}\,dy\big|_{\substack{x=8,\,y=9\\ dx=-0.03,\,dy=0.03}}$

$$= \tfrac{2}{3}(\tfrac{3}{2})(-0.03) + \tfrac{1}{2}(\tfrac{4}{3})(0.03) = -0.03 + 0.02 = -0.01.$$

23. $f_x = \dfrac{(x-y)-x(1)}{(x-y)^2} = -\dfrac{y}{(x-y)^2}, \quad f_y = \dfrac{(x-y)(0)-x(-1)}{(x-y)^2} = \dfrac{x}{(x-y)^2}.$

Therefore, $\Delta z \approx dz = \dfrac{-y\,dx + x\,dy}{(x-y)^2}$. Here $x = -3$, $y = -2$, $dx = -0.02$, and $dy = 0.02$.

So the approximate change in

$z = f(x,y)$ is $\Delta z \approx \dfrac{-(-2)(-0.02) + (-3)(0.02)}{(-1)^2} = -0.1.$

25. $\Delta z \approx dz = 2e^{-y}\,dx - 2xe^{-y}\,dy.$ With $x = 4$, $y = 0$, $dx = 0.03$ and $dy = 0.03$, we have
$$\Delta z \approx 2e^0(0.03) - 2(4)(e^0)(0.03) = -0.18.$$

27. $f(x,y) = xe^{xy} - y^2;\quad f_x = xye^{xy} + e^{xy} = e^{xy}(1+xy);\quad f_y = x^2 e^{xy} - 2y;$

and $\Delta z \approx dz = e^{xy}(1+xy)dx + (x^2 e^{xy} - 2y)dy.$ With
$x = -1$, $y = 0$, $dx = 0.03$, and $y = 0.03$, we have
$$\Delta z \approx e^0(1)(0.03) + (1)(e^0)(0.03) = 0.06.$$

29. $f(x,y) = x\ln x + y\ln x;\quad f_x = \ln x + x\left(\dfrac{1}{x}\right) + \dfrac{y}{x} = \ln x + 1 + \dfrac{y}{x},\ f_y = \ln x.$

So $\Delta z \approx dz = \left(\ln x + 1 + \dfrac{y}{x}\right)dx + \ln x\,dy.$ Using $x = 2$, $y = 3$, $dx = -0.02$, and

$dy = -0.11$, we have $\Delta z \approx (\ln 2 + 1 + 1.5)(-0.02) + (\ln 2)(-0.11) \approx -0.1401.$

31. $\Delta P \approx dP = (-0.04x + y + 39)dx + (-30y + x + 25)dy.$
With $x = 4000$, $y = 150$, $dx = 500$, and $dy = -10$, we have
$$\Delta P \approx [-0.04(4000) + 150 + 39](500) + [-30(150) + 4000 + 25](-10)$$
$$= 19{,}250 \text{ or } \$19{,}250 \text{ per month.}$$

33. $R(x,y) = -x^2 - 0.5y^2 + xy + 8x + 3y + 20$
$dR = -2x\,dx - y\,dy + y\,dx + x\,dy + 8dx + 3dy$
$\quad = (-2x + y + 8)\,dx = (-y + x + 3)\,dy.$
If $x = 10$, $y = 15$, $dx = 1$, and $dy = -1$,

$$dR = [-2(10)+15+8](1)+(-15+10+3)(-1) = 3(1)+(-2)(-1) = 5.$$

So the revenue is expected to increase by $5000 per month

35. $R = \dfrac{x}{y}; \ \Delta R \approx dR = \dfrac{1}{y}dx - \dfrac{x}{y^2}dy$

Therefore,

$$\Delta R = \frac{1}{4}(2) - \frac{60}{4^2}(-0.2) = \frac{1}{2} + \frac{15}{4}\left(\frac{1}{5}\right)$$

$$= \frac{1}{2} + \frac{3}{4} = \frac{5}{4} = 1.25 , \ \text{or } \$1.25.$$

37. $V = \pi r^2 h; \ \Delta V \approx dV = 2\pi r h\, dr + \pi r^2\, dh.$

If $r = 8$, $h = 20$, $dr = (\pm 0.1)$, then

$$dV = 2\pi(8)(20)(\pm 0.1) + \pi(8)^2(\pm 0.1) = 320\pi(\pm 0.1) + 64\pi(\pm 0.1) = \pm 38.4 \ \pi \ \text{cc} .$$

39. Differentiating implicitly, we have

$$-\frac{1}{R^2}dR = -\frac{1}{R_1^2}dR_1 - \frac{1}{R_2^2}dR_2 - \frac{1}{R_3^2}dR_3$$

or $|dR| \le \left(\dfrac{R}{R_1}\right)^2 |dR_1| + \left(\dfrac{R}{R_2}\right)^2 |dR_2| + \left(\dfrac{R}{R_3}\right)^2 |dR_3|.$

With $R_1 = 100$, $R_2 = 200$, $R_3 = 300$, $|dR_1| \le 1$, $|dR_2| \le 2$, and $|dR_3| \le 3$, we have

$$\frac{1}{R} = \frac{1}{100} + \frac{1}{200} + \frac{1}{300} = \frac{6+3+2}{600} = \frac{11}{600}.$$

$$|dR| \le \left(\frac{6}{11}\right)^2 (1) + \left(\frac{3}{11}\right)^2 (2) + \left(\frac{2}{11}\right)^2 (3) = 0.5455, \ \text{or } 0.55 \ \text{ohms}.$$

EXERCISES 8.7, page 608

1. $\displaystyle\int_1^2 \int_0^1 (y+2x)\,dy\,dx = \int_1^2 \frac{1}{2}y^2 + 2xy\Big|_{y=0}^{y=1} dx = \int_1^2 \left(\frac{1}{2}+2x\right) dx = \frac{1}{2}x + x^2\Big|_1^2 = 5 - \frac{3}{2} = \frac{7}{2}.$

3. $\displaystyle\int_{-1}^1 \int_0^1 xy^2\,dy\,dx = \int_{-1}^1 \frac{1}{3}xy^3\Big|_0^1 dx = \int_{-1}^1 \frac{1}{3}x\,dx = \frac{x^2}{6}\Big|_{-1}^1 = \frac{1}{6} - \left(\frac{1}{6}\right) = 0 .$

5. $\int_{-1}^{2}\int_{1}^{e^3}\dfrac{x}{y}\,dy\,dx = \int_{-1}^{2}x\ln y\Big|_{1}^{e^3}\,dx = \int_{-1}^{2}x\ln e^3\,dx = \int_{-1}^{2}3x\,dx = \tfrac{3}{2}x^2\Big|_{-1}^{2} = \tfrac{3}{2}(4)-\tfrac{3}{2}(1)=\tfrac{9}{2}.$

7. $\int_{-2}^{0}\int_{0}^{1}4xe^{2x^2+y}\,dx\,dy = \int_{-2}^{0}e^{2x^2+y}\Big|_{x=0}^{x=1}\,dy = \int_{-2}^{0}(e^{2+y}-e^{y})\,dy = (e^{2+y}-e^{y})\Big|_{-2}^{0}$
$$= [(e^2-1)-(e^0-e^{-2}) = e^2 - 2 + e^{-2} = (e^2-1)(1-e^{-2}).$$

9. $\int_{0}^{1}\int_{1}^{e}\ln y\,dy\,dx = \int_{0}^{1}y\ln y - y\Big|_{y=1}^{y=e}\,dx = \int_{0}^{1}dx = 1.$

11. $\int_{0}^{1}\int_{0}^{x}(x+2y)\,dy\,dx = \int_{0}^{1}(xy+y^2)\Big|_{y=0}^{y=x}\,dx = \int_{0}^{1}2x^2\,dx = \tfrac{2}{3}x^3\Big|_{0}^{1} = \tfrac{2}{3}.$

13. $\int_{1}^{3}\int_{0}^{x+1}(2x+4y)\,dy\,dx = \int_{1}^{3}2xy+2y^2\Big|_{y=0}^{y=x+1}\,dx = \int_{1}^{3}[2x(x+1)+2(x+1)^2]\,dx$
$$= \int_{1}^{3}(4x^2+6x+2)\,dx = (\tfrac{4}{3}x^3+3x^2+2x)\Big|_{1}^{3}$$
$$= (36+27+6)-(\tfrac{4}{3}+3+2) = \tfrac{188}{3}.$$

15. $\int_{0}^{4}\int_{0}^{\sqrt{y}}(x+y)\,dx\,dy = \int_{0}^{4}\tfrac{1}{2}x^2+xy\Big|_{x=0}^{x=\sqrt{y}}\,dy = \int_{0}^{4}(\tfrac{1}{2}y+y^{3/2})\,dy$
$$= (\tfrac{1}{4}y^2+\tfrac{2}{5}y^{5/2})\Big|_{0}^{4} = 4+\tfrac{64}{5} = \tfrac{84}{5}.$$

17. $\int_{0}^{2}\int_{0}^{\sqrt{4-y^2}}y\,dx\,dy = \int_{0}^{2}xy\Big|_{0}^{\sqrt{4-y^2}}\,dy = \int_{0}^{2}y\sqrt{4-y^2}\,dy = -\tfrac{1}{2}(\tfrac{2}{3})(4-y^2)^{3/2}\Big|_{0}^{2}$
$$= \tfrac{1}{3}(4^{3/2}) = \tfrac{8}{3}.$$

19. $\int_{0}^{1}\int_{0}^{x}2xe^{y}\,dy\,dx = \int_{0}^{1}2xe^{y}\Big|_{y=0}^{y=x}\,dx = \int_{0}^{1}(2xe^x-2x)\,dx = 2(x-1)e^x - x^2\Big|_{0}^{1} = (-1)+2 = 1.$

21. $\int_{0}^{1}\int_{x}^{\sqrt{x}}ye^x\,dy\,dx = -\int_{0}^{1}\int_{x}^{\sqrt{x}}ye^x\,dy\,dx = \int_{0}^{1}-\tfrac{1}{2}y^2e^x\Big|_{y=\sqrt{x}}^{y=x}\,dx = -\tfrac{1}{2}\int_{0}^{1}(x^2e^x-xe^x)\,dx$
$$= -\tfrac{1}{2}[x^2e^x\Big|_{0}^{1} - 2\int_{0}^{1}xe^x\,dx - \int_{0}^{1}xe^x\,dx] = -\tfrac{1}{2}[x^2e^x\Big|_{0}^{1} - 3\int_{0}^{1}xe^x\,dx]$$
$$= -\tfrac{1}{2}[x^2e^x - 3xe^x + 3e^x]\Big|_{0}^{1} = -\tfrac{1}{2}[e-3e+3e-3] = \tfrac{1}{2}(3-e).$$

23. $\displaystyle\int_0^1\int_{2x}^2 e^{y^2}\,dy\,dx = \int_0^2\int_0^{y/2} e^{y^2}\,dx\,dy = \int_0^2 xe^{y^2}\Big|_{x=0}^{x=y/2}\,dy = \int_0^2 \tfrac{1}{2}ye^{y^2}\,dy$

$\qquad\qquad = \tfrac{1}{4}e^{y^2}\Big|_0^2 = \tfrac{1}{4}(e^4-1).$

25. $\displaystyle\int_0^2\int_{y/2}^1 ye^{x^3}\,dx\,dy = \int_0^1\int_0^{2x} ye^{x^3}\,dy\,dx = \int_0^1 \tfrac{1}{2}y^2 e^{x^3}\Big|_{y=0}^{y=2x}\,dx = \int_0^1 2x^2 e^{x^3}\,dx$

$\qquad\qquad = \tfrac{2}{3}e^{x^3}\Big|_0^1 = \tfrac{2}{3}(e-1).$

27. False. Let $f(x,y) = \dfrac{x}{y-2}$, $a=0$, $b=3$, $c=0$, and $d=1$. Then $\displaystyle\iint_{R_1} f(x,y)\,dA$ is

defined on $R_1 = \{(x,y)\mid 0\le x\le 3,\, 0\le y\le 1\}$. But $\displaystyle\iint_{R_2} f(x,y)\,dA$ is not defined on

$R_2 = \{(x,y)\mid 0\le x\le 1,\, 0\le y\le 3\}$, because f is discontinuous on R_2 (where $y=2$).

EXERCISES 8.8, page 614

1. $\displaystyle V = \int_0^4\int_0^3 (4-x+\tfrac{1}{2}y)\,dx\,dy = \int_0^4 (4x-\tfrac{1}{2}x^2+\tfrac{1}{2}xy)\Big|_{x=0}^{x=3}\,dy$

$\qquad = \int_0^4 (\tfrac{15}{2}+\tfrac{3}{2}y)\,dy = \tfrac{15}{2}y+\tfrac{3}{4}y^2\Big|_0^4 = 42,$ or 42 cu units.

3. $\displaystyle V = \int_0^2\int_x^{4-x} 5\,dy\,dx = \int_0^2 5y\Big|_{y=x}^{y=4-x}\,dx = 5\int_0^2 (4-2x)\,dx = 5(4x-x^2)\Big|_0^2 = 20$

or 20 cu units.

5. $\displaystyle V = \int_0^2\int_0^{4-x^2} 4\,dy\,dx = \int_0^2 4y\Big|_{y=0}^{y=4-x^2}\,dx = \int_0^2 4(4-x^2)\,dx = 4\int_0^2 (4-x^2)\,dx$

$\qquad = 4(4x-\tfrac{1}{3}x^3)\Big|_0^2 = \tfrac{64}{3},$ or $21\tfrac{1}{3}$ cu units.

7. $\displaystyle V = \int_0^1\int_0^y \sqrt{1-y^2}\,dx\,dy = \int_0^1 x\sqrt{1-y^2}\Big|_{x=0}^{x=y}\,dy = \int_0^1 y\sqrt{1-y^2}\,dy$

$\qquad\qquad = (-\tfrac{1}{2})(\tfrac{2}{3})(1-y^2)^{3/2}\Big|_0^1 = \tfrac{1}{3},$ or $\tfrac{1}{3}$ cu units.

9. $V = \int_0^1 \int_0^2 (4-2x-y)\,dy\,dx = \int_0^1 [4y-2xy-\tfrac{1}{2}y^2]\Big|_{y=0}^{y=2}\,dx$

$= \int_0^1 (8-4x-2)\,dx = \int_0^1 (6-4x)\,dx = 6x-2x^2\Big|_0^1 = 6-2 = 4$ cu units.

11. $V = \int_0^1 \int_0^2 (x^2+y^2)\,dy\,dx = \int_0^1 x^2 y + \tfrac{1}{3}y^3\Big|_{y=0}^{y=2}\,dx = \int_0^1 (2x^2+\tfrac{8}{3})\,dx$

$= \int_0^1 (\tfrac{2}{3}x^2+\tfrac{8}{3})\,dx = (\tfrac{2}{3}x^3+\tfrac{8}{3}x)\Big|_0^1 = \tfrac{2}{3}+\tfrac{8}{3} = \tfrac{10}{3}$ cu units.

13. $V = \int_0^2 \int_x^2 2xe^y\,dy\,dx = \int_0^2 2xe^y\Big|_{y=x}^{y=2}\,dx = \int_0^2 (2xe^2-2xe^x)\,dx$

$= e^2 x^2 - 2(x-1)e^x\Big|_0^2$ (Integrating by parts.)

$= 4e^2 - 2e^2 - 2 = 2(e^2 - 1)$ cu units.

15. $V = \int_0^1 \int_{x^2}^{x} 2x^2 y\,dy\,dx = \int_0^1 x^2 y^2\Big|_{y=x^2}^{y=x}\,dx = \int_0^1 (x^4-x^6)\,dx = \tfrac{1}{5}-\tfrac{1}{7} = \tfrac{2}{35}$, or $\tfrac{2}{35}$ cu units.

17. $A = \tfrac{1}{6}\int_0^3 \int_0^2 6x^2 y^3\,dx\,dy = \int_0^3 \tfrac{1}{3}x^3 y^3\Big|_0^2\,dy = \tfrac{8}{3}\int_0^3 y^3\,dy = \tfrac{2}{3}y^4\Big|_0^3 = 54.$

19. The area of R is $\tfrac{1}{2}(2)(1) = 1.$ Therefore, the average value of f is

$\int_0^1 \int_y^{2-y} xy\,dx\,dy = \int_0^1 \tfrac{1}{2}x^2 y\Big|_{x=y}^{x=2-y}\,dy = \int_0^1 \tfrac{1}{2}(2-y)^2 y - \tfrac{1}{2}y^3]\,dy$

$= \int_0^1 (2y-2y^2)\,dy = (y^2 - \tfrac{2}{3}y^3)\Big|_0^1 = \tfrac{1}{3}.$

21. The area of R is $1/2$. Therefore, the average value of f is

$2\int_0^1 \int_0^x xe^y\,dy\,dx = 2\int_0^1 xe^y\Big|_0^x\,dx = 2\int_0^1 (xe^x - x)\,dx = 2(xe^x - e^x - \tfrac{1}{2}x^2)\Big|_0^1$

$= 2(e-e-\tfrac{1}{2}+1) = 1.$

23. The population is

$2\int_0^5 \int_{-2}^0 \frac{10{,}000e^y}{1+0.5x}\,dy\,dx = 20{,}000 \int_0^5 \frac{e^y}{1+0.5x}\Big|_{y=-2}^{y=0}\,dx$

$$= 20{,}000(1-e^{-2})\int_0^5 \frac{1}{1+0.5x}\,dx = 20{,}000(1-e^{-2})2\,\ln(1+0.5x)\Big|_0^5$$

$$= 40{,}000(1-e^{-2})\,\ln 3.5 \approx 43{,}329.$$

25. By symmetry, it suffices to compute the population in the first quadrant. In the first quadrant, $f(x,y) = \dfrac{50{,}000xy}{(x^2+20)(y^2+36)}$. Therefore, the population in R is given by

$$\iint_R f(x,y)\,dA = 4\int_0^{15}\left[\int_0^{20}\frac{50{,}000xy}{(x^2+20)(y^2+36)}\,dy\right]dx$$

$$= 4\int_0^{15}\left[\frac{50{,}000(x)(\tfrac{1}{2})\ln(y^2+36)}{(x^2+20)}\Bigg|_0^{20}\right]dx$$

$$= 100{,}000(\ln 436 - \ln 36)\int_0^{15}\frac{x}{x^2+20}\,dx$$

$$= 100{,}000(\ln 436 - \ln 36)(\tfrac{1}{2})\ln(x^2+20)\Big|_0^{15}$$

$$= 50{,}000(\ln 436 - \ln 36)(\ln 245 - \ln 20) \approx 312{,}439.08$$

or approximately 312,439 people.

27. The average price is

$$\tfrac{1}{2}\int_0^1\int_0^2 [200 - 10(x-\tfrac{1}{2})^2 - 15(y-1)^2]\,dy\,dx$$

$$= \tfrac{1}{2}\int_0^1 [200y - 10(x-\tfrac{1}{2})^2 y - 5(y-1)^3]\Big|_0^2\,dx$$

$$= \tfrac{1}{2}\int_0^1 [400 - 20(x-\tfrac{1}{2})^2 - 5 - 5]\,dx$$

$$= \tfrac{1}{2}\int_0^1 [390 - 20(x-\tfrac{1}{2})^2]\,dx = \tfrac{1}{2}[390x - \tfrac{20}{3}(x-\tfrac{1}{2})^3]\Big|_0^1$$

$$= \tfrac{1}{2}[390 - \tfrac{20}{3}(\tfrac{1}{8}) - \tfrac{20}{3}(\tfrac{1}{8})] \approx 194.17 \text{ or approximately \$194 per sq ft.}$$

29. True. The average value (AV) is given by

$$AV = \frac{\displaystyle\iint_R f(x,y)\,dA}{\displaystyle\iint_R dA} \qquad \text{and so } AV\iint_R dA = \iint_R f(x,y)\,dA.$$

The quantity on the left-hand side is the volume of such a cylinder.

CHAPTER 8 REVIEW EXERCISES, page 618

1. $f(0,1) = 0$; $f(1,0) = 0$; $f(1,1) = \dfrac{1}{1+1} = \dfrac{1}{2}$.

$f(0,0)$ does not exist because the point $(0,0)$ does not lie in the domain of f.

3. $h(1,1,0) = 1 + 1 = 2$; $h(-1,1,1) = -e - 1 = -(e+1)$;
$h(1,-1,1) = -e - 1 = -(e+1)$.

5. $D = \{(x,y)\,|\,y \neq -x\}$

7. The domain of f is the set of all ordered triplets (x,y,z) of real numbers such that $z \geq 0$ and $x \neq 1$, $y \neq 1$, and $z \neq 1$.

9. $z = y - x^2$

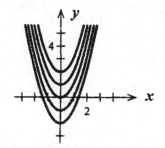

11. $z = e^{xy}$

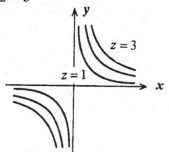

13. $f(x,y) = x\sqrt{y} + y\sqrt{x}$; $f_x = \sqrt{y} + \dfrac{y}{2\sqrt{x}}$; $f_y = \dfrac{x}{2\sqrt{y}} + \sqrt{x}$

15. $f(x,y) = \dfrac{x-y}{y+2x}$. $f_x = \dfrac{(y+2x)-(x-y)(2)}{(y+2x)^2} = \dfrac{3y}{(y+2x)^2}$.

$f_y = \dfrac{(y+2x)(-1)-(x-y)}{(y+2x)^2} = \dfrac{-3x}{(y+2x)^2}$.

17. $h(x,y) = (2xy+3y^2)^5$; $h_x = 10y(2xy+3y^2)^4$; $h_y = 10(x+3y)(2xy+3y^2)^4$.

19. $f(x,y) = (x^2+y^2)e^{x^2+y^2}$;
$f_x = 2xe^{x^2+y^2} + (x^2+y^2)(2x)e^{x^2+y^2} = 2x(x^2+y^2+1)e^{x^2+y^2}$.

$$f_y = 2ye^{x^2+y^2} + (x^2+y^2)(2y)e^{x^2+y^2} = 2y(x^2+y^2+1)e^{x^2+y^2}.$$

21. $f(x,y) = \ln\left(1+\dfrac{x^2}{y^2}\right). \; f_x = \dfrac{\frac{2x}{y^2}}{1+\frac{x^2}{y^2}} = \dfrac{2x}{x^2+y^2}; \; f_y = \dfrac{-\frac{2x^2}{y^3}}{1+\frac{x^2}{y^2}} = -\dfrac{2x^2}{y(x^2+y^2)}.$

23. $f(x,y) = x^4 + 2x^2y^2 - y^4; \; f_x = 4x^3 + 4xy^2; \; f_y = 4x^2y - 4y^3;$
$f_{xx} = 12x^2 + 4y^2, \; f_{xy} = 8xy = f_{yx}, \; f_{yy} = 4x^2 - 12y^2.$

25. $g(x,y) = \dfrac{x}{x+y^2}; \; g_x = \dfrac{(x+y^2)-x}{(x+y^2)^2} = \dfrac{y^2}{(x+y^2)^2}, \; g_y = \dfrac{-2xy}{(x+y^2)^2}.$

Therefore, $g_{xx} = -2y^2(x+y^2)^{-3} = -\dfrac{2y^2}{(x+y^2)^3},$

$$g_{yy} = \frac{(x+y^2)^2(-2x) + 2xy(2)(x+y^2)2y}{(x+y^2)^4} = \frac{2x(x^2+y^2)[-x-y^2+4y^2]}{(x+y^2)^4}$$

$$= \frac{2x(3y^2-x)}{(x+y^2)^3}.$$

and $g_{xy} = \dfrac{(x+y^2)2y - y^2(2)(x+y^2)2y}{(x+y^2)^4} = \dfrac{2(x+y^2)[xy+y^3-2y^3]}{(x+y^2)^4}$

$$= \frac{2y(x-y^2)}{(x+y^2)^3} = g_{yx}.$$

27. $h(s,t) = \ln\left(\dfrac{s}{t}\right).$ Write $h(s,t) = \ln s - \ln t.$ Then $h_s = \dfrac{1}{s}, \; h_t = -\dfrac{1}{t}.$

Therefore, $h_{ss} = -\dfrac{1}{s^2}, \; h_{st} = h_{ts} = 0, \; h_{tt} = \dfrac{1}{t^2}.$

29. $f(x,y) = 2x^2 + y^2 - 8x - 6y + 4;$ To find the critical points of f, we solve the

system $\begin{cases} f_x = 4x - 8 = 0 \\ f_y = 2y - 6 = 0 \end{cases}$ obtaining $x = 2$ and $y = 3$. Therefore, the sole critical

point of f is $(2,3)$. Next, $f_{xx} = 4, f_{xy} = 0, f_{yy} = 2$. Therefore,
$$D = f_{xx}(2,3)f_{yy}(2,3) - f_{xy}(2,3)^2 = 8 > 0.$$
Since $f_{xx}(2,3) > 0$, we see that $f(2,3) = -13$ is a relative minimum.

31. $f(x,y) = x^3 - 3xy + y^2$. We solve the system of equations $\begin{cases} f_x = 3x^2 - 3y = 0 \\ f_y = -3x + 2y = 0 \end{cases}$

obtaining $x^2 - y = 0$, or $y = x^2$. Then $-3x + 2x^2 = 0$, and $x(2x - 3) = 0$, and $x = 0$, or $x = 3/2$ and $y = 0$, or $y = 9/4$. Therefore, the critical points are $(0,0)$ and $(\frac{3}{2}, \frac{9}{4})$. Next, $f_{xx} = 6x, f_{xy} = -3$, and $f_{yy} = 2$ and $D(x,y) = 12x - 9 = 3(4x - 3)$. Therefore, $D(0,0) = -9$ so $(0,0)$ is a saddle point. $D(\frac{3}{2}, \frac{9}{4}) = 3(6-3) = 9 > 0$, and $f_{xx}(\frac{3}{2}, \frac{9}{4}) > 0$ and therefore, $f(\frac{3}{2}, \frac{9}{4}) = \frac{27}{8} - \frac{81}{8} + \frac{81}{16} = -\frac{27}{16}$ is the relative minimum value.

33. $f(x,y) = f(x,y) = e^{2x^2 + y^2}$. To find the critical points of f, we solve the system

$$\begin{cases} f_x = 4xe^{2x^2 + y^2} = 0 \\ f_y = 2ye^{2x^2 + y^2} = 0 \end{cases}$$

giving $(0,0)$ as the only critical point of f. Next,

$$f_{xx} = 4(e^{2x^2 + y^2} + 4x^2 e^{2x^2 + y^2}) = 4(1 + 4x^2)e^{2x^2 + y^2}$$

$$f_{xy} = 8xye^{2x^2 + y^2}$$

$$f_{yy} = 2(1 + 2y^2)e^{2x^2 + y^2}.$$

Therefore, $D = f_{xx}(0,0)f_{yy}(0,0) - f_{xy}^2(0,0) = (4)(2) - 0 = 8 > 0$, and so $(0,0)$ gives a relative minimum of f since $f_{xx}(0,0) > 0$. The minimum value of f is $f(0,0) = e^0 = 1$.

35. We form the Lagrangian function $F(x,y,\lambda) = -3x^2 - y^2 + 2xy + \lambda(2x + y - 4)$. Next, we solve the system

$$\begin{cases} F_x = 6x + 2y + 2\lambda = 0 \\ F_y = -2y + 2x + \lambda = 0. \\ F_\lambda = 2x + y - 4 = 0 \end{cases}$$

Multiplying the second equation by 2 and subtracting the resultant equation from the first equation yields $6y - 10x = 0$ so $y = 5x/3$. Substituting this value of y into the third equation of the system gives $2x + \frac{5}{3}x - 4 = 0$. So $x = \frac{12}{11}$ and consequently $y = \frac{20}{11}$. So $(\frac{12}{11}, \frac{20}{11})$ gives the maximum value for f subject to the given constraint.

37. The Lagrangian function is $F(x,y,\lambda) = 2x - 3y + 1 + \lambda(2x^2 + 3y^2 - 125)$. Next, we solve the system of equations

$$\begin{cases} F_x = 2 + 4\lambda x = 0 \\ F_y = -3 + 6\lambda y = 0 \\ F_\lambda = 2x^2 + 3y^2 - 125 = 0. \end{cases}$$

Solving the first equation for x gives $x = -1/2\lambda$. The second equation gives $y = 1/2\lambda$. Substituting these values of x and y into the third equation gives

$$2\left(-\frac{1}{2\lambda}\right)^2 + 3\left(\frac{1}{2\lambda}\right)^2 - 125 = 0$$

$$\frac{1}{2\lambda^2} + \frac{3}{4\lambda^2} - 125 = 0$$

$$2 + 3 - 500\lambda^2 = 0, \text{ or } \lambda = \pm\frac{1}{10}.$$

Therefore, $x = \pm 5$ and $y = \pm 5$ and so the critical points of f are $(-5,5)$ and $(5,-5)$. Next, we compute

$$f(-5,5) = 2(-5) - 3(5) + 1 = -24.$$
$$f(5,-5) = 2(5) - 3(-5) + 1 = 26.$$

So f has a maximum value of 26 at $(5,-5)$ and a minimum value of -24 at $(-5,5)$.

39. $df = \frac{3}{2}(x^2 + y^4)^{1/2}(2x)dx + \frac{3}{2}(x^2 + y^4)^{1/2}(4y^3)dy$.

At $(3,2)$, $df = 9(9 + 16)^{1/2}dx + 48(9 + 16)^{1/2}dy = 45dx + 240dy$.

41. $\Delta f \approx df = (4xy^3 + 6y^2x - 2y)dx + (6x^2y^2 + 6yx^2 - 2x)dy$.

With $x = 1$, $y = -1$, $dx = 0.02$, and $dy = 0.02$, we find
$\Delta f \approx (-4 + 6 + 2)(0.02) + (6 - 6 - 2)(0.02) = 0.04.$

43. $\int_{-1}^{2}\int_{2}^{4}(3x - 2y)dx\,dy = \int_{-1}^{2}\frac{3}{2}x^2 - 2xy\Big|_{x=2}^{x=4}dy = \int_{-1}^{2}[(24 - 8y) - (6 - 4y)]dy$

$$= \int_{-1}^{2}(18 - 4y)\,dy = (18y - 2y^2)\Big|_{-1}^{2} = (36 - 8) - (-18 - 2) = 48.$$

45. $\int_{0}^{1}\int_{x^3}^{x^2}2x^2y\,dy\,dx = \int_{0}^{1}x^2y^2\Big|_{y=x^3}^{y=x^2}dx = \int_{0}^{1}x^2(x^4 - x^6)dx$

$$= \int_{0}^{1}(x^6 - x^8)dx = \frac{x^7}{7} - \frac{x^9}{9}\Big|_{0}^{1} = \frac{1}{7} - \frac{1}{9} = \frac{2}{63}.$$

47. $\displaystyle\int_0^2 \int_0^1 (4x^2 + y^2)\,dy\,dx = \int_0^2 4x^2 y + \tfrac{1}{3}y^3 \Big|_{y=0}^{y=1} dx = \int_0^2 (4x^2 + \tfrac{1}{3})dx$

$\displaystyle = (\tfrac{4}{3}x^3 + \tfrac{1}{3}x)\Big|_0^2 = \tfrac{32}{3} + \tfrac{2}{3} = \tfrac{34}{3}.$

49. The area of R is

$\displaystyle\int_0^2 \int_{x^2}^{2x} dy\,dx = \int_0^2 y\Big|_{y=x^2}^{y=2x} dx = \int_0^2 (2x - x^2)\,dx = (x^2 - \tfrac{1}{3}x^3)\Big|_0^2 = \tfrac{4}{3}.$

Then

$\displaystyle AV = \frac{1}{4/3}\int_0^2 \int_{x^2}^{2x} (xy + 1)\,dy\,dx = \frac{3}{4}\int_0^2 \frac{xy^2}{2} + y\Big|_{x^2}^{2x} dx$

$\displaystyle = \tfrac{3}{4}\int_0^2 (-\tfrac{1}{2}x^5 + 2x^3 - x^2 + 2x)\,dx = \tfrac{3}{4}(-\tfrac{1}{12}x^6 + \tfrac{1}{2}x^4 - \tfrac{1}{3}x^3 + x^2)\Big|_0^2$

$\displaystyle = \tfrac{3}{4}(-\tfrac{16}{3} + 8 - \tfrac{8}{3} + 4) = 3.$

51. $f(p,q) = 900 - 9p - e^{0.4q}$; $g(p,q) = 20,000 - 3000q - 4p.$

We compute $\dfrac{\partial f}{\partial q} = -0.4e^{0.4q}$ and $\dfrac{\partial g}{\partial p} = -4.$ Since $\dfrac{\partial f}{\partial q} < 0$ and $\dfrac{\partial g}{\partial p} < 0$

for all $p > 0$ and $q > 0$, we conclude that compact disc players and audio discs are complementary commodities.

53. We first summarize the data:

x	y	x^2	xy
1	369	1	369
3	390	9	1170
5	396	25	1980
7	420	49	2940
9	436	81	3924
25	2011	165	10383

The normal equations are

$$5b + 25m = 2011$$

$$25b + 165m = 10383$$

The solutions are $b = 361.2$ and $m = 8.2$. Therefore, the least-square line has equation $y = 8.2x + 361.2.$

b. The average daily viewing time in 2002 ($x = 11$) wiill be
$y = 8.2(1) + 361.2 = 451.4 = 7.52$, or 7 hr 31 min.

55. Refer to the following diagram.

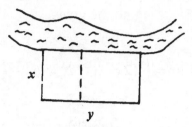

We want to minimize $C(x,y) = 3(2x) + 2(x) + 3y = 8x + 3y$ subject to
$xy = 303{,}750.$ The Lagrangian function is
$$F(x,y,\lambda) = 8x + 3y + \lambda(xy - 303{,}750).$$
Next, we solve the system
$$\begin{cases} F_x = 8 + \lambda y = 0 \\ F_y = 3 + \lambda x = 0 \\ F_\lambda = xy - 303{,}750 = 0 \end{cases}$$
Solving the first equation for y gives $y = -8/\lambda$. The second equation gives
$x = -3/\lambda$. Substituting this value into the third equation gives
$$\left(-\frac{3}{\lambda}\right)\left(-\frac{8}{\lambda}\right) = 303{,}750 \ \text{ or } \ \lambda^2 = \frac{24}{303{,}750} = \frac{4}{50{,}625},$$
or $\lambda = \pm\frac{2}{225}$. Therefore, $x = 337.5$ and $y = 900$ and so the required dimensions of
the pasture are 337.5 yd by 900 yd.

CHAPTER 9

EXERCISES 9.1, page 627

1. $y' = 2x$. Substituting this value into the given differential equation yields

    $$x(2x) + x^2 = 3x^2 \qquad [xy' + y = 3x^2]$$

    and the given differential equation is satisfied. Therefore $y = x^2$ is a solution of the differential equation.

3. $y = \frac{1}{2} + ce^{-x^2}$, $y' = -2cxe^{-x^2}$. Substituting this value into the differential equation gives

    $$-2cxe^{-x^2} + 2x(\tfrac{1}{2} + ce^{-x^2}) = x \qquad [y' + 2xy = x]$$

    and the differential equation is satisfied. So $y = \frac{1}{2} + ce^{-x^2}$ is a solution of the differential equation.

5. $y = e^{-2x}$, $y' = -2e^{-2x}$. $y'' = 4e^{-2x}$. Substituting these values into the differential equation yields

    $$4e^{-2x} - 2e^{-2x} - 2e^{-2x} = 0, \quad [y'' + y' - 2y = 0]$$

    and so the differential equation is satisfied. Therefore, $y = e^{-2x}$ is a solution of the differential equation.

7. $y = C_1 e^{-2x} + C_2 xe^{-2x}$.

    $y' = -2C_1 e^{-2x} + C_2 e^{-2x} - 2C_2 xe^{-2x}$

    $y'' = 4C_1 e^{-2x} - 2C_2 e^{-2x} - 2C_2 e^{-2x} + 4C_2 xe^{-2x} = 4C_1 e^{-2x} - 4C_2 e^{-2x} + 4C_2 xe^{-2x}$.

    Substituting these values into the differential equation, we find

    $4C_1 e^{-2x} - 4C_2 e^{-2x} + 4C_2 xe^{-2x} - 8C_1 e^{-2x} + 4C_2 e^{-2x} - 8C_2 xe^{-2x} + 4C_1 e^{-2x} + 4C_2 xe^{-2x}$

    $= 0$

    and so the equation is satisfied and $y = C_1 e^{-2x} + C_2 xe^{-2x}$ is a solution of the given equation.

9. $y = \dfrac{C_1}{x} + C_2 \dfrac{\ln x}{x} = C_1 x^{-1} + C_2 x^{-1} \ln x$

$y' = -C_1 x^{-2} + C_2(-x^{-2} \ln x + x^{-2}) = -C_1 x^{-2} + C_2 x^{-2}(1 - \ln x)$

$y'' = 2C_1 x^{-3} + C_2[-2x^{-3}(1 - \ln x) - x^{-3}] = 2C_1 x^{-3} + C_2 x^{-3}(2 \ln x - 3).$

Substituting these values into the differential equation yields

$x^2[2C_1 x^{-3} + C_2 x^{-3}(2 \ln x - 3)] + 3x[-C_1 x^{-2} + C_2 x^{-2}(1 - \ln x)] + C_1 x^{-1} + C_2 x^{-1} \ln x$

$\qquad = 2C_1 x^{-1} + C_2 x^{-1}(2 \ln x - 3) - 3C_1 x^{-1} + 3C_2 x^{-1}(1 - \ln x) + C_1 x^{-1} + C_2 x^{-1} \ln x$

$\qquad = 0,$

$$[x^2 y'' + 3xy' + y = 0]$$

and so the equation is satisfied. Therefore, $y = \dfrac{C_1}{x} + \dfrac{\ln x}{x}$ is a solution of the

differential equation.

11. $y = C - Ae^{-kt}; y' = kAe^{-kt}.$

   Substituting these values into the differential equation, we find
   $$kAe^{-kt} = k[C - (C - Ae^{-kt})] = kAe^{-kt}, \qquad\qquad [y' = k(C - y)]$$
   and so the equation is satisfied.

13. $y = Cx^2 - 2x, \; y' = 2Cx - 2.$

   Substituting these values into the differential equation gives
   $$2Cx - 2 - 2 \cdot \frac{1}{x}(cx^2 - 2x) = 2Cx - 2 - 2Cx + 4 = 2$$
   and so the equation is satisfied. Next,
   $\quad y(1) = 10$ implies $C - 2 = 10$, or $C = 12.$
   Therefore, a particular solution is $y = 12x^2 - 2x.$

15. $y = \dfrac{C}{x} = Cx^{-1}, \; y' = -Cx^{-2} = -\dfrac{C}{x^2}.$

   Substituting these values into the differential equation gives
   $$-\frac{C}{x^2} + \left(\frac{1}{x}\right)\left(\frac{C}{x}\right) = 0,$$
   and so the equation is satisfied. Next, $y(1) = 1$ implies $C = 1$ and so a particular

   solution is $y = \dfrac{1}{x}.$

17. $y = \dfrac{Ce^x}{x} + \dfrac{1}{2}xe^x = Cx^{-1}e^x + \dfrac{1}{2}xe^x.$

$y' = C(x^{-1}e^x - x^{-2}e^x) + \dfrac{1}{2}(e^x + xe^x).$

Substituting these values into the differential equation gives

$$C\left(\dfrac{e^x}{x} - \dfrac{e^x}{x^2}\right) + \dfrac{1}{2}e^x + \dfrac{1}{2}xe^x + \left(\dfrac{1}{x} - 1\right)\left(\dfrac{Ce^x}{x} + \dfrac{1}{2}xe^x\right) = e^x,$$

and so the equation is satisfied. Next,

$$y(1) = -\dfrac{1}{2}e \text{ implies } \dfrac{Ce}{1} + \dfrac{1}{2}e = -\dfrac{1}{2}e \text{ or } C = -1.$$

Therefore, a particular solution is $y = -\dfrac{e^x}{x} + \dfrac{1}{2}xe^x.$

19. Let $Q(t)$ denote the amount of the substance present at time t. Since the substance decays at a rate directly proportional to the amount present, we have

$$\dfrac{dQ}{dt} = -kQ,$$

where k (positive) is the constant of proportion. The side condition is $Q(0) = Q_0$.

21. Let $A(t)$ denote the total investment at time t. Then

$$\dfrac{dA}{dt} = k(C - A),$$

where k is the constant of proportion.

23. Since the rate of decrease of the concentration of the drug at time t is proportional to the concentration $C(t)$ at any time t, we have

$$\dfrac{dC}{dt} = -kC,$$

where k is the (positive) constant of proportion. The initial condition is $C(0) = C_0$.

25. The rate of decrease of the temperature is dy/dt. Since this is proportional to the difference between the temperature y and C, we have

$$\dfrac{dy}{dt} = -k(y - C),$$

where k is a constant of proportionality. Furthermore, the initial temperature of y_0 degrees translates into the condition $y(0) = y_0$.

27. Since the relative growth rate of one organ, $(dx/dt)/x$ is proportional to the relative growth rate of the other, $(dy/dt)/y$, we have

$$\frac{1}{x} \cdot \frac{dx}{dt} = k \frac{1}{y} \cdot \frac{dy}{dt},$$

where k is a constant of proportionality.

29. True. $y = x^2 + 2x + x^{-1}$, $y' = 2x + 2 - \dfrac{1}{x^2}$;

$$x\left(2x + 2 - \frac{1}{x^2}\right) + x^2 + 2x + \frac{1}{x} = 2x^2 + 2x - \frac{1}{x} + x^2 + 2x + \frac{1}{x} = 3x^2 + 4x$$

31. False. $y = 2 + ce^{-x^3}$, $y' = -3cx^2 e^{-x^3}$

$$-3cx^2 e^{-x^3} + 3x^2(2 + ce^{-x^3}) = -3cx^2 e^{-x^3} + 6x^2 + 3cx^2 e^{-x^3} = 6x^2 \neq x.$$

33. False. Consider the solution in Problem 29. Next,

$$y = 2f(x) = 2x^2 + 4x + 2x^{-1}.$$

Then $y' = 4x + 4 - \dfrac{2}{x^2}$. Substituting these values, we have

$$x\left(4x + 4 - \frac{2}{x^2}\right) + 2x^2 + 4x + \frac{2}{x} = 6x^2 + 8x \neq 3x^2 + 4x.$$

EXERCISES 9.2, page 633

1. $\dfrac{dy}{dx} = \dfrac{x+1}{y^2}$, $y^2\, dy = (x+1)\, dx$. So

$$\int y^2\, dy = \int (x+1)\, dx, \quad \tfrac{1}{3} y^3 = \tfrac{1}{2} x^2 + x + C.$$

3. $\dfrac{dy}{dx} = \dfrac{e^x}{y^2}$, $y^2\, dy = e^x\, dx$. So

$$\int y^2\, dy = \int e^x\, dx, \quad \tfrac{1}{3} y^3 = e^x + C.$$

5. $\dfrac{dy}{dx} = 2y$, $\dfrac{dy}{y} = 2\, dx$.

$\int \dfrac{dy}{y} = \int 2\,dx,\ \ln|y| = 2x + C_1$

so $y = e^{2x+C_1} = e^{C_1}e^{2x} = ce^{2x}$, where $c = e^{C_1}$.

$\dfrac{dy}{dx} = xy^2;\ \dfrac{dy}{y^2} = x\,dx.\ \ \int \dfrac{dy}{y^2} = \int x\,dx,\ -\dfrac{1}{y} = \dfrac{1}{2}x^2 + C.$

$\dfrac{dy}{dx} = -2(3y+4);\ \dfrac{dy}{3y+4} = -2\,dx;$

$\int \dfrac{dy}{3y+4} = \int -2\,dx,\ \tfrac{1}{3}\ln|3y+4| = -2x + c_1\,;\ 3y+4 = c_2e^{-6x}$, where $c_2 = e^{c_1}$

or $y = -\tfrac{4}{3} + ce^{-6x}$, where $c = \tfrac{1}{3}c_2$.

1. $\dfrac{dy}{dx} = \dfrac{x^2+1}{3y^2},\ 3y^2\,dy = (x^2+1)\,dx.\ \int 3y^2\,dy = \int (x^2+1)\,dx.$ Therefore,

$y^3 = \tfrac{1}{3}x^3 + x + C.$

3. $\dfrac{dy}{dx} = \sqrt{\dfrac{y}{x}} = \dfrac{y^{1/2}}{x^{1/2}},\ \dfrac{dy}{y^{1/2}} = \dfrac{dx}{x^{1/2}},\ \int y^{-1/2}\,dy = \int x^{-1/2}\,dx;\ 2y^{1/2} = 2x^{1/2} + C_1,$

or $y^{1/2} - x^{1/2} = C$, where $C = \tfrac{1}{2}C_1$.

5. $\dfrac{dy}{dx} = \dfrac{y\ln x}{x},\ \dfrac{dy}{y} = \dfrac{\ln x}{x}dx\,;\ \int \dfrac{dy}{y} = \int \dfrac{\ln x}{x}dx,\ \ln|y| = \tfrac{1}{2}(\ln x)^2 + C.$

7. $\dfrac{dy}{dx} = \dfrac{2x}{y},\ \int y\,dy = \int 2x\,dx,\ \dfrac{1}{2}y^2 = x^2 + C,$

$y(1) = -2$ implies $\tfrac{1}{2}(-2)^2 = 1^2 + C$, or $C = 1$. Therefore, the solution is

$y^2 = 2x^2 + 2.$

9. $\dfrac{dy}{dx} = 2 - y,\ \int \dfrac{dy}{2-y} = dx$

$-\ln|2-y| = x + C.$ The condition $y(0) = 3$ implies

$-\ln|2-3| = 0+C,$ or $C = -\ln 1 = 0.$ Therefore, the solution is

$-\ln|2-y| = x,$ $\ln|2-y| = -x,$ $2-y = e^{-x},$ or $y = 2-e^{-x}.$

21. $\dfrac{dy}{dx} = 3xy - 2x = x(3y-2)$

$\displaystyle\int \dfrac{dy}{3y-2} = \int x\, dx,\ \tfrac{1}{3}\ln|3y-2| = \tfrac{1}{2}x^2 + C$

The condition $y(0) = 1$ gives $\tfrac{1}{3}\ln 1 = 0+c,$ or $c = 0.$

Therefore, the solution is $\tfrac{1}{3}\ln|3y-2| = \tfrac{1}{2}x^2,$

$\ln|3y-2| = \tfrac{3}{2}x^2,\ 3y-2 = e^{(3/2)x^2},$ or $y = \tfrac{2}{3} + \tfrac{1}{3}e^{(3/2)x^2}.$

23. $\dfrac{dy}{dx} = \dfrac{xy}{x^2+1}.$ Separating variables and integrating, we have

$\displaystyle\int \dfrac{dy}{y} = \int \dfrac{x}{x^2+1}\,dx$

$\ln|y| - \ln(x^2+1)^{1/2} = \ln C$

$\ln \dfrac{|y|}{\sqrt{x^2+1}} = C$ and $y = C\sqrt{x^2+1}.$

Next, using the condition $y(0) = 1$ gives $y(0) = C = 1.$ Therefore, the solution is $y = \sqrt{x^2+1}.$

25. $\dfrac{dy}{dx} = xye^x,\ \displaystyle\int \dfrac{dy}{y} = \int xe^x\, dx$

$\ln|y| = (x-1)e^x + C.$

The condition $y(1) = 1$ gives $\ln 1 = 0 + C,$ or $C = 0.$ Therefore, the solution is $\ln|y| = (x-1)e^x.$

27. $\dfrac{dy}{dx} = 3x^2 e^{-y},\ e^y\, dy = 3x^2\, dx,\ e^y = x^3 + C,\ y = \ln(x^3 + C)$

$y(0) = 1$ implies that $1 = \ln C$ or $C = e.$ Therefore, $y = \ln(x^3 + e).$

29. Let $y = f(x).$ Then

$$\frac{dy}{dx} = \frac{3x^2}{2y}, \quad \int 2y \, dy = \int 3x^2 \, dx$$

$$y^2 = x^3 + C.$$

The given condition implies that $9 = 1 + C$, or $C = 8$. Therefore, a required equation defining the function is $y^2 = x^3 + 8$.

1. $\dfrac{dQ}{dt} = -kQ, \quad \int \dfrac{dQ}{Q} = \int -k \, dt.$

$\ln|Q| = -kt + C_1, \quad Q = e^{-kt+C_1} = Ce^{-kt}$, where $C = e^{C_1}$.

The condition $Q(0) = Q_0$ implies $Ce^0 = C = Q_0$. Therefore, $Q(t) = Q_0 e^{-kt}$.

3. $\dfrac{dA}{dt} = k\left(\frac{C}{k} - A\right), \quad \displaystyle\int \dfrac{dA}{\frac{C}{k} - A} = \int k \, dt; \quad -\ln\left|\frac{C}{k} - A\right| = kt + d \quad$ (Assume $\left(\frac{C}{k} - A\right) > 0$)

$\ln\left|\frac{C}{k} - A\right| = -kt - d_1.$

Therefore, $\frac{C}{k} - A = d_2 e^{-kt}$, or $A = \frac{C}{k} - d_2 e^{-kt}$ where $d_2 = e^{-d_1}$.

5. $\dfrac{dS}{dt} = k(D - S), \quad \displaystyle\int \dfrac{dS}{D - S} = \int k \, dt$

$-\ln(D - S) = kt + d_1 \quad$ (where $D - S > 0$ and d_1 is a constant).

$\ln(D - S) = -kt - d_1$

$\quad D - S = d_2 e^{-kt}$, where $d_2 = e^{-d_1}$

$\quad\quad S = D - d_2 e^{-kt}.$

The condition $S(0) = S_0$ gives $D - d_2 = S_0$, or $d_2 = D - S_0$. Therefore, $S(t) = D - (D - S_0)e^{-kt}.$

7. True. Rewrite it in the form $\dfrac{dy}{dx} = y(x-1) + 2(x-1) = (y+2)(x-1)$ and it is evident that the equation is separable.

8. True. The equation can be rewritten as $f(x)g(y)dx + F(x)G(y)dy = 0$ or

$$\frac{dy}{dx} = -\frac{f(x)g(y)}{F(x)G(y)} = \phi(x)\psi(y) \quad \text{where } \phi(x) = -\frac{f(x)}{F(x)} \quad \text{and} \quad \psi(y) = \frac{g(y)}{G(y)}$$

and is evidently separable.

41. False. It cannot be written in the form $\dfrac{dy}{dx} = f(x)g(y)$.

EXERCISES 9.3, page 641

1. $\dfrac{dy}{dx} = -ky,\ y = y_0 e^{-kt}$.

3. Solving $\dfrac{dQ}{dt} = kQ$, we find $Q = Q_0 e^{kt}$. Here $k = 0.02$ and $Q_0 = 4.5$. Therefore, $Q(t) = 4.5e^{0.02t}$. At the beginning of the year 2008, the population will be $Q(28) = 4.5e^{0.02(28)} \approx 7.9$, or 7.9 billion.

5. $\dfrac{dL}{L} = k\,dx$. Integrating, we have

$\ln L = kx + C_1$, or $L = L_0 e^{kx}$, where $L_0 = e^{C_1}$.

Using the given condition $L\left(\tfrac{1}{2}\right) = \tfrac{1}{2}L_0$, we have

$\tfrac{1}{2}L_0 = L_0 e^{k/2}$, $e^{k/2} = \tfrac{1}{2}$, $\tfrac{1}{2}k = \ln\tfrac{1}{2}$, $k = -2\ln 2 \approx -1.3863$.

Therefore, $L = L_0 e^{-1.3863x}$. We want to find x so that $L = \tfrac{1}{4}L_0$; that is, so that

$\tfrac{1}{4}L_0 = L_0 e^{-1.3863x}$, $e^{-1.3863x} = \tfrac{1}{4}$,

or $-1.3863x = -\ln 4 = -1.3863$. So $x = 1$. Therefore, 1/2 inch additional material is needed.

7. $\dfrac{dQ}{dt} = kQ^2$; $\displaystyle\int \frac{dQ}{Q^2} = \int k\,dt$ and $-\dfrac{1}{Q} = kt + C$. Therefore, $Q = -\dfrac{1}{kt + C}$.

Now, $Q = 50$ when $t = 0$, and, therefore, $50 = -\dfrac{1}{C}$, and $C = -\dfrac{1}{50}$. Next,

$Q = -\dfrac{1}{kt - \frac{1}{50}}$. Since $Q = 10$ when $t = 1$,

$$10 = -\frac{1}{k - \frac{1}{50}}, \; 10\left(k - \tfrac{1}{50}\right) = -1, \; 10k - \tfrac{1}{5} = -1.$$

$$10k = -1 + \tfrac{1}{5} = -\tfrac{4}{5}. \;\; k = -\tfrac{4}{50} = -\tfrac{2}{25}.$$

Therefore, $\; Q(t) = \dfrac{1}{\frac{2}{25}t + \frac{1}{50}} = \dfrac{1}{\frac{4t+1}{50}} = \dfrac{50}{4t+1}$,

and $\qquad Q(2) = \dfrac{50}{8+1} \approx 5.56$ grams.

9. Let C be the temperature of the surrounding medium and let T be the temperature at any time t. Then $\dfrac{dT}{dt} = k(C - T)$. The solution of the differential equation is

$$T = C - Ae^{-kt} \qquad \text{[see Equation (10)]}$$

With $C = 72$, we have $T = 72 - Ae^{-kt}$. Next, $T(0) = 212$ gives
   $212 = 72 - A$, or $A = -140$.
Therefore, $T = 72 + 140e^{-kt}$. Using the condition $T(2) = 140$, we have

$$72 + 140e^{-2k} = 140, \; e^{-2k} = \frac{68}{140}, \; \text{or } k = -\frac{1}{2}\ln\left(\frac{68}{140}\right) \approx 0.3611.$$

So $\; T(t) = 72 + 140e^{-0.3611t}$. When $T = 110$, we have
   $72 + 140e^{-0.3611t} = 110, \; 140e^{-0.3611t} = 38,$

or $\; e^{-0.3611t} = \dfrac{38}{140}, \; -0.3611t = \ln\left(\dfrac{38}{140}\right)$, and $t \approx 3.6$, or 3.6 minutes.

11. The differential equation is $\dfrac{dQ}{dt} = k(40 - Q)$, whose solution is

   $Q(t) = 40 - Ae^{-kt}$ (see Example 2). $Q(0) = 0$ gives $A = 40$ and so
   $Q(t) = 40(1 - e^{-kt})$. Next, $Q(2) = 10$ gives $40 = (1 - e^{-2k}) = 10$,
   $1 - e^{-2k} = 0.25, \; e^{-2k} = 0.75$ and $k = -\frac{1}{2}\ln 0.75 \approx 0.1438.$
   So $Q(t) = 40(1 - e^{-0.1438t})$.
   The number of claims the average trainee can process after six weeks is
   $\qquad Q(6) = 40(1 - e^{-0.1438(6)}) \approx 23$, or 23 claims.

13. $P(28) = -\dfrac{0.5}{0.008} + \left(289 + \dfrac{0.5}{0.008}\right)e^{0.008(28)} \approx 377.3$ or, 377.3 million.

15. Let $Q(t)$ denote the number of people who have heard the rumor. Then $\dfrac{dQ}{dt} = kQ(400 - Q)$. The solution is $Q(t) = \dfrac{400}{1 + Ae^{-400kt}}$. The condition $Q(0) = 10$ gives $10 = \dfrac{400}{1 + A}$, $10 + 10A = 400$, or $A = 39$. Therefore, $Q(t) = \dfrac{400}{1 + 39e^{-400kt}}$.

Next, the condition, $Q(2) = 80$ gives
$$\frac{400}{1 + 39e^{-400kt}} = 80, \quad 1 + 39e^{-900k} = 5,$$
$$39e^{-800k} = 4, \quad \text{and} \quad k = -\frac{1}{800}\ln\left(\frac{4}{39}\right) \approx 0.0028466.$$

Therefore, $Q(t) = \dfrac{400}{1 + 39e^{-1.1386t}}$. In particular, the number of people who wil have heard the rumor after a week is
$$Q(7) = \frac{400}{1 + 39e^{-1.1386(7)}} \approx 395, \quad \text{or 395 people.}$$

17. Separating variables and integrating, we obtain
$$\int \frac{dQ}{Q(C - \ln Q)} = \int k\, dt$$
$$-\ln|C - \ln Q| = kt + C_1, \quad C - \ln Q = C_2 e^{-kt} \qquad (C_2 = e^{-C_1})$$
$$\ln Q = C - C_2 e^{-kt}, \quad Q = e^C e^{-C_2 e^{-kt}}.$$
Using the condition $Q(0) = Q_0$, we have
$$Q_0 = e^C e^{-C_2}, \quad \ln Q_0 = C - C_2, \quad \text{or} \quad C_2 = C - \ln Q_0.$$
So $Q(t) = e^C e^{-(C - \ln Q_0)e^{-kt}}$.

19. The differential equation governing this process is
$$\frac{dx}{dt} = 6 - \frac{3x}{20}.$$
Separating varibles and integrating, we obtain

$$\frac{dx}{dt} = \frac{120 - 3x}{20} = \frac{3(40 - x)}{20},$$

$$\frac{dx}{40 - x} = \frac{3}{20} dt.$$

$$-\ln|40 - x| = \frac{3}{20} t + C_1.$$

$$40 - x = Ce^{-3t/20}.$$

Therefore, $x = 40 - Ce^{-3t/20}$. The initial condition $x(0) = 0$ implies $0 = 40 - C$, and $C = 40$. Therefore, $x(t) = 40(1 - e^{-3t/20})$. The amount of salt present at the end of 20 minutes is $x(20) = 40(1 - e^{-3}) \approx 38$ lbs. The amount of salt present in the long run is $\lim_{t \to \infty} 40(1 - e^{-3t/20}) = 40$ lbs.

EXERCISES 9.4, page 650

1. a. $x_0 = 0$, $b = 1$, and $n = 4$.

    Therefore, $h = \frac{1}{4}$ and $x_0 = 0$, $x_1 = \frac{1}{4}$, $x_2 = \frac{1}{2}$, $x_3 = \frac{3}{4}$, and $x_4 = b = 1$. Also

    $F(x, y) = x + y$ and $y_0 = y(0) = 1$.

    $\tilde{y}_0 = y_0 = 1$

    $\tilde{y}_1 = \tilde{y}_0 + hF(x_0, \tilde{y}_0) = 1 + \frac{1}{4}(0 + 1) = \frac{5}{4}$.

    $\tilde{y}_2 = \tilde{y}_1 + hF(x_1, \tilde{y}_1) = \frac{5}{4} + \frac{1}{4}\left(\frac{1}{4} + \frac{5}{4}\right) = \frac{13}{8}$

    $\tilde{y}_3 = \tilde{y}_2 + hF(x_2, \tilde{y}_2) = \frac{13}{8} + \frac{1}{4}\left(\frac{1}{2} + \frac{13}{8}\right) = \frac{69}{32}$.

    $\tilde{y}_4 = \tilde{y}_3 + hF(x_3, \tilde{y}_3) = \frac{69}{32} + \frac{1}{4}\left(\frac{3}{4} + \frac{69}{32}\right) = \frac{369}{128}$.

    Therefore, $y(1) = \frac{369}{128} = 2.8828$.

    b. $n = 6$. Therefore, $h = \frac{1}{6}$.

    So $x_0 = 0$, $x_1 = \frac{1}{6}$, $x_2 = \frac{2}{6}$, $x_3 = \frac{3}{6}$, $x_4 = \frac{4}{6}$, $x_5 = \frac{5}{6}$, $x_6 = 1$.

    Therefore,

    $\quad \tilde{y}_0 = y_0 = 1$

    $\quad \tilde{y}_1 = \tilde{y}_0 + hF(x_0, \tilde{y}_0) = 1 + \frac{1}{6}(0 + 1) = \frac{7}{6}$.

    $\quad \tilde{y}_2 = \tilde{y}_1 + hF(x_1, \tilde{y}_1) = \frac{7}{6} + \frac{1}{6}\left(\frac{1}{6} + \frac{7}{6}\right) = \frac{50}{36} = \frac{25}{18}$.

    $\quad \tilde{y}_3 = \tilde{y}_2 + hF(x_2, \tilde{y}_2) = \frac{25}{18} + \frac{1}{6}\left(\frac{2}{6} + \frac{25}{18}\right) = \frac{181}{108}$.

$$\tilde{y}_4 = \tilde{y}_3 + hF(x_3, \tilde{y}_3) = \frac{181}{108} + \frac{1}{6}\left(\frac{3}{6} + \frac{181}{108}\right) = \frac{1321}{648}.$$

$$\tilde{y}_5 = \tilde{y}_4 + hF(x_4, \tilde{y}_4) = \frac{1321}{648} + \frac{1}{6}\left(\frac{4}{6} + \frac{1321}{648}\right) = \frac{9679}{3888}.$$

$$\tilde{y}_6 = \tilde{y}_5 + hF(x_5, \tilde{y}_5) = \frac{9679}{3888} + \frac{1}{6}\left(\frac{5}{6} + \frac{9679}{3888}\right) = \frac{70993}{23328}.$$

Therefore, $y(1) = \frac{70993}{23328} \approx 3.043$.

3.   a. Here $x_0 = 0$ and $b = 2$. Taking $n = 4$, we have $h = \frac{2}{4} = \frac{1}{2}$ and

$$x_0 = 0, \ x_1 = \frac{1}{2}, \ x_2 = 1, \ x_3 = \frac{3}{2}, \text{ and } x_4 = 2.$$

Also, $F(x, y) = 2x - y + 1$ and $y(0) = y_0 = 2$. Therefore,

$$\tilde{y}_0 = y_0 = 2$$

$$\tilde{y}_1 = \tilde{y}_0 + hF(x_0, \tilde{y}_0) = 2 + \frac{1}{2}(0 - 2 + 1) = \frac{3}{2}$$

$$\tilde{y}_2 = \tilde{y}_1 + hF(x_1, \tilde{y}_1) = \frac{3}{2} + \frac{1}{2}\left(1 - \frac{3}{2} + 1\right) = \frac{7}{4}.$$

$$\tilde{y}_3 = \tilde{y}_2 + hF(x_2, \tilde{y}_2) = \frac{7}{4} + \frac{1}{2}\left(2 - \frac{7}{4} + 1\right) = \frac{19}{8}.$$

$$\tilde{y}_4 = \tilde{y}_3 + hF(x_3, \tilde{y}_3) = \frac{19}{8} + \frac{1}{2}\left(3 - \frac{19}{8} + 1\right) = \frac{51}{16}.$$

Therefore, $y(2) = \frac{51}{16} \approx 3.1875$.

b. $n = 6$. Therefore, $h = \frac{2}{6} = \frac{1}{3}$, and so

$$x_0 = 0, \ x_1 = \frac{1}{3}, \ x_2 = \frac{2}{3}, \ x_3 = 1, \ x_4 = \frac{4}{3}, \ x_5 = \frac{5}{3}, \ x_6 = 2.$$

Therefore,

$$\tilde{y}_0 = y_0 = 2$$

$$\tilde{y}_1 = \tilde{y}_0 + hF(x_0, \tilde{y}_0) = 2 + \frac{1}{3}(0 - 2 + 1) = \frac{5}{3}$$

$$\tilde{y}_2 = \tilde{y}_1 + hF(x_1, \tilde{y}_1) = \frac{5}{3} + \frac{1}{3}\left(\frac{2}{3} - \frac{5}{3} + 1\right) = \frac{5}{3}$$

$$\tilde{y}_3 = \tilde{y}_2 + hF(x_2, \tilde{y}_2) = \frac{5}{3} + \frac{1}{3}\left(\frac{4}{3} - \frac{5}{3} + 1\right) = \frac{17}{9}$$

$$\tilde{y}_4 = \tilde{y}_3 + hF(x_3, \tilde{y}_3) = \frac{17}{9} + \frac{1}{3}\left(2 - \frac{17}{9} + 1\right) = \frac{61}{27}.$$

$$\tilde{y}_5 = \tilde{y}_4 + hF(x_4, \tilde{y}_4) = \frac{61}{27} + \frac{1}{3}\left(\frac{8}{3} - \frac{61}{27} + 1\right) = \frac{221}{81}.$$

$$\tilde{y}_6 = \tilde{y}_5 + hF(x_5, \tilde{y}_5) = \frac{221}{81} + \frac{1}{3}\left(\frac{10}{3} - \frac{221}{81} + 1\right) = \frac{793}{243}.$$

Therefore, $y(2) = \frac{793}{243} \approx 3.2634$.

5.   a. Here $x_0 = 0$ and $b = 0.5$. Taking $n = 4$, we have $h = \dfrac{0.5}{4} \approx 0.125$.

Therefore, $x_0 = 0, \ x_1 = 0.125, \ x_2 = 0.25, \ x_3 = 0.375, \text{ and } x_4 = 0.5$. Also,
$F(x, y) = -2xy^2$ and $y(0) = y_0 = 1$. So

$\tilde{y}_0 = y_0 = 1$

$\tilde{y}_1 = \tilde{y}_0 + hF(x_0, \tilde{y}_0) = 1 + 0.125(0) = 1$

$\tilde{y}_2 = \tilde{y}_1 + hF(x_1, \tilde{y}_1) = 1 + 0.125[-2(0.125)(1)] = 0.96875.$

$\tilde{y}_3 = \tilde{y}_2 + hF(x_2, \tilde{y}_2) = 0.96875 + 0.125[-2(0.25)(0.96875)^2] \approx 0.910095.$

$\tilde{y}_4 = \tilde{y}_3 + hF(x_3, \tilde{y}_3) = 0.910095 + 0.125[-2(0.375)(0.910095)^2]$
$\qquad = 0.832445.$

Therefore, $y(0.5) = 0.8324.$

b. $n = 6.$ Therefore, $h = \dfrac{0.5}{6} \approx 0.83333.$ So

$\qquad x_0 = 0,\ x_1 = 0.083333,\ x_2 = 0.166667,\ x_3 = 0.245000,$ and $x_4 = 0.333333,$

$\qquad x_5 = 0.416666,$ and $x_5 = 0.5.$

Therefore,

$\tilde{y}_0 = y_0 = 1$

$\tilde{y}_1 = \tilde{y}_0 + hF(x_0, \tilde{y}_0) = 1 + 0.083333(0) = 1$

$\tilde{y}_2 = \tilde{y}_1 + hF(x_1, \tilde{y}_1) = 1 + 0.083333[-2(0.083333)(1) \approx 0.986111.$

$\tilde{y}_3 = \tilde{y}_2 + hF(x_2, \tilde{y}_2) = 0.986111 + 0.083333[-2(0.1666670(0.986111)^2] \approx 0.920772$

$\tilde{y}_4 = \tilde{y}_3 + hF(x_3, \tilde{y}_3) = 0.95100 - 0.083333[-2(0.245000)(0.959100)^2] \approx 0.959100.$

$\tilde{y}_5 = \tilde{y}_4 + hF(x_4, \tilde{y}_4) = 0.921539 + 0.08333[-2(0.333333)(0.921539)^2] \approx 0.873671.$

$\tilde{y}_6 = \tilde{y}_5 + hF(x_5, \tilde{y}_5) = 0.874360 + 0.083333[-2(0.416666)(0.874360)^2] \approx 0.820665.$

Therefore, $y(0.5) \approx 0.8207.$

7. a. $x_0 = 1,\ b = 1.5,\ n = 4,$ and $h = 0.125.$ Therefore,

$\qquad\qquad x_0 = 1,\ x_1 = 1.125,\ x_2 = 1.25,\ x_3 = 1.375,\ x_4 = 1.5.$

Also, $F(x, y) = \sqrt{x + y},\ y(1) = 1.$ Therefore,

$\qquad \tilde{y}_0 = y_0 = 1$

$\qquad \tilde{y}_1 = \tilde{y}_0 + hF(x_0, \tilde{y}_0) = 1 + 0.125\sqrt{1 + 1} \approx 1.7667767.$

$\qquad \tilde{y}_2 = \tilde{y}_1 + hF(x_1, \tilde{y}_1) = 1.1767767 + 0.125\sqrt{1.125 + 1.767767}$
$\qquad\quad \approx 1.3664218.$

$$\tilde{y}_3 = \tilde{y}_2 + hF(x_2, \tilde{y}_2) = 1.3664218 + 0.125\sqrt{1.25 + 1.3664218}$$
$$\approx 1.5686138.$$

$$\tilde{y}_4 = \tilde{y}_3 + hF(x_3, \tilde{y}_3) = 1.5686138 + 0.125\sqrt{1.375 + 1.5686138}$$
$$\approx 1.7830758.$$

Therefore, $y(1.5) \approx 1.7831$.

b. $n = 6$. Therefore, $h = \dfrac{0.5}{6} \approx 0.0833333$. So

$x_0 = 1$, $x_1 = 1.0833333$, $x_2 = 1.1666666$, $x_3 = 1.25$, $x_4 = 1.3333332$, $x_5 = 1.4166665$, and $x_5 = 1.5$. So

$$\tilde{y}_0 = y_0 = 1$$

$$\tilde{y}_1 = \tilde{y}_0 + hF(x_0, \tilde{y}_0) = 1 + 0.0833333\sqrt{1+1} \approx 1.1178511.$$

$$\tilde{y}_2 = \tilde{y}_1 + hF(x_1, \tilde{y}_1) = 1.1178511 + 0.0833333\sqrt{1.0833333 + 1.1178511} \approx 1.2414876.$$

$$\tilde{y}_3 = \tilde{y}_2 + hF(x_2, \tilde{y}_2) = 1.2414876 + 0.0833333\sqrt{1.166666 + 1.2414876} \approx 1.3708061.$$

$$\tilde{y}_4 = \tilde{y}_3 + hF(x_3, \tilde{y}_3) = 1.3708061 + 0.0833333\sqrt{1.25 + 1.3708061} \approx 1.5057136.$$

$$\tilde{y}_5 = \tilde{y}_4 + hF(x_4, \tilde{y}_4) = 1.5057136 + 0.0833333\sqrt{1.333332 + 1.5057136} \approx 1.6461258.$$

$$\tilde{y}_6 = \tilde{y}_5 + hF(x_5, \tilde{y}_5) = 1.6461258 + 0.0833333\sqrt{1.4166665 + 1.6461258} \approx 1.791966.$$

Therefore $y(1.5) \approx 1.7920$.

9. a. Here $x_0 = 0$ and $b = 1$. Taking $n = 4$, we have $h = 0.25$. Therefore,
$x_0 = 0$, $x_1 = 0.25$, $x_2 = 0.5$, $x_3 = 0.75$, and $x_4 = 1$.

Also, $F(x, y) = \dfrac{x}{y}$ and $y_0 = y(0) = 1$. Therefore,

$$\tilde{y}_0 = y_0 = 1$$
$$\tilde{y}_1 = \tilde{y}_0 + hF(x_0, \tilde{y}_0) = 1 + 0.25(0) = 1$$
$$\tilde{y}_2 = \tilde{y}_1 + hF(x_1, \tilde{y}_1) = 1 + \left(\frac{0.25}{1}\right) \approx 1.0625$$

$$\tilde{y}_3 = \tilde{y}_2 + hF(x_2, \tilde{y}_2) = 1.0625 + 0.25\left(\frac{0.5}{1.0625}\right) \approx 1.180147.$$

$$\tilde{y}_4 = \tilde{y}_3 + hF(x_3, \tilde{y}_3) = 1.180147 + 0.25\left(\frac{0.75}{1.180147}\right) \approx 1.339026.$$

Therefore, $y(1) \approx 1.3390$.

b. $n = 6$. Therefore, $h = \frac{1}{6} \approx 0.166667$ and

$x_0 = 0, \; x_1 \approx 0.166667, \; x_2 \approx 0.333333, \; x_3 \approx 0.500000, x_4 \approx 0.666667,$
$x_5 \approx 0.833333,$ and $x_6 = 1.$

Therefore,

$\tilde{y}_0 = y_0 = 1$

$\tilde{y}_1 = \tilde{y}_0 + hF(x_0, \tilde{y}_0) = 1 + 0.1666671(0) = 1.$

$\tilde{y}_2 = \tilde{y}_1 + hF(x_1, \tilde{y}_1) = 1 + 0.166667\left(\dfrac{0.166667}{1}\right) \approx 1.027778.$

$\tilde{y}_3 = \tilde{y}_2 + hF(x_2, \tilde{y}_2) = 1.0277778 + 0.166667\left(\dfrac{0.333333}{1.027778}\right) \approx 1.081832.$

$\tilde{y}_4 = \tilde{y}_3 + hF(x_3, \tilde{y}_3) = 1.081832 + 0.166667\left(\dfrac{0.500000}{1.081832}\right) \approx 1.158862.$

$\tilde{y}_5 = \tilde{y}_4 + hF(x_4, \tilde{y}_4) = 1.158862 + 0.166667\left(\dfrac{0.666667}{1.158862}\right) \approx 1.254742.$

$\tilde{y}_6 = \tilde{y}_5 + hF(x_5, \tilde{y}_5) = 1.254742 + 0.166667\left(\dfrac{0.833333}{1.254742}\right) \approx 1.365433.$

Therefore, $y(1) \approx 1.3654.$

11. Here $x_0 = 0, \; b = 1.$ With $n = 5$, we have $h = 1.5 = 0.2$ and

$x_0 = 0, \; x_1 = 0.2, \; x_2 = 0.4, \; x_3 = 0.6, \; x_4 = 0.8,$ and $x_5 = 1.$
$F(x, y) = \frac{1}{2}xy$ and $y(0) = 1.$ Therefore,

$\tilde{y}_0 = y_0 = 1$

$\tilde{y}_1 = \tilde{y}_0 + hF(x_0, \tilde{y}_0) = 1 + 0.2[0.5(0)(1)] = 1.$

$\tilde{y}_2 = \tilde{y}_1 + hF(x_1, \tilde{y}_1) = 1 + 0.2[0.5(0.2)(1)] = 1.02$

$\tilde{y}_3 = \tilde{y}_2 + hF(x_2, \tilde{y}_2) = 1.02 + 0.2[0.5(0.4)(1.02)] = 1.0608.$

$\tilde{y}_4 = \tilde{y}_3 + hF(x_3, \tilde{y}_3) = 1.0608 + 0.2[0.5(0.6)(1.0608)] = 1.124448.$

$\tilde{y}_5 = \tilde{y}_4 + hF(x_4, \tilde{y}_4) = 1.124448 + 0.2[0.5(0.8)(1.124448)] = 1.21440384.$

The solutions are summarized below.

x	0	0.2	0.4	0.6	0.8	1
$\tilde{y}_n$	1	1	1.02	1.0608	1.1245	1.2144

13. $x_0 = 0$, $b = 1$, $n = 5$, and, therefore, $h = 0.2$. So

$x_0 = 0$, $x_1 = 0.2$, $x_2 = 0.4$, $x_3 = 0.6$, $x_4 = 0.8$, $x_5 = 1$ and

$F(x, y) = 2x - y + 1$, $y(0) = 2$

$\tilde{y}_0 = y_0 = 2$

$\tilde{y}_1 = \tilde{y}_0 + hF(x_0, \tilde{y}_0) = 2 + 0.2[2(0) - 2 + 1] = 1.8.$

$\tilde{y}_2 = \tilde{y}_1 + hF(x_1, \tilde{y}_1) = 1.8 + 0.2[2(0.2) - 1.8 + 1] = 1.72$

$\tilde{y}_3 = \tilde{y}_2 + hF(x_2, \tilde{y}_2) = 1.72 + 0.2[2(0.4) - 1.72 + 1] = 1.736.$

$\tilde{y}_4 = \tilde{y}_3 + hF(x_3, \tilde{y}_3) = 1.736 + 0.2[2(0.6) - 1.736 + 1] = 1.8288.$

$\tilde{y}_5 = \tilde{y}_4 + hF(x_4, \tilde{y}_4) = 1.8288 + 0.2[2(0.8) - 1.8288 + 1] = 1.98304$

$\tilde{y}_6 = \tilde{y}_5 + hF(x_5, \tilde{y}_5) =$

The solutions are summarized below.

x	0	0.2	0.4	0.6	0.8	1
$\tilde{y}_n$	2	1.8	1.72	1.736	1.8288	1.9830

15. Here $x_0 = 0$, $b = 0.5$, and with $n = 5$, we have $h = 0.1$. So

$x_0 = 0$, $x_1 = 0.1$, $x_2 = 0.2$, $x_3 = 0.3$, $x_4 = 0.4$, and $x_5 = 0.5$.

$F(x, y) = x^2 + y$ and $y(0) = 1$. Therefore,

$\tilde{y}_0 = y_0 = 1$

$\tilde{y}_1 = \tilde{y}_0 + hF(x_0, \tilde{y}_0) = 1 + 0.1(0 + 1) = 1.1$

$y_2 = \tilde{y}_1 + hF(x_1, \tilde{y}_1) = 1.1 + 0.1(0.1)^2 + 1.1] \approx 1.211.$

$\tilde{y}_3 = \tilde{y}_2 + hF(x_2, \tilde{y}_2) = 1.211 + 0.1[(0.2)^2 + 1.211] \approx 1.3361.$

$\tilde{y}_4 = \tilde{y}_3 + hF(x_3, \tilde{y}_3) = 1.3361 + 0.1[(0.3)^2 + 1.3361] \approx 1.47871.$

$\tilde{y}_5 = \tilde{y}_4 + hF(x_4, \tilde{y}_4) = 1.47871 + 0.1[(0.4)^2 + 1.573481] \approx 1.64258.$

The solutions are summarized below.

x	0	0.1	0.2	0.3	0.4	0.5
$\tilde{y}_n$	1	1.1	1.211	1.3361	1.4787	1.64258

CHAPTER 9, REVIEW EXERCISES, page 651

1. $y = C_1 e^{2x} + C_2 e^{-3x}$; $y' = 2C_1 e^{2x} - 3C_2 e^{-3x}$; $y'' = 4C_1 e^{2x} + 9C_2 e^{-3x}$.
   Substituting these values into the differential equation, we have
   $$4C_1 e^{2x} + 9C_2 e^{-3x} + 2C_2 e^{2x} - 3C_2 e^{-3x} - 6(C_1 e^{2x} + C_2 e^{-3x}) = 0$$
   $$[y'' + y' - 6y = 0]$$
   and the differential equation is satisfied.

3. $y = Cx^{-4/3}$. So $y' = -\frac{4}{3} Cx^{-7/3}$. Substituting these values into the given differential
   equation, which can be written in the form $\dfrac{dy}{dx} = -\dfrac{4xy^3}{3x^2 y^2} = -\dfrac{4y}{3x}$, we find

   $$-\frac{4}{3} Cx^{-7/3} = -\frac{4(Cx^{-4/3})}{3x} = -\frac{4}{3} Cx^{-7/3}$$

   we see that y is a solution of the differential equation.

5. $y = (9x + C)^{-1/3}$. So $y' = -\dfrac{1}{3}(9x + C)^{-4/3}(9) = -3(9x + C)^{-4/3}$. Substituting into the
   differential equation, we have
   $$-3(9x + C)^{-4/3} = -3[(9x + C)^{-1/3}]^{-4} = -3(9x + C)^{-4/3}$$
   $$[y' = -3y^4]$$
   and it is satisfied. Therefore, y is indeed a solution. Next, using the side condition,
   we find $y(0) = C^{-1/3} = \frac{1}{2}$, or $C = 8$.
   Therefore, the required solution is $y = (9x + 8)^{-1/3}$.

7. $\dfrac{dy}{4-y} = 2\,dt$ implies that $-\ln|4 - y| = 2t + C_1$.
   $4 - y = Ce^{-2t}$, or $y = 4 - Ce^{-2t}$, where $C = e^{-C_1}$.

9. We have $\dfrac{dy}{dx} = 3x^2 y + y^2 = y^2(3x^2 + 1)$. Separating variables and integrating, we
   have

$$\int y^{-2} dy = \int (3x^2 + 1) dx$$

$$-\frac{1}{y} = x^3 + x + C.$$

Using the side condition, we find $\frac{1}{2} = C$. So the solution is

$$y = -\frac{1}{x^3 + x + \frac{1}{2}} = -\frac{2}{2x^3 + 2x + 1}.$$

11. We have $\dfrac{dy}{dx} = -\frac{3}{2} x^2 y$. Separating variables and integrating, we have

$$\int \frac{dy}{y} = \int -\frac{3}{2} x^2 \, dx$$

$$\ln|y| = -\frac{1}{2} x^3 + C$$

Using the condition $y(0) = 3$, we have $\ln 3 = C$. Therefore,

$$\ln|y| = -\frac{1}{2} x^3 + \ln 3 \quad \text{or} \quad y = e^{-(1/2)x^3 + \ln 3} = 3e^{-x^3/2}.$$

13. a. $x_0 = 0$, $b = 1$, and $n = 4$. Therefore, $h = 0.25$. So
   $x_0 = 0$, $x_1 = 0.25$, $x_2 = 0.5$, $x_3 = 0.75$, $x_4 = 1$.
   Also $F(x, y) = x + y^2$, $y(0) = 0$. Therefore,
   $\tilde{y}_0 = y_0 = 0$
   $\tilde{y}_1 = \tilde{y}_0 + hF(x_0, \tilde{y}_0) = 0 + 0.25(0) = 0$
   $\tilde{y}_2 = \tilde{y}_1 + hF(x_1, \tilde{y}_1) = 0 + 0.25(0.25 + 0) = 0.0625$.
   $\tilde{y}_3 = \tilde{y}_2 + hF(x_2, \tilde{y}_2) = 0.0625 + 0.25(0.5 + 0.0625^2) = 0.1884766$.
   $\tilde{y}_4 = \tilde{y}_3 + hF(x_3, \tilde{y}_3) = 0.1884766 + 0.25(0.75 + 0.884766^2) = 0.3848575$.
   Therefore, $y(1) = 0.3849$.
   b. $n = 6$. Therefore, $h = 0.1666667$.
   So $x_0 = 0$, $x_1 = 1.666667$, $x_2 = 0.3333334$, $x_3 = 0.5$, $x_4 = 0.6666667$,
      $x_5 = 0.8333334$, and $x_6 = 1$.
   So
   $\tilde{y}_0 = y_0 = 0$
   $\tilde{y}_1 = \tilde{y}_0 + hF(x_0, \tilde{y}_0) = 0 + 0.1666667(0 + 0) = 0$

$\tilde{y}_2 = \tilde{y}_1 + hF(x_1, \tilde{y}_1) = 1 + 0.1666667[0.1666667 + 0^2] = 0.0277778.$

$\tilde{y}_3 = \tilde{y}_2 + hF(x_2, \tilde{y}_2) = 0.0277778 + 0.1666667(0.3333334 + 0.0277778^2) = 0.0834619.$

$\tilde{y}_4 = \tilde{y}_3 + hF(x_3, \tilde{y}_3) = 0.0834619 + 0.1666667(0.5 + 0.0834619^2)$

$\approx 0.1679562.$

$\tilde{y}_5 = \tilde{y}_4 + hF(x_4, \tilde{y}_4) = 0.1679562 + 0.1666667(0.666667 + 0.1679562^2)$

$\approx 0.2837689.$

$\tilde{y}_6 = \tilde{y}_5 + hF(x_5, \tilde{y}_5) = 0.2837689 + 0.1666667(0.8333334 + 0.2837689^2)$

$\approx 0.4360785.$

Therefore, $y(1) \approx 0.4361.$

15. a. Here $x_0 = 0$, $b = 1$, and $n = 4$. So $h = 1/4 = 0.25$. So
$x_1 = 0$, $x_1 = 0.25$, $x_2 = 0.5$, $x_3 = 0.75$, and $x_4 = 1$.
$F(x, y) = 1 + 2xy^2$ and $y(0) = 0$. So
$\tilde{y}_0 = y_0 = 0$
$\tilde{y}_1 = \tilde{y}_0 + hF(x_0, \tilde{y}_0) = 0 + 0.25[1 + 0] = 0.25.$
$\tilde{y}_2 = \tilde{y}_1 + hF(x_1, \tilde{y}_1) = 0.25 + 0.25[1 + 2(0.05)(0.25)^2] = 0.507812.$
$\tilde{y}_3 = \tilde{y}_2 + hF(x_2, \tilde{y}_2) = 0.507813 + 0.25[1 + 2(0.5)(0.507813)^2] \approx 0.822281.$
$\tilde{y}_4 = \tilde{y}_3 + hF(x_3, \tilde{y}_3) = 0.822281 + 0.25[1 + 2(0.75)(0.822281)^2] \approx 1.32584.$
Therefore, $y(1) = 1.3258.$
b. $n = 6$. Therefore, $h = 1/6 = 0.1666667$.
So
$x_0 = 0$, $x_1 = 1.666667$, $x_2 = 0.3333334$, $x_3 = 0.5$, $x_4 = 0.6666667$, $x_5 = 0.8333334$,
and $x_6 = 1$. So
$\tilde{y}_0 = y_0 = 0$
$\tilde{y}_1 = \tilde{y}_0 + hF(x_0, \tilde{y}_0) = 0 + 0.1666667(1 + 0) = 0.166667.$
$\tilde{y}_2 = \tilde{y}_1 + hF(x_1, \tilde{y}_1) = 1 + 0.1666667[1 + 2(0.166667)(0.166667)^2] = 0.334877.$
$\tilde{y}_3 = \tilde{y}_2 + hF(x_2, \tilde{y}_2) = 0.334877 + 0.1666667[1 + 2(0.333333)(0.334877)^2]$

$\approx 0.514003.$

$\tilde{y}_4 = \tilde{y}_3 + hF(x_3, \tilde{y}_3) = 0.514004 + 0.1666667[1 + 2(0.50000)(0.515005)^2]$

$\approx 0.724703.$

$$\tilde{y}_5 = \tilde{y}_4 + hF(x_4, \tilde{y}_4) = 0.724704 + 0.1666667(1 + 2(0.666667)(0.724704)^2]$$
$$\approx 1.008081.$$
$$\tilde{y}_6 = \tilde{y}_5 + hF(x_5, \tilde{y}_5) = 1.008081 + 0.1666667[1 + 2(0.833333)(1.008081)^2]$$
$$\approx 1.457034.$$
Therefore, $y(1) = 1.4570.$

17. Here $x_0 = 0$, $b = 1$, so with $n = 5$, and $h = 1/5 = 0.2$ Then
$x_0 = 0$, $x_1 = 0.2$, $x_2 = 0.4$, $x_3 = 0.6$, $x_4 = 0.8$, and $x_5 = 1$.
Also, $F(x, y) = 2xy$ and $y_0 = y(0) = 1$.
Therefore,
$$\tilde{y}_0 = y_0 = 1$$
$$\tilde{y}_1 = \tilde{y}_0 + hF(x_0, \tilde{y}_0) = 1 + (0.2)(2)(0)(1) = 1$$
$$\tilde{y} = \tilde{y}_1 + hF(x_1, \tilde{y}_1) = 1 + (0.2)(2)(0.2)(1) = 1.08.$$
$$\tilde{y}_3 = \tilde{y}_2 + hF(x_2, \tilde{y}_2) = 1.08 + (0.2)(2)(0.4)(1.08) = 1.2528.$$
$$\tilde{y}_4 = \tilde{y}_3 + hF(x_3, \tilde{y}_3) = 1.2528 + (0.2)(2)(0.6)(1.2528) = 1.553472.$$
$$\tilde{y}_5 = \tilde{y}_4 + hF(x_4, \tilde{y}_4) = 1.553472 + (0.2)(2)(0.8)(1.553472) = 2.05058.$$
Therefore, $y(1) = 2.0506.$

19. $\dfrac{dS}{dT} = -kS;$ Thus, $S(t) = S_0 e^{-kt}$, $S(t) = 50,000 e^{-kt}$.

$$S(2) = 32,000 = 50,000 e^{-2k}, \ e^{-2k} = \frac{32}{50} = 0.64.$$
$-2k \ln e \approx \ln 0.64$,   or $k = 0.223144.$
a. Therefore, $S = 50,000 e^{-0.223144t} = 50,000(0.8)^t$.
b. $S(5) = 50,000(0.8)^5 = 16,384$,   or \$16,384.

21. Separating variables and integrating, we have
$$\int \frac{dA}{rA + P} = \int dt$$

$$\frac{1}{r} \ln(rA + P) = t + C_1$$
$$\ln(rA + P) = rt + C_2 \quad (C_2 = C_1 r)$$
$$rA + P = Ce^{rt}.$$

So $A = \dfrac{1}{r}(Ce^{rt} - P)$. Using the condition $A(0) = 0$, we have

$$0 = \dfrac{1}{r}(C - P), \text{ or } C = P.$$

So $A = \dfrac{P}{R}(e^{rt} - 1)$. The size of the fund after five years is

$$A = \dfrac{50,000}{0.12}(e^{(0.12)(5)} - 1) \approx 342,549.50, \text{ or approximately } \$342,549.50.$$

23. According to Newton's Law of cooling,

$$\dfrac{dT}{dt} = k(350 - T)$$

where T is the temperature of the roast. We also have the conditions $T(0) = 68$ and $T(2) = 118$. Separating the variables in the differential equation and integrating, we have

$$\dfrac{dT}{350 - T} = k\,dt$$

$$\ln|350 - T| = kt + C_1$$

$$350 - T = Ce^{kt},$$

or
$$T = 350 - Ce^{kt}.$$
Using the condition $T(0) = 68$, we have
$$350 - C = 68, \text{ or } C = 282.$$
So
$$T = 350 - 282e^{kt}.$$
Next, we use the condition $T(2) = 118$ to find
$$118 = 350 - 282e^{2k}$$

$$e^{2k} = 0.822695035.$$

So
$$k = \tfrac{1}{2}\ln 0.822695035 \approx -0.097584.$$
Therefore, $T = 350 - 282e^{-0.097584t}$.
We want to find t when $T = 150$. So we solve
$$150 = 350 - 282e^{-0.097584t}$$
$$e^{-0.097584t} = 0.709219858$$
$$-0.097584t = \ln 0.709219858 = -0.343589704,$$
or
$$t = 3.52096, \text{ or approximately } 3.5 \text{ hours.}$$
So the roast would have been $150°\,F$ at approximately 7:30 P.M.

25. $N = \dfrac{200}{1 + 49e^{-200kt}}$. When $t = 2$, $N = 40$, and

$$40 = \frac{200}{1 + 49e^{-400k}}$$

$$40 + 1960e^{-400k} = 200$$

$$e^{-400k} = \frac{160}{1960}(0.0816327)$$

$$-400 \ln k = \ln 0.0816327$$

$$k = 0.00626.$$

Therefore, $N(5) = \dfrac{200}{1 + 49e^{-200(5)(0.00626)}} \approx 183$, or 183 families.

CHAPTER 10

EXERCISES 10.1 , page 663

1. $f(x) = \dfrac{1}{3} > 0$ on [3,6]. Next, $\displaystyle\int_3^6 \frac{1}{3}\,dx = \frac{1}{3}x\Big|_3^6 = \frac{1}{3}(6-3) = 1.$

3. $f(x) \geq 0$ on [2,6]. Next $\displaystyle\int_2^6 \frac{2}{32}x\,dx = \frac{1}{32}x^2\Big|_2^6 = \frac{1}{32}(36-4) = 1,$
   and so f is a probability density function on [2,6].

5. $f(x) = \frac{2}{9}(3x - x^2)$ is nonnegative on [0,3] since both the factors x
   and 3 - x are nonnegative there. Next, we compute
   $$\int_0^3 \frac{2}{9}(3x - x^2)\,dx = \frac{2}{9}\left(\frac{3}{2}x^2 - \frac{1}{3}x^3\right)\Big|_0^3 = \frac{2}{9}\left(\frac{27}{2} - 9\right) = 1,$$
   and so f is a probability density function.

7. $f(x) = \dfrac{12 - x}{72}$ is nonnegative on [0,12]. Next, we see that
   $$\int_0^{12} \frac{12-x}{72}\,dx = \int_0^{12}\left(\frac{1}{6} - \frac{x}{72}\right)dx = \frac{1}{6}x - \frac{1}{144}x^2\Big|_0^{12} = 2 - 1 = 1,$$
   and conclude that f is a probability function.

9. $f(x) = \dfrac{8}{7x^2}$ is nonnegative on [1,8]. Next, we compute
   $$\int_1^8 \frac{8}{7x^2}\,dx = -\frac{8}{7x}\Big|_1^8 = -\frac{8}{7}\left(\frac{1}{8} - 1\right) = 1,$$
   and so f is a probability density function.

11. First $f(x) \geq 0$ on [0,∞) is self-evident. Next,
    $$\int_0^\infty \frac{x}{(x^2+1)^{3/2}}\,dx = \lim_{b\to\infty}\int_0^b x(x^2+1)^{-3/2}\,dx$$
    Let $I = \int x(x^2+1)^{-3/2}\,dx$. Integrate I by the method of substitution letting
    $u = x^2 + 1$ so that $du = 2x\,dx,$ and so

$$I = \frac{1}{2}\int u^{-3/2} du = \frac{1}{2}(-2u^{-1/2}) + C = -\frac{1}{\sqrt{u}} + C = -\frac{1}{\sqrt{x^2+1}} + C.$$

Therefore, $\displaystyle\int_0^\infty \frac{x}{(x^2+1)^{3/2}} dx = \lim_{b\to\infty}\left[-\frac{1}{\sqrt{x^2+1}}\Big|_0^b\right] = \lim_{b\to\infty}\left(-\frac{1}{\sqrt{b^2+1}}+1\right) = 1,$

and this completes the proof.

13. $\displaystyle\int_1^4 k\,dx = kx\Big|_1^4 = 3k = 1$ implies $k = \frac{1}{3}$.

15. $\displaystyle\int_0^4 k(4-x)\,dx = k\int_0^4 (4-x)\,dx = k(4x - \frac{1}{2}x^2)\Big|_0^4 = k(16-8) = 8k = 1$
implies that $k = 1/8$.

17. $\displaystyle\int_0^4 kx^{1/2}\,dx = \frac{2}{3}kx^{3/2}\Big|_0^4 = \frac{16}{3}k = 1$ implies $k = \frac{3}{16}$.

19. $\displaystyle\int_1^\infty \frac{k}{x^3}\,dx = \lim_{b\to\infty}\int_1^b kx^{-3}\,dx = \lim_{b\to\infty}-\frac{k}{2x^2}\Big|_1^b = \lim_{b\to\infty}\left(-\frac{k}{2b^2}+\frac{k}{2}\right) = \frac{k}{2} = 1$
implies $k = 2$.

21. a. $\displaystyle P(2 \le x \le 4) = \int_2^4 \frac{1}{12}x\,dx = \frac{1}{24}x^2\Big|_2^4 = \frac{1}{24}(16-4) = \frac{1}{2}.$

    b. $\displaystyle P(1 \le x \le 4) = \int_1^4 \frac{1}{12}x\,dx = \frac{1}{24}x^2\Big|_1^4 = \frac{1}{24}(16-1) = \frac{5}{8}.$

    c. $\displaystyle P(x \ge 2) = \int_2^5 \frac{1}{12}x\,dx = \frac{1}{24}x^2\Big|_2^5 = \frac{1}{24}(25-4) = \frac{7}{8}.$

    d. $\displaystyle P(x = 2) = \int_2^2 \frac{1}{12}x\,dx = 0.$

23. a. $\displaystyle P(-1 \le x \le 1) = \int_{-1}^1 \frac{3}{32}(4-x^2)\,dx = \frac{3}{32}(4x - \frac{1}{3}x^3)\Big|_{-1}^1 = \frac{3}{32}[(4-\frac{1}{3})-(-4+\frac{1}{3})] = \frac{11}{16}.$

    b. $\displaystyle P(x \le 0) = \int_{-2}^0 \frac{3}{32}(4-x^2)\,dx = \frac{3}{32}(4x - \frac{1}{3}x^3)\Big|_{-2}^0 = \frac{3}{32}[0-(-8+\frac{8}{3})] = \frac{1}{2}.$

    c. $\displaystyle P(x > -1) = \int_{-1}^2 \frac{3}{32}(4-x^2)\,dx = \frac{3}{32}(4x - \frac{1}{3}x^3)\Big|_{-1}^2 = \frac{3}{32}[(8-\frac{8}{3})-(-4+\frac{1}{3})] = \frac{27}{32}.$

    d. $\displaystyle P(x = 0) = \int_0^0 \frac{3}{32}(4-x^2)\,dx = 0.$

25.   a. $P(x \geq 4) = \int_4^9 \frac{1}{4} x^{-1/2} \, dx = \frac{1}{2} x^{1/2} \Big|_4^9 = \frac{1}{2}(3-2) = \frac{1}{2}.$

b. $P(1 \leq x \leq 8) = \int_1^8 \frac{1}{4} x^{-1/2} \, dx = \frac{1}{2} x^{1/2} \Big|_1^8 = \frac{1}{2}(2\sqrt{2} - 1) \approx 0.9142.$

c. $P(x = 3) = \int_3^3 \frac{1}{4} x^{-1/2} \, dx = 0.$

d. $P(x \leq 4) = \int_1^4 \frac{1}{4} x^{-1/2} \, dx = \frac{1}{2} x^{1/2} \Big|_1^4 = \frac{1}{2}(2-1) = \frac{1}{2}.$

27.   a. $P(0 \leq x \leq 4) = \int_0^4 4xe^{-2x^2} \, dx = -e^{-2x^2} \Big|_0^4 = -e^{-32} + 1 \approx 1.$

b. $P(x \geq 1) = \int_1^\infty 4xe^{-2x^2} \, dx = \lim_{b \to \infty} \int_1^b 4xe^{-2x^2} \, dx$

$$= \lim_{b \to \infty} \left[-e^{-2x^2} \Big|_1^b \right] = \lim_{b \to \infty}(-e^{-2b^2} + e^{-2}) = e^{-2} \approx 0.1353.$$

29.   a. $P(\frac{1}{2} \leq x \leq 1) = \int_{1/2}^1 x \, dx = \frac{1}{2} x^2 \Big|_{1/2}^1 = \frac{1}{2}(1) - \frac{1}{2}\left(\frac{1}{4}\right) = \frac{1}{2} - \frac{1}{8} = \frac{3}{8}.$

b. $P(\frac{1}{2} \leq x \leq \frac{3}{2}) = \int_{1/2}^1 x \, dx + \int_1^{3/2} (2-x) \, dx = \frac{3}{8} + \left(2x - \frac{1}{2} x^2 \right) \Big|_1^{3/2}$

(Using the results from (9))

$$= \frac{3}{8} + \frac{1}{2} x(4-x) \Big|_1^{3/2} = \frac{3}{8} + \left[\frac{1}{2}\left(\frac{3}{2}\right)\left(4 - \frac{3}{2}\right) - \frac{1}{2}(3) \right]$$

$$= \frac{3}{8} + \frac{3}{4}\left(\frac{5}{2}\right) - \frac{3}{2} = \frac{3}{4}.$$

c. $P(x \geq 1) = \int_1^2 (2-x) \, dx = 2x - \frac{1}{2} x^2 \Big|_1^2 = (4-2) - \left(2 - \frac{1}{2}\right) = \frac{1}{2}.$

d. $P(x \leq \frac{3}{2}) = \int_0^1 x \, dx + \int_1^{3/2} (2-x) \, dx = \frac{1}{2} x^2 \Big|_0^1 + \left(2x - \frac{1}{2} x^2 \right) \Big|_1^{3/2}$

$$= \frac{1}{2} + \left[2\left(\frac{3}{2}\right) - \frac{1}{2}\left(\frac{9}{4}\right) \right] - \left(2 - \frac{1}{2} \right) = \frac{1}{2} + \left(3 - \frac{9}{8} \right) - \frac{3}{2} = \frac{1}{2} + \frac{15}{8} - \frac{3}{2} = \frac{7}{8}.$$

31. f is nonnegative on D. $\int_1^3 \int_0^2 \frac{1}{4} xy\, dx\, dy = \int_1^3 \frac{1}{4} x^2 \Big|_0^2 \, dy = \int_1^3 \frac{1}{2} dy = \frac{1}{2} y \Big|_1^3 = \frac{3}{4} - \frac{1}{2} = 1.$

33. f is nonnegative on D.

$$\frac{1}{3} \int_1^2 \int_0^2 xy\, dx\, dy = \frac{1}{3} \int_1^2 \left[\frac{1}{2} x^2 y \Big|_0^2 \right] dy = \frac{1}{3} \int_1^2 2y\, dy = y^2 \Big|_1^2 = \frac{1}{3}(4-1) = \frac{1}{3} \cdot 3 = 1.$$

35. $\quad k \int_1^2 \int_0^1 x^2 y\, dx\, dy = k \int_1^2 \left[\frac{1}{3} x^3 y \Big|_0^1 \, dy \right] = k \int_1^2 \frac{1}{3} y\, dy = k \frac{1}{6} y^2 \Big|_1^2 = \frac{3k}{6} = \frac{k}{2}.$

Therefore, $k = 2$.

37. $\quad k \int_1^\infty \int_0^1 (x - x^2) e^{-2y}\, dx\, dy = k \int_1^\infty \frac{1}{2} x^2 - \frac{1}{3} x^3 \Big|_0^1 \, e^{-2y}\, dy = k \left(\frac{1}{6} \right) \left(-\frac{1}{2} e^{-2y} \right) \Big|_1^\infty$

$$= -\frac{k}{12} e^{-2y} \Big|_1^\infty = \frac{ke^{-2}}{12}.$$

Therefore, $k = 12e^2$.

39. a. $P(0 \le x \le 1, 0 \le y \le 1) = \int_0^1 \int_0^1 x\, y\, dx\, dy = \int_0^1 \left[\frac{1}{2} x^2 y \Big|_0^1 \right] dy = \int_0^1 \frac{1}{2} y\, dy = \frac{1}{4} y^2 \Big|_0^1 = \frac{1}{4}.$

b. $P(\{x, y\} | x + 2y \le 1) = \int_0^1 \int_0^{(1-x)/2} xy\, dy\, dx = \int_0^1 \left[\frac{1}{2} xy^2 \Big|_0^{(1-x)/2} \right] dx = \frac{1}{8} \int_0^1 (x - 2x^2 + x^3)\, dx$

$$= \frac{1}{8} \left(\frac{1}{2} x^2 - \frac{2}{3} x^3 + \frac{1}{4} x^4 \right) \Big|_0^1 = \frac{1}{8} \left(\frac{1}{2} - \frac{2}{3} + \frac{1}{4} \right) = \frac{1}{96} \ .$$

41. a. $P(0 \le x \le 2, 0 \le y \le 1) = \dfrac{9}{224} \int_0^1 \int_1^2 x^{1/2} y^{1/2}\, dx\, dy = \dfrac{9}{224} \int_0^1 \frac{2}{3} x^{3/2} y^{1/2} \Big|_1^2 \, dy$

$$= \frac{3}{112} \int_0^1 (2\sqrt{2} y^{3/2} - y^{3/2})\, dy = \frac{3}{112} (2\sqrt{2} - 1) \int_0^1 y^{3/2} dy$$

$$= \frac{3}{112} (2\sqrt{2} - 1) \left[\frac{2}{3} y^{3/2} \Big|_0^1 \right] = \frac{1}{56} (2\sqrt{2} - 1)$$

b. $P(\{(x, y) | 1 \le x \le 4; 0 \le y \le \sqrt{x}) = \dfrac{9}{224} \int_1^4 \int_0^{\sqrt{x}} x^{1/2} y^{1/2}\, dy\, dx = \dfrac{9}{224} \int_1^4 \frac{2}{3} x^{1/2} y^{3/2} \Big|_0^{\sqrt{x}} \, dx$

$$= \frac{3}{112} \int_1^4 x^{1/2} (x^{1/2})^{3/2}\, dx = \frac{3}{112} \int_1^4 x^{5/4}\, dx$$

$$= \frac{3}{112} \cdot \frac{4}{9} x^{9/4} \Big|_1^4 = \frac{4}{336}(4^{9/4} - 1) \approx .2575$$

43.   a. $P(x \le 100) = \int_0^{100} \frac{1}{100} e^{-x/100}\, dx = -e^{-x/100}\Big|_0^{100} = -e^{-1} + 1 \approx 0.6321.$

     b. $P(x \ge 120) = \int_{120}^{\infty} \frac{1}{100} e^{-x/100} dx = \lim_{b \to \infty} \int_{120}^{b} \frac{1}{100} e^{-x/100} dx$

$$= \lim_{b \to \infty} -e^{-x/100}\Big|_{120}^{b} = \lim_{b \to \infty}(-e^{-b/100} + e^{-(120/100)}) = e^{-1.2} \approx 0.3012.$$

     c. $P(60 < x < 140) = \int_{60}^{140} \frac{1}{100} e^{-x/100} dx = -e^{-x/100}\Big|_{60}^{140} = -e^{-1.4} + e^{-0.6} \approx 0.3022.$

45.   a. $P(600 \le t \le 800) = \int_{600}^{800} 0.001 e^{-0.001t}\, dt = -e^{-0.001t}\Big|_{600}^{800} = -e^{-0.8} + e^{-0.6}$

$$\approx 0.0995.$$

     b. $P(t \ge 1200) = \int_{1200}^{\infty} 0.001 e^{-0.001t} dt = \lim_{b \to \infty} \int_{1200}^{b} 0.001 e^{-0.001t} dt$

$$= \lim_{b \to \infty} -e^{-0.001t}\Big|_{1200}^{b} = \lim_{b \to \infty}(-e^{-0.001b} + e^{-1.2}) = e^{-1.2} \approx 0.3012.$$

47. $P(t \ge 2) = \int_2^{\infty} \frac{1}{30} e^{-t/30}\, dt = \lim_{b \to \infty} \int_2^{b} \frac{1}{30} e^{-t/30}\, dt = \lim_{b \to \infty} -e^{-t/30}\Big|_2^{b}$

$$= \lim_{b \to \infty}(-e^{-b/30} + e^{-1/15}) = e^{-1/15} \approx 0.9355.$$

49. $P(1 \le x \le 2) = \int_1^2 \frac{2}{9} x(3-x) dx = \frac{2}{9}(\frac{3}{2}x^2 - \frac{1}{3}x^3)\Big|_1^2 = \frac{2}{9}[(6 - \frac{8}{3}) - (\frac{3}{2} - \frac{1}{3})] = \frac{13}{27} \approx 0.4815.$

     $P(x \ge 1) = \int_1^3 \frac{2}{9} x(3-x) dx = \frac{2}{9}(\frac{3}{2}x^2 - \frac{1}{3}x^3)\Big|_1^3 = \frac{2}{9}[(\frac{27}{2} - 9) - (\frac{3}{2} - \frac{1}{3})] = \frac{20}{27} \approx 0.740741.$

51. $P(t > 4) = \int_4^{\infty} 9(9 + t^2)^{-3/2}\, dt = \lim_{b \to \infty} \int_4^{b} 9(9 + t^2)^{-3/2}\, dt$

$$= \lim_{b \to \infty} \frac{t}{\sqrt{9 + t^2}}\Big|_4^{b} = \lim_{b \to \infty}\left(\frac{b}{\sqrt{9 + b^2}} - \frac{4}{5}\right) = 1 - \frac{4}{5} = \frac{1}{5}.$$

53. $P(2 \le x \le 2.5, 1 \le y \le 2) = \dfrac{9}{4000} \displaystyle\int_{2}^{2.5} \int_{1}^{2} xy(25 - x^2)^{1/2}(4 - y)\,dx\,dx$

$$= \dfrac{9}{4000} \int_{2}^{2.5} \left[(2y^2 - \dfrac{y^3}{3}) x(25 - x^2)^{1/2} \Big|_{1}^{2} \right] dy = \dfrac{9}{4000} \cdot \dfrac{11}{3} \int_{2}^{2.5} x(25 - x^2)^{1/2}\,dx$$

$$= -\dfrac{11}{4000}(25 - x^2)^{3/2} \Big|_{2}^{2.5} = -\dfrac{11}{4000} \left\{ \left[24 - \left(\dfrac{5}{2} \right)^2 \right]^{3/2} - [25 - 4]^{3/2} \right\} \approx .041372$$

55. False. f must be nonnegative on $[a, b]$ as well.

USING TECHNOLOGY EXERCISES 10.1, page 665

1.

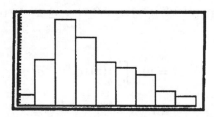

3. a.

x	0	1	2	3	4	5
$P(X=x)$	0.017	0.067	0.033	0.117	0.233	0.133

x	6	7	8	9	10
$P(X=x)$	0.167	0.1	0.05	0.067	0.017

   b.

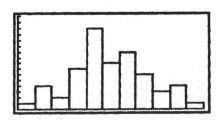

1. $\mu = \int_3^6 \frac{1}{3}x\,dx = \frac{1}{6}x^2\Big|_3^6 = \frac{1}{6}(36-9) = \frac{9}{2}.$

$\text{Var}(x) = \int_3^6 \frac{1}{3}x^2\,dx - \frac{81}{4} = \frac{1}{9}x^3\Big|_3^6 - \frac{81}{4} = \frac{1}{9}(216-27) - \frac{81}{4} = \frac{3}{4}.\ \sigma = \sqrt{\frac{3}{4}} = \frac{\sqrt{3}}{2} \approx 0.8660.$

3. $\mu = \int_0^5 \frac{3}{125}x^3\,dx = \frac{3}{500}x^4\Big|_0^5 = \frac{15}{4}.$

$\text{Var}(x) = \int_0^5 \frac{3}{125}x^4\,dx - \frac{225}{16} = \frac{3}{625}x^5\Big|_0^5 - \frac{225}{16} = 15 - \frac{225}{16} = \frac{15}{16}.\ \sigma = \sqrt{\frac{15}{16}} = \frac{\sqrt{15}}{4} \approx 0.9682.$

5. $\mu = \int_1^5 \frac{3}{32}x(x-1)(5-x)\,dx = \frac{3}{32}\int_1^5 (-x^3 + 6x^2 - 5x)\,dx$

$= \frac{3}{32}\left(-\frac{1}{4}x^4 + 2x^3 - \frac{5}{2}x^2\right)\Big|_1^5 = \frac{3}{32}\left[\left(-\frac{625}{4} + 250 - \frac{125}{2}\right) - \left(-\frac{1}{4} + 2 - \frac{5}{2}\right)\right] = 3.$

$\text{Var}(x) = \int_1^5 \frac{3}{32}\left[-\frac{1}{4}x^4 + 6x^3 - 5x^2\right)dx - 9 = \frac{3}{32}\left(-\frac{1}{5}x^5 + \frac{3}{2}x^4 - \frac{5}{3}x^3\right)\Big|_1^5 - 9$

$= \frac{3}{32}\left[(-625 + \frac{1875}{2} - \frac{625}{3}) - (-\frac{1}{5} + \frac{3}{2} - \frac{5}{3})\right] - 9 \approx 0.8..$

$\sigma = \sqrt{0.8} \approx 0.8944.$

7. $\mu = \int_1^8 \frac{8}{7x}\,dx = \frac{8}{7}\ln x\Big|_1^8 = \frac{8}{7}\ln 9 \approx 2.3765.$

$\text{Var}(x) = \int_0^8 \frac{8}{7}\,dx - 5.64777 = \frac{8}{7}x\Big|_1^8 - 5.64777 = 8 - 5.647777 \approx 2.3522.$

$\sigma = \sqrt{2.3522} \approx 1.5337.$

9. $\mu = \int_1^4 \frac{3}{14}x^{3/2}\,dx = \frac{3}{35}x^{5/2}\Big|_1^4 = \frac{3}{35}(32-1) = \frac{93}{35}.$

$\text{Var}(x) = \int_1^4 \frac{3}{14}x^{5/2}\,dx - \frac{9}{25} = \frac{3}{49}x^{7/2}\Big|_1^4 - \frac{9}{25} = \frac{3}{49}(128-1) - \frac{9}{25} \approx 0.715102.$

$\sigma = \sqrt{7.4155} \approx 0.8456.$

11. $\mu = \int_1^\infty \dfrac{3}{x^3} dx = \lim_{b \to \infty} \int_1^b 3x^{-3} dx = \lim_{b \to \infty} -\dfrac{3}{2x^2}\Big|_1^b = \lim_{b \to \infty}\left(-\dfrac{3}{2b^2} + \dfrac{3}{2}\right) = \dfrac{3}{2}.$

$\mathrm{Var}(x) = \int_1^\infty \dfrac{3}{x^2} dx - \dfrac{9}{4} = \lim_{b \to \infty} \int_1^b 3x^{-2} dx - \dfrac{9}{4} = \lim_{b \to \infty} -\dfrac{3}{x}\Big|_1^b - \dfrac{9}{4}$

$= \lim_{b \to \infty}\left(-\dfrac{3}{b} + 3\right) - \dfrac{9}{4} = \dfrac{3}{4}.$

$\sigma = \sqrt{\dfrac{3}{4}} = \dfrac{\sqrt{3}}{2} \approx 0.8660.$

13. $\mu = \int_0^\infty \tfrac{1}{4} x e^{-x/4} dx = \lim_{b \to \infty} \int_0^b \tfrac{1}{4} x e^{-x/4} dx = \lim_{b \to \infty} 4(-\tfrac{1}{4} - 1)e^{-x/4}\Big|_0^b$

$= \lim_{b \to \infty}[4(-\tfrac{1}{4}b - 1)e^{-b/4} + 4] = 4.$

$\mathrm{Var}(x) = \int_0^\infty \tfrac{1}{4} x^2 e^{-x/4} dx - 16 = \lim_{b \to \infty} \int_0^b \tfrac{1}{4} x^2 e^{-x/4} dx - 16$

$= \lim_{b \to \infty} x^2 e^{-x/4} - 32(-\tfrac{1}{4}x - 1)e^{-x/4}\Big|_0^b - 16$

$= \lim_{b \to \infty}\left\{[b^2 e^{-b/4} - 32(-\tfrac{1}{4}b - 1)e^{-b/4}] + 32\right\} - 16 = 16.$

$\sigma = \sqrt{16} = 4.$

15. $\mu = \int_0^\infty x \cdot \dfrac{1}{100} e^{-x/100} dx = \lim_{b \to \infty} \int_0^b \tfrac{1}{100} x e^{-x/100} dx = \lim_{b \to \infty}\left[100\left(-\dfrac{x}{100} - 1\right)e^{-x/100}\right]\Big|_0^b$

$= \lim_{b \to \infty}\left[100\left(-\dfrac{b}{100} - 1\right)e^{-b/100} + 100\right] = 100.$

So a plant of this species is expected to live 100 days.

17. $\mu = \int_0^5 t \cdot \dfrac{2}{25} t \, dt = \dfrac{2}{25} \int_0^5 t^2 \, dt = \dfrac{2}{75} t^3\Big|_0^5 = \dfrac{2}{75}(125) = 3\dfrac{1}{3}.$

So a shopper is expected to spend $3\tfrac{1}{3}$ minutes in the magazine section.

19. $\mu = \int_0^3 x \cdot \dfrac{2}{9} x(3 - x) \, dx = \dfrac{2}{9} \int_0^3 (3x^2 - x^3) dx = \dfrac{2}{9}\left(x^3 - \dfrac{1}{4}x^4\right)\Big|_0^3$

$= \dfrac{2}{9}\left(27 - \dfrac{81}{4}\right) = 1.5.$

So the expected amount of snowfall is 1.5 ft.

21. $\mu = \int_0^5 x \cdot \frac{6}{125} x(5-x)\, dx = \frac{6}{125}\int_0^5 (5x^2 - x^3)\, dx = \frac{6}{125}\left(\frac{5}{3}x^3 - \frac{1}{4}x^4\right)\Big|_0^5$

$= \frac{6}{125}\left(\frac{625}{3} - \frac{625}{4}\right) = 2.5.$

So the expected demand is 2500 lb/wk.

23. $P(x \le m) = \int_2^m \frac{1}{6}\, dx = \frac{1}{6}x\Big|_2^m = \frac{1}{6}(m-2) = \frac{1}{2}.$ $m-2 = 3$, or $m = 5$.

25. $P(x \le m) = \int_0^m \frac{3}{16} x^{1/2}\, dx = \frac{1}{8}x^{3/2}\Big|_0^m = \frac{1}{8}m^{3/2}$. Next, we solve

$\frac{1}{8}m^{3/2} = \frac{1}{2}$, $m^{3/2} = 4$, or $m = 4^{2/3} \approx 2.5198.$

27. $P(x \le m) = \int_1^m 3x^{-2}\, dx = -\frac{3}{x}\Big|_1^m = -\frac{3}{m} + 3.$

$-\frac{3}{m} + 3 = \frac{1}{2}$ gives $-\frac{3}{m} = -\frac{5}{2}$, or $-5m = -6$, or $m = \frac{6}{5}.$

29. False. The expected value of x is $\int_a^b x f(x)\, dx$.

USING TECHNOLOGY EXERCISES 10.2, page 681

1. a.

   b. $\mu = 4$ and $\sigma = 1.40$

3. a. X gives the minimum age requirement for a regular driver's license.

b.

x	15	16	17	18	19	21
P(X=x)	0.02	0.30	0.08	0.56	0.02	0.02

c.

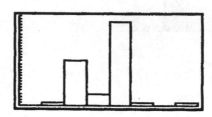

d. $\mu = 17.34$ and $\sigma = 1.11$

5.   a. Let X denote the random variable that gives the weight of a carton of sugar.
b. The probability distribution for the random variable X is

x	4.96	4.97	4.98	4.99	5.00	5.01	5.02	5.03
P(X = x)	$\dfrac{3}{30}$	$\dfrac{4}{30}$	$\dfrac{4}{30}$	$\dfrac{1}{30}$	$\dfrac{1}{30}$	$\dfrac{5}{30}$	$\dfrac{3}{30}$	$\dfrac{3}{30}$

x	5.04	5.05	5.06
P(X = x)	$\dfrac{4}{30}$	$\dfrac{1}{30}$	$\dfrac{1}{30}$

c. $\mu = 5.00467 \approx 5.00$; $V(X) = 0.0009$; $\sigma = \sqrt{0.0009} = 0.03$

EXERCISES 10.3, page 692

1. $P(Z < 1.45) = 0.9265$.   3. $P(Z < -1.75) = 0.0401$.

5. $P(-1.32 < Z < 1.74) = P(Z < 1.74) - P(Z < -1.32) = 0.9591 - 0.0934 = 0.8657$.

7. $P(Z < 1.37) = 0.9147$.

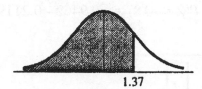

9. $P(Z < -0.65) = 0.2578$.

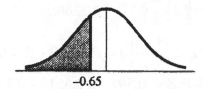

11. $P(Z > -1.25) = 1 - P(Z < -1.25) = 1 - 0.1056 = 0.8944$

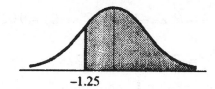

13. $P(0.68 < Z < 2.02) = P(Z < 2.02) - P(Z < 0.68) = 0.9783 - 0.7517 = 0.2266$.

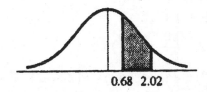

15.   a. Referring to Table 4, we see that $P(Z < z) = 0.8907$ implies that $z = 1.23$.
      b. Referring to Table 4, we see that $P(Z < z) = 0.2090$ implies that $z = -0.81$.

17.   a. $P(Z > -z) = 1 - P(Z < -z) = 1 - 0.9713 = 0.0287$ implies $z = 1.9$.
      b. $P(Z < -z) = 0.9713$ implies that $z = -1.9$.

19.   a. $P(X < 60) = P(Z < \dfrac{60-50}{5}) = P(Z < 2) = 0.9772$.

      b. $P(X > 43) = P(Z > \dfrac{43-50}{5}) = P(Z > -1.4) = P(Z < 1.4) = 0.9192$.

c. $P(46 < X < 58) = P(\frac{46-50}{5} < Z < \frac{58-50}{5}) = P(-0.8 < Z < 1.6)$

$\qquad\qquad = P(Z < 1.6) - P(Z < -0.8) = 0.9452 - 0.2119 = 0.7333.$

21. $\mu = 20$ and $\sigma = 2.6$.

a. $P(X > 22) = P(Z > \frac{22-20}{2.6}) = P(Z > 0.77) = P(Z < -0.77) = 0.2206.$

b. $P(X < 18) = P(Z < \frac{18-20}{2.6}) = P(Z < -0.77) = 0.2206.$

c. $P(19 < X < 21) = P(\frac{19-20}{2.6} < Z < \frac{21-20}{2.6}) = P(-0.39 < Z < 0.39)$

$\qquad\qquad = P(Z < 0.39) - P(Z < -0.39) = 0.6517 - 0.3483 = 0.3034.$

23. $\mu = 750$ and $\sigma = 75$.

a. $P(X > 900) = P(Z > \frac{900-750}{75}) = P(Z > 2) = P(Z < -2) = 0.0228.$

b. $P(X < 600) = P(Z < \frac{600-750}{75}) = P(Z < -2) = 0.0228.$

c. $P(750 < X < 900) = P(Z < \frac{750-750}{75} < Z < \frac{900-750}{75})$

$\qquad\qquad = P(0 < Z < 2) = P(Z < 2) - P(Z < 0)$

$\qquad\qquad = 0.9772 - 0.5000 = 0.4772.$

d. $P(600 < X < 800) = P(\frac{600-750}{75} < Z < \frac{800-750}{75})$

$\qquad\qquad = P(-2 < Z < .667) = P(Z < .667) - P(Z < -2)$

$\qquad\qquad = 0.7486 - 0.0228 = 0.7258.$

25. $\mu = 100$ and $\sigma = 15$.

a. $P(X > 140) = P(Z > \frac{140-100}{15}) = P(Z > 2.667) = P(Z < -2.667)$

$\qquad\qquad = 0.0038.$

b. $P(X > 120) = P(Z > \frac{120-100}{15}) = P(Z > 1.33) = P(Z < -1.33) = 0.0918.$

c. $P(100 < X < 120) = P(\frac{100-100}{15} < Z < \frac{120-100}{15}) = P(0 < Z < 1.333)$

$$= P(Z < 0) - P(Z < 1.333) = 0.9082 - 0.5000 = 0.4082.$$

d. $P(X < 90) = P(Z < \dfrac{90 - 100}{15}) = P(Z < -0.667) = 0.2514.$

27. Here $\mu = 475$ and $\sigma = 50$.

$$P(450 < X < 550) = P(\dfrac{450 - 475}{50} < Z < \dfrac{550 - 475}{50}) = P(-0.5 < Z < 1.5)$$

$$= P(Z < 1.5) - P(Z < -0.5) = 0.9332 - 0.3085 = 0.6247.$$

29. Here $\mu = 22$ and $\sigma = 4$.

$$P(X < 12) = P(Z < \dfrac{12 - 22}{4}) = P(Z < -2.5) = 0.0062, \text{ or } 0.62 \text{ percent.}$$

31. $\mu = 70$ and $\sigma = 10$.

To find the cut-off point for an A, we solve $P(Y < y) = 0.85$ for y. Now

$$P(Y < y) = P\left(Z < \dfrac{y - 70}{10}\right) = 0.85 \text{ implies } \dfrac{y - 70}{10} = 1.04$$

or $y = 80.4 \approx 80$.

For a B: $P(Y < y) = P\left(Z < \dfrac{y - 70}{10}\right) = 0.75; \dfrac{y - 70}{10} = .67, \text{ or } y \approx 77.$

For a C: $P(Y < y) = P\left(Z < \dfrac{y - 70}{10}\right) = 0.60 \text{ implies } \dfrac{y - 70}{10} = 0.25, \text{ or } y \approx 73.$

For a D: $P(Y < y) = P\left(Z < \dfrac{y - 70}{10}\right) = 0.15 \text{ implies } \dfrac{y - 70}{10} = -0.84 \text{ or } y \approx 62.$

For a F: $P(Y < y) = P\left(Z < \dfrac{y - 70}{10}\right) = 0.05 \text{ implies } \dfrac{y - 70}{10} = -1.65, \text{ or } y \approx 54.$

CHAPTER 10, REVIEW EXERCISES, page 695

. First $f(x) = \frac{1}{28}(2x + 3)$ is nonnegative on $[0,4]$. Next,

$$\int_0^4 \tfrac{1}{28}(2x + 3)dx = \tfrac{1}{28}(x^2 + 3x)\Big|_0^4 = \tfrac{1}{28}(16 + 12) = 1.$$

. First, $f(x) = \frac{1}{4} > 0$ on $[7,11]$. Next, $\int_7^{11} \tfrac{1}{4}dx = \tfrac{1}{4}x\Big|_7^{11} = \tfrac{1}{4}(11 - 7) = 1.$

10 Probability and Calculus

5. $\int_0^9 kx^2\,dx = \dfrac{k}{3}x^3\Big|_0^9 = \dfrac{k}{3}(729) = 1$. Therefore, $k = \dfrac{1}{243}$.

7. $\int_1^3 kx^{-2}\,dx = \dfrac{2k}{3} = 1$. Therefore, $k = \dfrac{3}{2}$.

9. a. $\int_2^4 \tfrac{2}{21}x\,dx = \tfrac{1}{21}x^2\Big|_2^4 = \tfrac{16}{21} - \tfrac{4}{21} = \tfrac{12}{21} = \tfrac{4}{7}$.

   b. $\int_4^4 \tfrac{2}{21}x\,dx = \tfrac{1}{21}x^2\Big|_4^4 = 0$.

   c. $\int_3^4 \tfrac{2}{21}x\,dx = \tfrac{1}{21}x^2\Big|_3^4 = \tfrac{16}{21} - \tfrac{9}{21} = \tfrac{7}{21} = \tfrac{1}{3}$.

11. a. $P(1 \le x \le 3) = \tfrac{3}{16}\int_1^3 x^{1/2}\,dx = \tfrac{1}{8}x^{3/2}\Big|_1^3 = \tfrac{1}{8}(3\sqrt{3}-1) \approx 0.52$.

    b. $P(x \le 3) = \tfrac{3}{16}\int_0^3 x^{1/2}\,dx = \tfrac{1}{8}x^{3/2}\Big|_0^3 = \tfrac{1}{8}(3\sqrt{3}-0) \approx 0.65$.

    c. $P(x = 2) = \tfrac{3}{16}\int_2^2 x^{1/2}\,dx = 0$.

13. $\mu = \tfrac{1}{5}\int_2^7 x\,dx = \tfrac{1}{10}x^2\Big|_2^7 = \tfrac{1}{10}(49-4) = \tfrac{9}{2}$.

    $\mathrm{Var}(x) = \tfrac{1}{5}\int_2^7 x^2\,dx - \left(\tfrac{9}{2}\right)^2 = \tfrac{1}{15}x^3\Big|_2^7 - (4.5)^2 = \tfrac{1}{15}(343-8) - (4.5)^2 \approx 2.083$.

    $\sigma = \sqrt{2.083} \approx 1.44$.

15. $\mu = \tfrac{1}{4}\int_{-1}^1 x(3x^2+1)\,dx = \tfrac{1}{4}\int_{-1}^1 (3x^3+x)\,dx = \tfrac{1}{4}\left(\tfrac{3}{4}x^4 + \tfrac{1}{2}x^2\right)\Big|_{-1}^1 = 0$.

    $\mathrm{Var}(x) = \int_{-1}^1 x^2(3x^2-1)\,dx - 0 = \tfrac{1}{4}\int_{-1}^1 (3x^4+x^2)\,dx = \tfrac{1}{4}\left(\tfrac{3}{5}x^5 + \tfrac{1}{3}x^3\right)\Big|_{-1}^1 = \tfrac{7}{15}$.

    $\sigma = \sqrt{\tfrac{7}{15}} \approx 0.6831$.

17. $P(Z < 2.24) = 0.9875$.

19. $P(0.24 \le Z \le 1.28) = P(Z \le 1.28) - P(Z \le 0.24) = 0.8997 - 0.5948 = 0.3049$.

21. f is nonnegative in D. Next,

$$\iint_D f(x,y)\,dA = \int_0^8 \int_0^4 \tfrac{1}{64} x^{1/2} y^{1/3}\,dx\,dy = \tfrac{1}{64}\int_0^8 \left[\tfrac{2}{3}x^{3/2}y^{1/3}\Big|_0^4\right]dy = \tfrac{1}{64}\cdot\tfrac{2}{3}\cdot 8\int_0^8 y^{1/3}\,dy$$

$$= \tfrac{1}{12}\left[\tfrac{3}{4}y^{4/3}\Big|_0^8\right] = \tfrac{1}{16}\cdot 16 = 1$$

and this proves that f is a joint probability density function on D.

23.   a. $P(X \le 84) = P\left(Z \le \dfrac{84-80}{8}\right) = P(Z \le 0.5) = 0.6915.$

     b. $P(X \ge 70) = P\left(\dfrac{70-80}{8}\right) = P(Z \ge -1.25) = P(Z \le 1.25) = 0.8944.$

     c. $P(75 \le X \le 85) = P\left(\dfrac{75-80}{8} \le Z \le \dfrac{85-80}{8}\right) = P(-0.625 \le Z \le 0.625)$

$$= P(Z \le 0.625) - P(Z \le -0.625) = 0.7341 - 0.2660 \approx 0.4681.$$

25.   a. $P(t > 6) = \displaystyle\int_6^\infty \tfrac{1}{4}e^{-t/4}\,dt = \lim_{b\to\infty}\int_6^b -e^{-t/4}\,dt = \lim_{b\to\infty}-e^{-t/4}\Big|_6^b$

$$= \lim_{b\to\infty}-e^{-b/4} + e^{-6/4} = 0.22313.$$

     b. $P(t < 2) = \displaystyle\int_0^2 \tfrac{1}{4}e^{-t/4}\,dt = -e^{-t/4}\Big|_0^2 = -e^{-1/2} + e^0 = -0.6065 + 1 = 0.39347.$

     c. $\mu = \displaystyle\int_0^\infty \tfrac{1}{4}te^{-t/4}\,dt = \lim_{b\to\infty}\int_0^b \tfrac{1}{4}te^{-t/4}\,dt = \lim_{b\to\infty} 4(-\tfrac{1}{4}t - 1)e^{-t/4}\Big|_0^b = 0 + 4 = 4.$

CHAPTER 11

EXERCISES 11.1, page 707

1. $f(x) = e^{-x}$, $f'(x) = -e^{-x}$, $f''(x) = e^{-x}$, and $f'''(x) = -e^{-x}$.
    So $f(0) = 1$, $f'(0) = -1$, $f''(0) = 1$, and $f'''(0) = -1$.
    Therefore,
    $$P_1(x) = f(0) + f'(0)x = 1 - x$$
    $$P_2(x) = f(0) = f'(0)x + \frac{f''(0)}{2!}x^2 = 1 - x + \tfrac{1}{2}x^2$$
    $$P_3(x) = f(0) + f'(0)x + \frac{f''(0)}{2!}x^2 + \frac{f'''(0)}{3!}x^3 = 1 - x + \tfrac{1}{2}x^2 - \tfrac{1}{6}x^3.$$

3. $f(x) = \dfrac{1}{x+1} = (x+1)^{-1}$, $f'(x) = -(x+1)^{-2}$, $f''(x) = 2(x+1)^{-3}$,
    and $f'''(x) = -6(x+1)^{-4}$. So
    $f(0) = 1$, $f'(0) = -1$, $f''(0) = 2$, and $f'''(0) = -6$. Therefore,
    $$P_1(x) = 1 - x$$
    $$P_2(x) = 1 - x + \frac{2}{2!}x^2 = 1 - x + x^2$$
    $$P_3(x) = 1 - x + \frac{2}{2!}x^2 - \frac{6}{3!}x^3 = 1 - x + x^2 - x^3.$$

5. $f(x) = \dfrac{1}{x} = x^{-2}$; $f'(x) = -x^{-2} = -\dfrac{1}{x^2}$, $f''(x) = (-1)(-2)x^{-3} = \dfrac{2}{x^3}$.
    $$f'''(x) = -\frac{6}{x^4}.$$
    So $f(1) = 1$, $f'(1) = -1$, $f''(1) = 2$, $f'''(1) = -6$. Therefore,
    $$P_1(x) = f(1) + f'(x)(x-1) = 1 - (x-1), \text{ or } P_1 = 1 - (x-1)$$
    $$P_2(x) = f(1) + f'(1)(x-1) + \frac{f''(1)}{2!}(x-1)^2 = 1 - (x-1) + (x-1)^2$$
    $$P_3(x) = f(1) + f'(1)(x-1) + \frac{f''(1)}{2!}(x-1)^2 + \frac{f'''(1)}{3!}(x-1)^3$$
    $$= 1 - (x-1) + (x-1)^2 - (x-1)^3.$$

7. $f(x) = (1-x)^{1/2}$, $f'(x) = -\dfrac{1}{2}(1-x)^{-1/2}$, $f''(x) = -\dfrac{1}{4}(1-x)^{-3/2}$,

$f'''(x) = -\dfrac{3}{8}(1-x)^{-5/2}$.

$f(0) = 1$, $f'(0) = -\dfrac{1}{2}$, $f''(0) = -\dfrac{1}{4}$, $f'''(0) = -\dfrac{3}{8}$.

$P_1 = 1 - \dfrac{1}{2}x$; $P_2(x) = 1 - \dfrac{1}{2}x + \dfrac{\left(-\frac{1}{4}\right)}{2!}x^2 = 1 - \dfrac{1}{2}x - \dfrac{1}{8}x^2$

$P_3 = 1 - \dfrac{1}{2}x - \dfrac{1}{8}x^2 + \dfrac{\left(-\frac{3}{8}\right)}{3!}x^3 = 1 - \dfrac{1}{2}x - \dfrac{1}{8}x^2 - \dfrac{1}{16}x^3$.

9. $f(x) = \ln(1-x)$, $f'(x) = -\dfrac{1}{1-x} = -(1-x)^{-1}$, $f''(x) = -(1-x)^{-2}$

$f'''(x) = -2(1-x)^{-3}$.
$f(0) = 0$, $f'(0) = -1$, $f''(0) = -1$, $f'''(0) = -2$.

$P_1(x) = -x$, $P_2(x) = -x + \dfrac{(-1)}{2!}x^2 = -x - \dfrac{1}{2}x^2$

$P_3(x) = -x - \dfrac{1}{2}x^2 + \dfrac{(-2)}{3!}x^3 = -x - \dfrac{1}{2}x^2 - \dfrac{1}{3}x^3$.

11. $f(x) = x^4$, $f'(x) = 4x^3$, $f''(x) = 12x^2$.
$f(2) = 16$, $f'(2) = 32$, $f''(2) = 48$

$P_2(x) = 16 + 32(x-2) + \dfrac{48}{2!}(x-2)^2 = 16 + 32(x-2) + 24(x-2)^2$.

13. $f(x) = \ln x$; $f'(x) = \dfrac{1}{x}$, $f''(x) = -\dfrac{1}{x^2}$, $f'''(x) = \dfrac{2}{x^3}$, $f^{(4)} = -\dfrac{6}{x^4}$.
$f(1) = 0$, $f'(1) = 1$, $f''(1) = -1$, $f'''(1) = 2$, $f^{(4)} = -6$.
$P_4(x) = (x-1) + \dfrac{(-1)}{2!}(x-1)^2 + \dfrac{2}{3!}(x-1)^3 + \dfrac{(-6)}{4!}(x-1)^4$

$= (x-1) - \dfrac{1}{2}(x-1)^2 + \dfrac{1}{3}(x-1)^3 - \dfrac{1}{4}(x-1)^4$.

15. $f(x) = e^x$, $f'(x) = f''(x) = f'''(x) = f^{(4)}(x) = e^x$

$$f(1) = f'(1) = f''(1) = f'''(x) = f^{(4)}(x) = e$$

$$P_4(x) = e + e(x-1) + \frac{e}{2!}(x-1)^2 + \frac{e}{3!}(x-1)^3 + \frac{e}{4!}(x-1)^4$$

17. $f(x) = (1-x)^{1/2}$, $f'(x) = -\frac{1}{2}(1-x)^{-1/2}$, $f''(x) = -\frac{1}{4}(1-x)^{-3/2}$,

$$f'''(x) = -\frac{3}{8}(1-x)^{-5/2}$$

$$f(0) = 1,\ f'(0) = -\frac{1}{2},\ f''(0) = -\frac{1}{4},\ f'''(0) = -\frac{3}{8}.$$

Therefore, $P_3(x) = 1 - \frac{1}{2}x + \frac{\left(-\frac{1}{4}\right)}{2!}x^2 + \frac{\left(-\frac{3}{8}\right)}{3!}x^3 = 1 - \frac{1}{2}x - \frac{1}{8}x^2 - \frac{1}{16}x^3.$

19. $f(x) = \dfrac{1}{2x+3} = (2x+3)^{-1}$, $f'(x) = -2(2x+3)^{-2}$, $f''(x) = 8(2x+3)^{-3}$

$$f'''(x) = -48(2x+3)^{-4}.$$

So $f(0) = \dfrac{1}{3}$, $f'(0) = -\dfrac{2}{9}$, $f''(0) = \dfrac{8}{27}$, $f'''(0) = -\dfrac{16}{27}$.   Therefore,

$$P_3(x) = \frac{1}{3} - \frac{2}{9}x + \frac{\left(\frac{8}{27}\right)}{2!}x^2 + \frac{\left(-\frac{16}{27}\right)}{3!}x^3 = \frac{1}{3} - \frac{2}{9}x + \frac{4}{27}x^2 - \frac{8}{81}x^3.$$

21. $f(x) = \dfrac{1}{1+x} = (1+x)^{-1}$, $f'(x) = -(1+x)^{-2}$, $f''(x) = 2(1+x)^{-3}$,

$f'''(x) = -6(1+x)^{-4}$, $f^{(4)}(x) = 24(1+x)^{-5}$. So

$f(0) = 1,\ f'(0) = -1,\ f''(0) = 2,\ f'''(0) = -6,\ f^{(4)} = 24.$

Therefore, $P_n(x) = 1 - x + x^2 - x^3 + \cdots + (-1)^n x^n$. In particular,

$P_4(x) = 1 - x + x^2 - x^3 + x^4$ and so $P_4(0.1) = 1 - 0.1 + 0.01 - 0.001 + 0.0001 = 0.9091$.

$$f(0.1) = \frac{1}{1.1} \approx 0.909090\ldots\ .$$

23. $f(x) = e^{-x/2}$, $f'(x) = -\frac{1}{2}e^{-x/2}$, $f''(x) = \frac{1}{4}e^{-x/2}$, $f'''(x) = -\frac{1}{8}e^{-x/2}$

$f^{(4)}(x) = \frac{1}{16}e^{-x/2}.$

$f(0) = 1,\ f'(0) = -\frac{1}{2},\ f''(0) = \frac{1}{4},\ f'''(0) = -\frac{1}{8},\ f^{(4)}(0) = \frac{1}{16}.$

Therefore, $P_4(x) = 1 - \frac{1}{2}x + \frac{1}{8}x^2 - \frac{1}{48}x^3 + \frac{1}{384}x^4.$

Since $f(x) = e^{-x/2}$, we let $x = 0.02$ to obtain $e^{-0.1}$. So
$$e^{-0.1} \approx P(0.2) = 1 - \tfrac{1}{2}(0.2) + \tfrac{1}{8}(0.2)^2 - \tfrac{1}{48}(0.2)^3 + \tfrac{1}{384}(0.2)^4 \approx 0.90484.$$

25. $f(x) = x^{1/2}$, $f'(x) = \tfrac{1}{2}x^{-1/2}$, $f''(x) = -\tfrac{1}{4}x^{-3/2}$.
Therefore, $f(16) = 4$, $f'(16) = \tfrac{1}{8}$, $f''(16) = -\tfrac{1}{256}$. Therefore,
$P_2(x) = 4 + \tfrac{1}{8}(x - 16) - \tfrac{1}{512}(x - 16)^2$. $\sqrt{15.6} \approx f(15.6) \approx 3.94969.$

27. $f(x) = \ln(x + 1)$.
We first obtain $P_3(x) = x - \tfrac{1}{2}x^2 + \tfrac{1}{3}x^3$.
Next, $\displaystyle\int_0^{1/2} \ln(x+1)\,dx = \int_0^{1/2}\left(x - \tfrac{1}{2}x^2 + \tfrac{1}{3}x^3\right)dx = \tfrac{1}{2}x^2 - \tfrac{1}{6}x^3 + \tfrac{1}{12}x^4\Big|_0^{1/2}$
$$= \tfrac{1}{8} - \tfrac{1}{48} + \tfrac{1}{192} \approx 0.109.$$
Actual value: $\displaystyle\int_0^{1/2} \ln(x+1)\,dx = (x+1)\ln(x+1) - (x+1)\Big|_0^{1/2}$
$$= \tfrac{3}{2}\ln\tfrac{3}{2} - \tfrac{3}{2} + 1 = \tfrac{3}{2}\ln\tfrac{3}{2} - \tfrac{1}{2} \approx 0.108.$$

29. $f(x) = x^{1/2}$, $f'(x) = \tfrac{1}{2}x^{-1/2}$, $f''(x) = -\tfrac{1}{4}x^{-3/2}$, $f'''(x) = \tfrac{3}{8}x^{-5/2}$
$f(4) = 2$, $f'(4) = \tfrac{1}{4}$, $f''(4) = -\tfrac{1}{32}$.
$P_2(x) = 2 + \tfrac{1}{4}(x - 4) - \tfrac{1}{64}(x - 4)^2$.
$\sqrt{4.06} = f(4.06) = P_2(4.06) = 2 + \tfrac{1}{4}(0.06) - \tfrac{1}{64}(0.06)^2 = 2.01494375.$

To find a bound for the approximation, observe that $f'''(x) = \dfrac{3}{8x^{5/2}}$ is
decreasing on $[4, 4.06]$. Therefore, f''' attains the absolute maximum value at
$x = 4$. So $|f'''(x)| \le \dfrac{3}{8(4)^{5/2}} = \dfrac{3}{256}$. Therefore,
$$\left|R_2(4.06)\right| \le \frac{\tfrac{3}{256}}{3!}(0.06)^3 \approx 0.000000421.$$

31. $f(x) = (1 - x)^{-1}$, $f'(x) = (1 - x)^{-2}$, $f''(x) = 2(1 - x)^{-3}$,

$f'''(x) = 6(1 - x)^{-4}$, $f^{(4)}(x) = 24(1 - x)^{-5}$, $f^{(5)} = 120(1 - x)^{-6} > 0$
So $f(0) = 1$, $f'(0) = 1$, $f''(0) = 2$, $f'''(0) = 6$, $f^{(4)}(0) = 24$, and
$P_3(x) = 1 + x + x^2 + x^3$. Therefore,
$$f(0.2) \approx P_3(0.2) = 1 + 0.2 + 0.04 + 0.008 = 1.248.$$

Since $f^{(5)}(x) > 0$, $f^{(4)}(x)$ is increasing and the maximum is

$M = \dfrac{24}{(1-0.2)^5} = 73.24 < 74$. Then the error bound is $\dfrac{74}{4!}(0.2)^4 = 0.00493$.

The exact value of $f(0.2)$ is $f(0.2) = \dfrac{1}{1-0.2} = 1.25$.

33. $f(x) = \ln x$, $f'(x) = \dfrac{1}{x}$, $f''(x) = -\dfrac{1}{x^2}$, $f'''(x) = \dfrac{2}{x^3}$, $f^{(4)} = -\dfrac{6}{x^4}$.

$f(1) = 0$, $f'(1) = 1$, $f''(1) = -1$ and $P_2(x) = x - 1 - \frac{1}{2}(x-1)^2$.

Therefore, $f(1.1) = 1.1 - 1 - \frac{1}{2}(0.1)^2 = 0.095$. Since $f^{(4)}(x) < 0$, $f^{(4)}(x)$ is

decreasing and $M = \dfrac{2}{x^3}\bigg|_{x=1}$, we see that $M = 2$. Therefore, the error bound is

$\dfrac{2}{3!}(0.1)^3 = 0.00033$.

35. a. $f(x) = e^{-x}$, $f'(x) = -e^{-x}$, $f''(x) = e^{-x}$, $f'''(x) = -e^{-x}$, $f^{(4)}(x) = e^{-x}$.
   Since $f^{(5)}(x) = -e^{-x} < 0$, $f^{(4)}(x)$ is decreasing and the maximum value of
   $f^{(4)}(x)$ is attained at $x = 0$. So, we may take
   $$M = \max_{0 \le x \le 1} \left|f^{(4)}(x)\right| = e^0 = 1.$$

   Therefore, a bound on the approximation of $f(x)$ by $P_3(x)$ is
   $$\tfrac{1}{4!}(1-0)^4 = 0.04167.$$

   b. $\displaystyle\int_0^1 e^{-x}\,dx \approx \int_0^1 P_3(x)\,dx = \int_0^1 (1 - x + \tfrac{1}{2}x^2 - \tfrac{1}{6}x^3)\,dx$ (See Problem 1 for $P_3(x)$]

   $= x - \tfrac{1}{2}x^2 + \tfrac{1}{6}x^3 - \tfrac{1}{24}x^4\Big|_0^1 = 1 - \tfrac{1}{2} + \tfrac{1}{6} - \tfrac{1}{24} = 0.625$.

   c. The error bound on the approximation in (b) is
   $$\int_0^1 0.04167\,dx = 0.04167x\Big|_0^1 = 0.04167 \quad \text{(using the result of (a))}$$

   d. Since $\displaystyle\int_0^1 e^{-x}\,dx = -e^{-x}\Big|_0^1 = -e^{-1} + 1 \approx 0.632121$,
   the actual error is $0.632121 - 0.625 = 0.007121$.

37. The required area is $A = \displaystyle\int_0^{0.5} e^{-x^2/2}\,dx$.

Now $e^u = 1 + u + \frac{1}{2}u^2$ is the second Taylor polynomial at $u = 0$.

Therefore, $e^{-x^2/2} \approx 1 - \frac{1}{2}x^2 + \frac{1}{2}(-\frac{x^2}{2})^2 = 1 - \frac{1}{2}x^2 + \frac{1}{8}x^4$

is the fourth Taylor polynomial at $x = 0$. So

$$A = \int_0^{0.5} e^{-x^2/2}\,dx = \int_0^{0.5}(1 - \frac{1}{2}x^2 + \frac{1}{8}x^4)\,dx = (x - \frac{1}{6}x^3 + \frac{1}{40}x^5)\Big|_0^{0.5}$$

$$= 0.5 - \frac{1}{6}(0.5)^3 + \frac{1}{40}(0.5)^5 \approx 0.47995, \text{ or approximately } 0.48 \text{ square units.}$$

39. The percentage of the nonfarm work force in the service industries t decades from now is given by $P(t) = \int 6e^{1/(2t+1)}\,dt$.

Let $u = \dfrac{1}{2t+1}$, so that $2t + 1 = \dfrac{1}{u}$ and $t = \dfrac{1}{2}\left(\dfrac{1}{u} - 1\right)$. Therefore, $dt = -\dfrac{1}{2u^2}\,du$.

So we have

$$\tilde{P}(u) = 6\int e^u\left(\frac{du}{-2u^2}\right) = -3\int \frac{e^u}{u^2}\,du = -3\int \frac{1}{u^2}\left(1 + u + \frac{u^2}{2} + \frac{u^3}{6} + \frac{u^4}{24}\right)du$$

$$= -3\int\left(\frac{1}{u^2} + \frac{1}{u} + \frac{1}{2} + \frac{u}{6} + \frac{u^2}{24}\right)du = -3\left(-\frac{1}{u} + \ln u + \frac{1}{2}u + \frac{u^2}{12} + \frac{u^3}{72}\right) + C.$$

Therefore,

$$P(t) = -3\left[-(2t+1) + \ln\left(\frac{1}{2t+1}\right) + \frac{1}{2(2t+1)} + \frac{1}{12(2t+1)^2} + \frac{1}{72(2t+1)^3}\right] + C.$$

Using the condition $P(0) = 30$, we find $P(0) = -3(-1 + \ln 1 + \frac{1}{2} + \frac{1}{12} + \frac{1}{72}) = 30$,

or $C = 28.79$. So

$$P(t) = -3\left[-2(2t+1) + \ln\left(\frac{1}{2t+1}\right) + \frac{1}{2(2t+1)} + \frac{1}{12(2t+1)^2} + \frac{1}{72(2t+1)^3}\right] + 28.79.$$

Two decades from now, the percentage of nonfarm workers will be

$$P(2) \approx -3\left[-5 + \ln(\tfrac{1}{5}) + \tfrac{1}{10} + \tfrac{1}{300} + \tfrac{1}{9000}\right] + 28.79 \approx 48.31,$$

or approximately 48 percent.

41. The average enrollment between $t = 0$ and $t = 2$ is given by

$$A = \frac{1}{2}\int_0^2\left[-\frac{20,000}{\sqrt{1+0.2t}} + 21,000\right]dt = -10,000\int_0^2(1+0.2t)^{-1/2}\,dt + \frac{1}{2}(21,000t)\Big|_0^2$$

$$= -10,000\int_0^2(1+0.2t)^{-1/2}\,dt + 21,000.$$

11 Taylor Polynomials

To evaluate the above integral, we approximate the integrand by the second Taylor polynomial at $t = 0$. We compute

$$f(t) = (1+0.2t)^{-1/2}, \quad f'(t) = -0.1(1+0.2t)^{-3/2},$$

$$f''(t) = 0.03(1+0.2t)^{-5/2},$$

$$f(0) = 1, \quad f'(0) = -0.1, \quad f''(0) = 0.03.$$

$$P_2(t) = 1 - 0.1t + 0.015t^2.$$

Therefore,

$$A = -10{,}000 \int_0^2 (1 - 0.1t + 0.015t^2)\,dt + 21{,}000$$

$$= -10{,}000(t - 0.05t^2 + 0.005t^3)\Big|_0^2 + 21{,}000$$

$$= -10{,}000(2 - 0.2 + 0.04) + 21{,}000 = 2600.$$

So the average enrollment between $t = 0$ and $t = 2$ is 2600.

43. False. The fourth Taylor polynomial $P_4(x)$ coincides with $f(x)$ for all values of x in this case. To see this, note that $R_4(x) = \dfrac{f^{(5)}(c)}{5!}(x-a)^5 = 0$ since $f^{(5)}(x) = 0$.

45. True. Since $f^{(n+1)}(x) = 0$ for all values of x, it follows that

$$R_n(x) = \frac{f^{(n+1)}(c)}{(n+1)!}(x-a)^{n+1} = 0 \text{ for all values of } x.$$

EXERCISES 11.2, page 716

1. 1, 2, 4, 8, 16.

3. $\dfrac{1-1}{1+1}, \dfrac{2-1}{2+1}, \dfrac{3-1}{3+1}, \dfrac{4-1}{4+1}, \dfrac{5-1}{5+1}$, or $0, \dfrac{1}{3}, \dfrac{2}{4}, \dfrac{3}{5}, \dfrac{4}{6}$.

5. $a_1 = \dfrac{2^0}{1!} = 1, \ a_2 = \dfrac{2^1}{2!} = 1, \ a_3 = \dfrac{2^2}{3!} = \dfrac{4}{6}, \ a_4 = \dfrac{2^3}{4!} = \dfrac{8}{24}, \ a_5 = \dfrac{2^4}{5!} = \dfrac{16}{120}.$

7. $a_1 = e, \ a_2 = \dfrac{e^2}{8}, \ a_3 = \dfrac{e^3}{27}, \ a_4 = \dfrac{e^4}{64}, \ a_5 = \dfrac{e^5}{125}.$

9. $a_1 = \dfrac{3-1+1}{2+1} = 1$, $a_2 = \dfrac{12-2+1}{8+1} = \dfrac{11}{9}$, $a_3 = \dfrac{27-3+1}{18+1} = \dfrac{25}{19}$,

   $a_4 = \dfrac{48-4+1}{32+1} = \dfrac{45}{33} = \dfrac{15}{11}$, $a_5 = \dfrac{75-5+1}{50+1} = \dfrac{71}{51}$.

11. $a_n = 3n - 2$

13. $a_n = \dfrac{1}{n^3}$

15. $a_n = 2\left(\dfrac{4}{5}\right)^{n-1} = \dfrac{2^{2n-1}}{5^{n-1}}$.

17. $a_n = \dfrac{(-1)^{n+1}}{2^{n-1}}$

19. $a_n = \dfrac{n}{(n+1)(n+2)}$

21. $a_n = \dfrac{e^{n-1}}{(n-1)!}$

23.

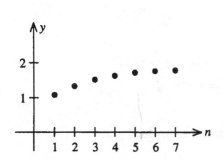

25.

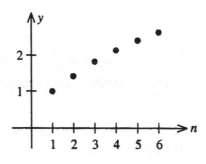

27.

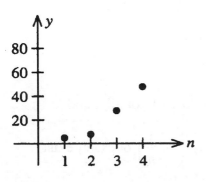

29.

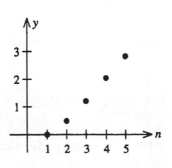

31. $\lim\limits_{n\to\infty} \dfrac{n+1}{2n} = \lim\limits_{n\to\infty} \dfrac{1+\frac{1}{n}}{2} = \dfrac{1}{2}$.

33. $\lim\limits_{n\to\infty} \dfrac{(-1)^n}{\sqrt{n}} = 0$.

35. $\lim\limits_{n\to\infty} \dfrac{\sqrt{n}-1}{\sqrt{n}+1} = \lim\limits_{n\to\infty} \dfrac{1-\frac{1}{\sqrt{n}}}{1+\frac{1}{\sqrt{n}}} = 1$.

37. $\lim\limits_{n\to\infty} \dfrac{2n^3-1}{n^3+2n+1} = \lim\limits_{n\to\infty} \dfrac{2-\frac{1}{n^3}}{1+\frac{2}{n^2}+\frac{1}{n^3}} = 2$.

39. $\lim\limits_{n\to\infty}\left(2-\dfrac{1}{2^n}\right) = \lim\limits_{n\to\infty} 2 - \lim\limits_{n\to\infty} \dfrac{1}{2^n} = 2-0 = 2$.

41. $\lim\limits_{n\to\infty} \dfrac{2^n}{3^n} = \lim\limits_{n\to\infty}\left(\dfrac{2}{3}\right)^n = 0$.

43. $\lim\limits_{n\to\infty} \dfrac{n}{\sqrt{2n^2+3}} = \lim\limits_{n\to\infty} \dfrac{1}{\sqrt{2+\frac{3}{n^2}}} = \dfrac{1}{\sqrt{2}} = \dfrac{\sqrt{2}}{2}$.

45. a. $a_1 = 0.015$, $a_{10} = 0.140$, $a_{100} = 0.77939$, $a_{1000} = 0.999999727$

   b. $\lim\limits_{n\to\infty} a_n = \lim\limits_{n\to\infty}[1-(0.985)^n] = 1 - \lim\limits_{n\to\infty}(0.985)^n = 1-0 = 1$.

47. a. The result follows from the compound interest formula .

   b. $a_n = 100(1.01)^n$

   c. $a_{24} = 100(1.01)^{24} = 126.97$, or \$126.97. Therefore, the accumulated amount at the end of two years is \$126.97.

49. True. $\lim\limits_{n\to\infty} a_n b_n = \left(\lim\limits_{n\to\infty} a_n\right)\left(\lim\limits_{n\to\infty} b_n\right) = (L)(0) = 0$.

51. False. Let $\lim\limits_{n\to\infty} a_n = (-1)^n$ and $b_n = 1+\dfrac{1}{n}$. Then $|a_n| \le 1$ for all n and

   $\lim\limits_{n\to\infty} b_n = \lim\limits_{n\to\infty}\left(1+\dfrac{1}{n}\right) = 1$. But $\lim\limits_{n\to\infty} a_n b_n = \lim\limits_{n\to\infty}(-1)^n\left(1+\dfrac{1}{n}\right)$ does not exist since if n is very large, $a_n b_n$ is close to 1 if n is even, but close to -1 if n is odd.

Therefore, $a_n b_n$ cannot approach a specific number as n approaches infinity.

EXERCISES 11.3, page 727

1. $S_1 = -2$, $S_2 = -2 + 4 = 2$, $S_3 = -2 + 4 - 8 = -6$, $S_4 = -6 + 16 = 10$, and

so $\lim_{N \to \infty} S_N$ does not exist and the series $\sum_{n=1}^{\infty} (-2)^n$ is divergent.

3. $S_N = \sum_{n=1}^{N} \frac{1}{n^2 + 3n + 2} = \sum_{n=1}^{N} \left(\frac{1}{n+1} - \frac{1}{n+2} \right)$

$= \left(\frac{1}{2} - \frac{1}{3} \right) + \left(\frac{1}{3} - \frac{1}{4} \right) + \cdots + \left(\frac{1}{N} - \frac{1}{N+1} \right) + \left(\frac{1}{N+1} - \frac{1}{N+2} \right) = \frac{1}{2} - \frac{1}{N+2}$.

So $\lim_{N \to \infty} S_N = \lim_{N \to \infty} \left(\frac{1}{2} - \frac{1}{N+2} \right) = \frac{1}{2}$ and $\sum_{n=1}^{\infty} \frac{1}{n^2 + 3n + 2} = \frac{1}{2}$.

5. $\sum_{n=0}^{\infty} \left(\frac{1}{3} \right)^n = 1 + \frac{1}{3} + \left(\frac{1}{3} \right)^2 + \cdots = \frac{1}{1 - \frac{1}{3}} = \frac{1}{\frac{2}{3}} = \frac{3}{2}$.

7. This is a geometric series with $r = 1.01 > 1$ and so it diverges.

9. $\sum_{n=0}^{\infty} \frac{(-2)^n}{3^n} = \sum_{n=0}^{\infty} \left(-\frac{2}{3} \right)^n$ is a geometric series with $|r| = \left| -\frac{2}{3} \right| = \frac{2}{3} < 1$,

and so it converges. The sum is $\dfrac{1}{1 - \left(-\frac{2}{3} \right)} = \dfrac{1}{1 + \frac{2}{3}} = \dfrac{3}{5}$.

11. $\sum_{n=0}^{\infty} \frac{2^n}{3^{n+2}} = \sum_{n=0}^{\infty} \frac{1}{9} \left(\frac{2}{3} \right)^n = \frac{1}{9} \cdot \frac{1}{1 - \frac{2}{3}} = \left(\frac{1}{9} \right)(3) = \frac{1}{3}$.

13. $\sum_{n=0}^{\infty} e^{-0.2n} = \sum_{n=0}^{\infty} \left(e^{-0.2} \right)^n = \sum_{n=0}^{\infty} \left(\frac{1}{e^{0.2}} \right)^n = \dfrac{1}{1 - \dfrac{1}{e^{0.2}}} \approx 5.52$.

15. Here $r = \left(-\dfrac{3}{\pi}\right)$ and since $|r| = \left|-\dfrac{3}{\pi}\right| < 1$, the series converges. In fact,

$$\sum_{n=0}^{\infty} \left(-\frac{3}{\pi}\right)^n = \frac{1}{1 - \left(-\dfrac{3}{\pi}\right)} = \frac{1}{1 + \dfrac{3}{\pi}} = \frac{1}{\dfrac{\pi + 3}{\pi}} = \frac{\pi}{\pi + 3}.$$

17. $8 + 4 + \dfrac{1}{2}\left[1 + \dfrac{1}{2} + \left(\dfrac{1}{2}\right)^2 + \cdots\right] = 12 + \dfrac{1}{2}\left(\dfrac{1}{1 - \frac{1}{2}}\right) = 13.$

19. $3 - \dfrac{1}{3}\left[1 - \dfrac{1}{3} + \left(-\dfrac{1}{3}\right)^2 - \cdots\right] = 3 - \dfrac{1}{3} \cdot \dfrac{1}{1 + \dfrac{1}{3}} = 3 - \dfrac{1}{3} \cdot \dfrac{3}{4} = 3 - \dfrac{1}{4} = \dfrac{11}{4}.$

21. $\displaystyle\sum_{n=0}^{\infty} \frac{3 + 2^n}{3^n} = 3\sum_{n=0}^{\infty}\left(\frac{1}{3^n}\right) + \sum_{n=0}^{\infty}\left(\frac{2}{3}\right)^n = 3 \cdot \frac{1}{1 - \frac{1}{3}} + \frac{1}{1 - \frac{2}{3}} = \frac{9}{2} + 3 = \frac{15}{2}.$

23. $\displaystyle\sum_{n=0}^{\infty} \frac{3 \cdot 2^n + 4^n}{3^n} = \sum_{n=0}^{\infty}\left[3\left(\tfrac{2}{3}\right)^n + \left(\tfrac{4}{3}\right)^n\right]$ diverges because the series

$\displaystyle\sum_{n=0}^{\infty}\left(\tfrac{4}{3}\right)^n$ is geometric with constant ratio $\tfrac{4}{3} > 1$.

25. Since $\dfrac{e}{\pi} < 1$, $\displaystyle\sum_{n=1}^{\infty}\left(\frac{e}{\pi}\right)^n$ converges. Also, $\dfrac{\pi}{e^2} < 1$ and so $\displaystyle\sum_{n=1}^{\infty}\left(\frac{\pi}{e^2}\right)^n$ converges.

Therefore, $\displaystyle\sum_{n=1}^{\infty}\left[\left(\frac{e}{\pi}\right)^n + \left(\frac{\pi}{e^2}\right)^n\right] = \sum_{n=1}^{\infty}\left(\frac{e}{\pi}\right)^n + \sum_{n=1}^{\infty}\left(\frac{\pi}{e^2}\right)^n$

$$= \frac{e}{\pi} \cdot \frac{1}{1 - \dfrac{e}{\pi}} + \frac{\pi}{e^2} \cdot \frac{1}{1 - \dfrac{\pi}{e^2}} = \frac{e}{\pi - e} + \frac{\pi}{e^2 - \pi} = \frac{e}{\pi} \cdot \frac{\pi}{\pi - e} + \frac{\pi}{e^2} \cdot \frac{e^2}{e^2 - \pi}$$

$$= \frac{e^3 - \pi e + \pi^2 - \pi e}{(\pi - e)(e^2 - \pi)} = \frac{e^3 - 2\pi e + \pi^2}{(\pi - e)(e^2 - \pi)}.$$

27. $0.3333\ldots = 0.3 + 0.03 + 0.003 + \cdots = \dfrac{3}{10} + \dfrac{3}{10^2} + \dfrac{3}{10^3} + \cdots$

$$= \frac{3}{10} \sum_{n=0}^{\infty} \left(\frac{1}{10} \right)^n = \frac{\frac{3}{10}}{1 - \frac{1}{10}} = \frac{3}{10} \cdot \frac{10}{9} = \frac{1}{3}.$$

29. $1.213213213 = 1 + 0.213 \left[1 + (0.0001) + (0.001)^2 + \cdots \right]$

$$= 1 + \frac{213}{1000} \cdot \frac{1}{1 - 0.001} = 1 + \frac{213}{1000} \cdot \frac{1000}{999} = 1 + \frac{213}{999} = \frac{1212}{999} = \frac{404}{333}.$$

31. The given series is geometric with $r = (-x)^n$. Therefore, it converges if $|-x| < 1$,

$|x| < 1,$ or $-1 < x < 1.$ Furthermore, $\displaystyle\sum_{n=0}^{\infty} (-x)^n = \frac{1}{1 - (-x)} = \frac{1}{1 + x}.$

33. The given series is geometric with $r = 2(x - 1)$ and so it converges provided

$|2(x-1)| < 1,\ |x-1| < \dfrac{1}{2},\ -\dfrac{1}{2} < x - 1 < \dfrac{1}{2},$ or $\dfrac{1}{2} < x < \dfrac{3}{2}.$ Next,

$$\sum_{n=1}^{\infty} 2^n (x-1)^n = 2(x-1) \cdot \frac{1}{1 - 2(x-1)} = 2(x-1) \cdot \frac{1}{3 - 2x} = \frac{2(x-1)}{3 - 2x}.$$

35. The additional spending generated by the proposed tax cut will be

$(0.91)(30) + (0.91)^2 (30) + (0.91)^3 (30) + \cdots$

$$= (0.91)(30)[1 + 0.91 + (0.91)^2 + (0.91)^3 + \cdots]$$

$$= 27.3 \left[\frac{1}{1 - 0.91} \right] = 303.33, \ \text{or approximately } \$303 \text{ billion.}$$

37. $p = \dfrac{1}{6} + \left(\dfrac{1}{6} \right)\left(\dfrac{5}{6} \right)^2 + \dfrac{1}{6}\left(\dfrac{5}{6} \right)^4 + \cdots = \dfrac{1}{6}\left\{ 1 + \left(\dfrac{5}{6} \right)^2 + \left[\left(\dfrac{5}{6} \right)^2 \right]^2 + \cdots \right\}$ $\left(r = \left(\dfrac{5}{6} \right)^2 \right)$

$$= \frac{1}{6} \cdot \frac{1}{1 - \left(\dfrac{5}{6} \right)^2} = \frac{1}{6} \cdot \frac{1}{1 - \dfrac{25}{36}} = \frac{1}{6} \cdot \frac{36}{36 - 25} = \frac{6}{11}.$$

39. The required upper bound is no larger than

$B = a_1 + aNa_1 + aNa_2 + aNa_3 + \cdots$

$= a_1 + aNa_1 + aN(ra_1) + aN(r^2 a_1) + \cdots.$ (a geometric series)

11 Taylor Polynomials

$$= a_1 + aNa_1(1 + r + r^2 + \cdots)$$

$$= a_1 + \frac{aNa_1}{1-r} = \frac{a_1 - a_1(1 + aN - b) + aNa_1}{1 - (1 + aN - b)} = \frac{a_1 b}{b - aN}.$$

41. $A = Pe^{-r} + Pe^{-2r} + Pe^{-3r} + \cdots$

$$= Pe^{-r}(1 + e^{-r} + e^{-2r} + \cdots) = \frac{Pe^{-r}}{1 - e^{-r}} = \frac{P}{e^r - 1}.$$

43. b. $C + R \leq S$;

$$C + \frac{Ce^{-kt}}{1 - e^{-kt}} \leq S, \; \frac{C - Ce^{-kt} + Ce^{-kt}}{1 - e^{-kt}} \leq S, \; \frac{C}{1 - e^{-kt}} \leq S$$

$$1 - e^{-kt} \geq \frac{C}{S}, \; e^{-kt} \leq 1 - \frac{C}{S} = \frac{S - C}{S}.$$

$$-kt \leq \ln \frac{S - C}{S}, \; kt \geq -\ln \frac{S - C}{S} = \ln \frac{S}{S - C} \quad \text{or} \quad t \geq \frac{1}{k} \ln \frac{S}{S - C}.$$

Therefore, the minimum time between dosages should be $\dfrac{1}{k} \ln \dfrac{S}{S - C}$ hours.

45. False. Take $a_n = -n$ and $b_n = n$. Then both $\displaystyle\sum_{n=0}^{\infty} a_n$ and $\displaystyle\sum_{n=0}^{\infty} b_n$ diverge but

$\displaystyle\sum_{n=0}^{\infty}(a_n + b_n) = \sum_{n=0}^{\infty} 0$ clearly converges to zero.

47. True. This is a convergent geometric series with common ratio $|r|$. If $|r| < 1$, then

$$\sum_{n=0}^{\infty} |r|^n = \frac{1}{1 - |r|}.$$

EXERCISES 11.4, page 738

1. Here $a_n = \dfrac{n}{n+1}$ and since $\displaystyle\lim_{n\to\infty} \dfrac{n}{n+1} = 1 \neq 0$, the series is divergent.

3. Here $a_n = \dfrac{2n}{3n+1}$ and $\displaystyle\lim_{n\to\infty} \dfrac{2n}{3n+1} = \dfrac{2}{3} \neq 0$, and so the series diverges.

5. Here $a_n = 2(1.5)^n$ and $\lim\limits_{n\to\infty} 2(1.5)^n = \infty$, and so the series diverges.

7. Here $a_n = \dfrac{1}{2+3^{-n}}$, and $\lim\limits_{n\to\infty}\dfrac{1}{2+3^{-n}} = \dfrac{1}{2} \neq 0$, and so the series diverges.

9. Here $a_n = \left(-\dfrac{\pi}{3}\right)^n = (-1)^n\left(\dfrac{\pi}{3}\right)^n$. Because $\dfrac{\pi}{3} > 1$, we see that $\lim\limits_{n\to\infty} a_n$ does not exist, and so the series is divergent.

11. Take $f(x) = \dfrac{1}{x+1}$ and note that it is nonnegative and decreasing for $x \geq 1$. Next,

$$\int_1^\infty \frac{1}{x+1}\,dx = \lim_{b\to\infty}\int_1^b \frac{1}{x+1}\,dx = \lim_{b\to\infty}\left[\ln(x+1)\big|_1^b\right] = \lim_{b\to\infty}\left[\ln(b+1) - \ln 2\right] = \infty$$

and so the series is divergent.

13. First $f(x) = \dfrac{x}{2x^2+1}$ is nonnegative and decreasing for $x \geq 1$. Next,

$$\int_1^\infty \frac{x}{2x^2+1}\,dx = \lim_{b\to\infty}\int_1^b \frac{x}{2x^2+1}\,dx = \lim_{b\to\infty}\left[\frac{1}{4}\ln(2x^2+1)\bigg|_1^b\right]$$

$$= \lim_{b\to\infty}\left[\frac{1}{4}\ln(2b^2+1) - \frac{1}{4}\ln 3\right] = \infty$$

and so the series is divergent.

15. Here $f(x) = xe^{-x}$ and nonnegative and decreasing for $x \geq 1$ (study its derivative).
Next, $\displaystyle\int_1^\infty xe^{-x}\,dx = \lim_{b\to\infty}\int_1^b xe^{-x}\,dx = \lim_{b\to\infty}\left[-(x+1)e^{-x}\Big|_1^b\right]$ (Integrate by parts)

$$= \lim_{b\to\infty}\left[-(b+1)e^{-b} + 2e^{-b}\right] = 0$$

[You can verify that be^{-b} approaches 0 by graphing it.]. So the integral converges.

17. Let $f(x) = \dfrac{x}{(x^2+1)^{3/2}}$. Then f is nonnegative and decreasing on $[1,\infty)$. Next,

$$\int_1^\infty \frac{x}{(x^2+1)^{3/2}}\,dx = \lim_{b\to\infty}\int_1^b x(x^2+1)^{-3/2}\,dx$$

$$= \lim_{b \to \infty} \left[\left(-\frac{1}{\sqrt{x^2+1}} \right) \Big|_1^b \right] \quad \text{[Use substitution.]}$$

$$= \lim_{b \to \infty} \left(-\frac{1}{\sqrt{b^2+1}} + \frac{1}{\sqrt{2}} \right) = \frac{1}{\sqrt{2}}$$

which converges and so the series converges.

19. Let $f(x) = \dfrac{1}{x \ln^3 x}$ which is nonnegative and decreasing on $[9, \infty)$. Next,

$$\int_9^\infty \frac{1}{x(\ln x)^3} \, dx = \lim_{b \to \infty} \int_9^b \frac{(\ln x)^{-3}}{x} \, dx$$

$$= \lim_{b \to \infty} \left[-\frac{1}{2(\ln x)^2} \Big|_9^b \right] \quad \text{[Use substitution with } u = \ln x.\text{]}$$

$$= \lim_{b \to \infty} \left[-\frac{1}{2(\ln b)^2} + \frac{1}{2(\ln 9)^2} \right] = \frac{1}{2(\ln 9)^2}$$

which is convergent, and so the series is convergent.

21. Here $p = 3 > 1$ and so the series is convergent.

23. Here $p = 1.01 > 1$, and so the series is convergent.

25. Here $p = \pi > 1$, and so the series is convergent.

27. Let $a_n = \dfrac{1}{2n^2+1}$. Then $a_n = \dfrac{1}{2n^2+1} < \dfrac{1}{2n^2} < \dfrac{1}{n^2} = b_n$. Since $\sum b_n$ is a convergent

p-series, $\displaystyle\sum_{n=1}^\infty \frac{1}{2n^2+1}$ converges by the comparison test.

29. Let $a_n = \dfrac{1}{n-2}$. Then $a_n = \dfrac{1}{n-2} > \dfrac{1}{n} = b_n$.

Since $\sum b_n$ diverges, the comparison test implies that $\displaystyle\sum_{n=3}^\infty \frac{1}{n-2}$ diverges as well.

31. Let $a_n = \dfrac{1}{\sqrt{n^2-1}}$. Then $a_n = \dfrac{1}{\sqrt{n^2-1}} > \dfrac{1}{\sqrt{n^2}} = \dfrac{1}{n} = b_n$. Since the harmonic series $\sum b_n$ diverges, the comparison test shows that the given series also diverges.

33. Let $a_n = \dfrac{2^n}{3^n+1} < \dfrac{2^n}{3^n} = \left(\dfrac{2}{3}\right)^n = b_n$. Since $\displaystyle\sum_{n=0}^{\infty}\left(\dfrac{2}{3}\right)^n$ is a convergent geometric series, $\displaystyle\sum_{n=0}^{\infty} a_n$ converges by the comparison test.

35. Let $a_n = \dfrac{\ln n}{n}$. Since $\ln n > 1$ if $n > 3$, we see that $a_n = \dfrac{\ln n}{n} > \dfrac{1}{n} = b_n$. But $\sum b_n$ is the divergent harmonic series, and the comparison test implies that $\displaystyle\sum_{n=2}^{\infty}\dfrac{\ln n}{n}$ diverges as well.

37. We use the Integral Test with $f(x) = \dfrac{1}{\sqrt{x+1}}$ which is nonnegative and decreasing on $[0,\infty)$. We find

$$\int_0^{\infty} \dfrac{1}{\sqrt{x+1}}\,dx = \lim_{b\to\infty}\int_0^b (x+1)^{-1/2}\,dx = \lim_{b\to\infty}\left[2(x+1)^{1/2}\Big|_0^b\right]$$
$$= \lim_{b\to\infty}[2(b+1)^{1/2} - 2] = \infty$$

and so the series diverges.

39. Let $a_n = \dfrac{1}{n(\sqrt{n^2+1}}$. Observe that if n is large, n^2+1 behaves like n^2 and this suggests we compare $\sum a_n$ with $\sum b_n$ where $b_n = \dfrac{1}{n\sqrt{n^2}} = \dfrac{1}{n^2}$. Now $\sum b_n$ is a convergent p-series with $p = 2$. Since $0 < \dfrac{1}{n\sqrt{n^2+1}} < \dfrac{1}{n^2}$ the comparison test implies that the given series converges.

41. $\displaystyle\sum_{n=1}^{\infty}\dfrac{1}{n\sqrt{n}} = \sum_{n=1}^{\infty}\dfrac{1}{n^{3/2}}$ is a convergent p-series with $p = \dfrac{3}{2}$. Next, $\displaystyle\sum_{n=1}^{\infty}\dfrac{2}{n^2}$ is a

convergent p-series with $p = 2$. Therefore, $\displaystyle\sum_{n=1}^{\infty}\left(\frac{1}{n\sqrt{n}} + \frac{2}{n^2}\right)$ is convergent.

43. We know that $\ln n > 1$ if $n > 3$ and so $a_n = \dfrac{\ln n}{\sqrt{n}} > \dfrac{1}{\sqrt{n}} = b_n$ if $n > 3$. But

$\displaystyle\sum_{n=2}^{\infty} b_n = \sum_{n=2}^{\infty}\frac{1}{n^{1/2}}$ is a divergent p-series with $p = \dfrac{1}{2}$. Therefore, by the comparison

test, the given series is divergent.

45. We use the Integral Test, letting $f(x) = \dfrac{1}{x(\ln x)^2}$. Observe that f is nonnegative

and decreasing on $[2, \infty)$. Next, we compute

$$\int_2^{\infty}\frac{1}{x(\ln x)^2}\,dx = \lim_{b\to\infty}\int_2^b \frac{(\ln x)^{-2}}{x}\,dx$$

$$= \lim_{b\to\infty}\left[-\frac{1}{\ln x}\Big|_2^b\right] \qquad \text{(Use substitution with } u = \ln x\text{)}$$

$$= \lim_{b\to\infty}\left(-\frac{1}{\ln b} + \frac{1}{\ln 2}\right) = \frac{1}{\ln 2},$$

which converges, and so the given series converges.

47. We use the comparison test. Since $a_n = \dfrac{1}{\sqrt{n}+4} > \dfrac{1}{3\sqrt{n}} = b_n$ and $\displaystyle\sum_{n=1}^{\infty} b_n$ is

divergent, we conclude that $\displaystyle\sum_{n=1}^{\infty}\frac{1}{\sqrt{n}+4}$ is also divergent.

49. We use the Integral Test with $f(x) = \dfrac{1}{x(\ln x)^p}$. Observe that f is nonnegative and

decreasing on $[2, \infty)$. Next, we compute

$$\int_2^{\infty}\frac{1}{x(\ln x)^p}\,dx = \lim_{b\to\infty}\int_2^b \frac{(\ln x)^{-p}}{x}\,dx$$

$$= \lim_{b \to \infty} \left[\frac{(\ln x)^{-p+1}}{1-p} \Big|_2^b \right] \qquad \text{(Use substitution with } u = \ln x \text{)}$$

$$= \lim_{b \to \infty} \left[\frac{(\ln b)^{1-p}}{1-p} - \frac{(\ln 2)^{1-p}}{1-p} \right]$$

$$= \frac{(\ln 2)^{1-p}}{p-1} \qquad \text{if } p > 1.$$

If $p \le 1$, then the improper integral diverges. Therefore, $\displaystyle\sum_{n=2}^{\infty} \frac{1}{n(\ln n)^p}$ converges

for $p > 1$.

51. Denoting the Nth partial sum of the series by S_N, we have

$$S_N = \sum_{n=1}^{N} \left(\frac{a}{n+1} - \frac{1}{n+2} \right) = \left(\frac{a}{2} - \frac{1}{3} \right) + \left(\frac{a}{3} - \frac{1}{4} \right) + \cdots + \left(\frac{a}{N+1} - \frac{1}{N+2} \right)$$

$$= \frac{a}{2} + \frac{a-1}{3} + \frac{a-1}{4} + \cdots + \frac{a-1}{N+1} - \frac{1}{N+2}.$$

If $a = 1$, then $S_N = \dfrac{1}{2} - \dfrac{1}{N+2}$ is the Nth partial sum of a telescoping series that

converges to ½. If $a \ne 0$, then S_N is the Nth partial sum of a series akin to the

harmonic series $\displaystyle\sum_{n=1}^{\infty} \frac{1}{n}$ and in this case, the series diverges. Therefore, the series

converges only for $a = 1$.

53. $\displaystyle\int_1^{\infty} \frac{1}{x^p} dx = \lim_{b \to \infty} \int_1^b x^{-p} dx = \lim_{b \to \infty} \left[\frac{x^{1-p}}{1-p} \Big|_1^b \right] = \lim_{b \to \infty} \left[\frac{b^{1-p}}{1-p} - \frac{1}{1-p} \right] = \frac{1}{p-1}$ if $p > 1$

and diverges to infinity if $p < 1$. If $p = 1$, then we have

$$\int_1^{\infty} \frac{1}{x} dx = \lim_{b \to \infty} \int_1^b \frac{1}{x} dx = \lim_{b \to \infty} \left[\ln |x| \Big|_1^b \right] = \lim_{b \to \infty} (\ln b - \ln 1) = \infty$$

and so the integral diverges in this case as well.

55. True. Compare it with the divergent harmonic series $\displaystyle\sum_{n=1}^{\infty} \frac{1}{n}$. In fact, if $\displaystyle\sum_{n=1}^{\infty} \frac{x}{n}$

converges for $x \neq 0$. Then $\displaystyle\sum_{n=1}^{\infty} \frac{x}{n} = x \sum_{n=1}^{\infty} \frac{1}{n}$ converges, a contradiction.

57. False. The harmonic series $\displaystyle\sum_{n=1}^{\infty} a_n$ with $a_n = \dfrac{1}{n}$ $(n = 1, 2, 3, \ldots)$ is divergent, but

$$\lim_{n\to\infty} a_n = 0.$$

59. False. Let $a_n = \dfrac{1}{n^2}$ and $b_n = \dfrac{2}{n^2}$. Clearly $b_n \geq a_n$ $(n = 1, 2, 3, \ldots)$ but both

$\displaystyle\sum a_n$ and $\displaystyle\sum b_n$ converge.

EXERCISES 11.5, page 747

1. $R = \lim\limits_{n\to\infty} \left| \dfrac{a_n}{a_{n+1}} \right| = \lim\limits_{n\to\infty} 1 = 1.$ Therefore, $R = 1$ and the interval of convergence is $(0, 2)$.

3. $R = \lim\limits_{n\to\infty} \left| \dfrac{a_n}{a_{n+1}} \right| = \lim\limits_{n\to\infty} \dfrac{n^2}{(n+1)^2} = \lim\limits_{n\to\infty} \dfrac{n^2}{n^2 + 2n + 1} = 1;\quad (-1,1)$

5. $R = \lim\limits_{n\to\infty} \left| \dfrac{a_n}{a_{n+1}} \right| = \lim\limits_{n\to\infty} \dfrac{\dfrac{1}{4^n}}{\dfrac{1}{4^{n+1}}} = \lim\limits_{n\to\infty} 4 = 4;\quad (-4,4)$

7. $R = \lim\limits_{n\to\infty} \left| \dfrac{a_n}{a_{n+1}} \right| = \lim\limits_{n\to\infty} \dfrac{1}{n!2^n}(n+1)!2^{n+1} = \lim\limits_{n\to\infty} 2(n+1) = \infty.$

Therefore, $R = \infty$ and the interval of convergence is $(-\infty,\infty)$.

9. $R = \lim\limits_{n\to\infty} \left| \dfrac{a_n}{a_{n+1}} \right| = \lim\limits_{n\to\infty} \dfrac{\dfrac{n!}{2^n}}{\dfrac{(n+1)!}{2^{n+1}}} = \lim\limits_{n\to\infty} \dfrac{2}{n+1} = 0;\quad x = -2$

11. $R = \lim\limits_{n\to\infty}\left|\dfrac{a_n}{a_{n+1}}\right| = \lim\limits_{n\to\infty}\dfrac{\dfrac{1}{(n+1)^2}}{\dfrac{1}{(n+2)^2}} = \lim\limits_{n\to\infty}\dfrac{n^2+4n+4}{n^2+2n+1} = 1; \quad (-4,-2)$

13. $R = \lim\limits_{n\to\infty}\left|\dfrac{a_n}{a_{n+1}}\right| = \lim\limits_{n\to\infty}\dfrac{2n(n+2)!}{(n+1)!(2n+2)} = \infty; \quad (-\infty,\infty)$

15. $R = \lim\limits_{n\to\infty}\left|\dfrac{a_n}{a_{n+1}}\right| = \lim\limits_{n\to\infty}\dfrac{\dfrac{n}{n+1}}{\dfrac{n+1}{n+2}} = \lim\limits_{n\to\infty}\dfrac{n(n+2)}{(n+1)^2} = \lim\limits_{n\to\infty}\dfrac{n^2+2n}{n^2+2n+1} = 1; \quad (-\tfrac{1}{2},\tfrac{1}{2})$

17. $R = \lim\limits_{n\to\infty}\left|\dfrac{a_n}{a_{n+1}}\right| = \lim\limits_{n\to\infty}\dfrac{\dfrac{n!}{3^n}}{\dfrac{(n+1)!}{3^{n+1}}} = \lim\limits_{n\to\infty}\dfrac{3}{n+1} = 0; \quad x = -1$

19. $R = \lim\limits_{n\to\infty}\left|\dfrac{a_n}{a_{n+1}}\right| = \lim\limits_{n\to\infty}\dfrac{n^3(3^{n+1})}{3^n(n+1)^3} = \lim\limits_{n\to\infty}\dfrac{3}{\left(\dfrac{n+1}{n}\right)^3} = 3; \quad (0,6)$

21. $f(x) = x^{-1}$, $f'(x) = -x^{-2}$, $f''(x) = 2x^{-3}$, $f'''(x) = -3\cdot 2x^{-4}$, ...,
$f^{(n)}(x) = (-1)^n n! x^{-n-1}$.
Therefore,
$$f(1) = 1,\ f'(1) = -1,\ f''(x) = 2,\ f'''(1) = -3!,\ ...,\ f^{(n)}(1) = (-1)^n n!,$$
and so
$$f(x) = 1 - (x-1) + \frac{2}{2!}(x-1)^2 + \cdots + \frac{(-1)^n n!}{n!}(x-1)^n + \cdots$$

$$= \sum_{n=0}^{\infty}(-1)^n(x-1)^n; \quad R = 1$$
and the interval of convergence is (0,2).

23. $f(x) = (x+1)^{-1}$, $f'(x) = -(x+1)^{-2}$, $f''(x) = 2(x+1)^{-3}$, $f'''(x) = -3!(x+1)^{-4}$, ...,
$f^{(n)}(x) = (-1)^n n!(x+1)^{-n-1}$.

Therefore, $f(2) = \dfrac{1}{3}$, $f'(2) = -\dfrac{1}{3^2}$, $f''(2) = \dfrac{2}{3^3}$, ..., $f^{(n)} = \dfrac{(-1)^n n!}{3^{n+1}}$,

and so

$$f(x) = \frac{1}{3} - \frac{1}{3^2}(x-2) + \frac{1}{3^3}(x-2)^2 + \cdots + \frac{(-1)^n}{3^{n+1}}(x-2)^n + \cdots$$

$$= \sum_{n=0}^{\infty} (-1)^n \frac{(x-2)^n}{3^{n+1}}.$$

$R = 3$ and the interval of convergence is (-1,5).

25. $f(x) = (1-x)^{-1}$, $f'(x) = (1-x)^{-2}$, $f''(x) = 2(1-x)^{-3}$, $f'(x) = 6(1-x)^{-4}$, ...,
$f^{(n)}(x) = n!(1-x)^{-n-1}$.

So

$$f(2) = -1, \; \jmath \;\; (2) = 1, \; f''(2) = -2, \; f'''(2) = 6, \ldots,$$
$$f^{(n)}(2) = (-1)^{n+1} n!$$

Therefore,

$$f(x) = -1 + (x-2) - (x-2)^2 + \cdots + (-1)^{n+1}(x-2)^n + \cdots$$

$$= \sum_{n=0}^{\infty} (-1)^{n+1}(x-2)^n.$$

$R = 1$ and the interval of convergence is (1,3).

27. $f(x) = x^{1/2}$, $f'(x) = \dfrac{1}{2}x^{-1/2}$, $f''(x) = -\dfrac{1}{4}x^{-3/2}$, $f'''(x) = \dfrac{3}{8}x^{-5/2}$,

$f^{(4)}(x) = -\dfrac{3 \cdot 5}{16}x^{-7/2}, \ldots$. So

$$f(1) = 1, \; f'(1) = \frac{1}{2}, \; f''(1) = -\frac{1}{2^2}, \; f'''(1) = \frac{1 \cdot 3}{2^3}, \ldots,$$

$$f^{(n)}(1) = (-1)^{n+1}\frac{1 \cdot 3 \cdot 5 \cdots (2n-3)}{2^n} \qquad (n \geq 2)$$

So $f(x) = 1 + \dfrac{1}{2}(x-1) + \displaystyle\sum_{n=2}^{\infty} (-1)^{n+1} \dfrac{1 \cdot 3 \cdot 5 \cdots (2n-3)}{n!2^n}(x-1)^n$

$R = 1$ and the interval of convergence is (0,2).

29. $f(x) = e^{2x}$, $f'(x) = 2e^{2x}$, $f''(x) = 2^2 e^{2x}$, ..., $f^{(n)}(x) = 2^n e^{2x}$.

$f(0) = 1$, $f'(0) = 2$, $f''(0) = 2^2$, ..., $f^{(n)}(0) = 2^n$.

Therefore,

$$f(x) = 1 + 2x + 2^2 x^2 + \cdots + \frac{2^n}{n!} x^n + \cdots = \sum_{n=0}^{\infty} \frac{2^n}{n!} x^n.$$

$R = \infty$ and the interval of convergence is $(-\infty, \infty)$.

31. $f(x) = (x+1)^{-1/2}$, $f'(x) = -\frac{1}{2}(x+1)^{-3/2}$, $f''(x) = \frac{3}{4}(x+1)^{-5/2}$.

$f'''(x) = -\frac{1 \cdot 3 \cdot 5}{2^3} x^{-7/2}$, ..., $f^{(n)}(x) = (-1)^n \frac{1 \cdot 3 \cdot 5 \cdots (2n-1)}{2^n} x^{-(2n+1)/2}$.

So $\displaystyle\sum_{n=0}^{\infty} (-1)^n \frac{1 \cdot 3 \cdot 5 \cdots (2n-1)}{n! 2^n} x^n$.

$R = 1$ and the interval of convergence is $(-1, 1)$.

33. For a Taylor series, $S_n(x) = P_n(x)$. Furthermore, $P_n(x) = f(x) - R_n(x)$. Therefore,

$$\lim_{n \to \infty} S_n(x) = \lim_{n \to \infty} P_n(x) = \lim_{n \to \infty}[f(x) - R_n(x)] = f(x) - \lim_{n \to \infty} R_n(x).$$

In other words for fixed x, the sequence of partial sums of the Taylor series of f converges to f if and only if $\lim_{n \to \infty} R_n(x) = 0$.

35. True. This follows from Theorem 9.

EXERCISES 11.6, page 756

1. $f(x) = \dfrac{1}{1-x} = \dfrac{1}{-1-(x-2)} = -\dfrac{1}{1+(x-2)}$.

Now, use the fact that

$$\frac{1}{1+u} = 1 - u + u^2 - u^3 + \cdots = \sum_{n=0}^{\infty} (-1)^n u^n \qquad |u| < 1$$

with $u = (x - 2)$ to obtain

$$f(x) = \frac{1}{1-x} = \sum_{n=0}^{\infty} (-1)^{n+1}(x-2)^n$$

with an interval of convergence of $(1, 3)$.

3.   Let $u = 3x$ in the series

$$\frac{1}{1+u} = 1 - u + u^2 - u^3 + \cdots = \sum_{n=0}^{\infty} (-1)^n u^n \qquad |u| < 1$$

and we have

$$f(x) = \frac{1}{1+3x} = 1 - 3x + (3x)^2 - (3x)^3 + \cdots = \sum_{n=0}^{\infty} (-1)^n 3^n x^n \qquad |3x| < 1$$

with an interval of convergence of $(-\frac{1}{3}, \frac{1}{3})$.

5. $f(x) = \dfrac{1}{4-3x} = \dfrac{1}{4(1-\frac{3x}{4})} = \dfrac{1}{4}\sum_{n=0}^{\infty}\left(\dfrac{3x}{4}\right)^n = \sum_{n=0}^{\infty} \dfrac{3^n}{4^{n+1}} x^n \qquad \left|\dfrac{3x}{4}\right| < 1$

with an interval of convergence of $(-\frac{4}{3}, \frac{4}{3})$.

7. $f(x) = \dfrac{1}{1-x^2} = 1 + (x^2) + (x^2)^2 + (x^2)^3 + \cdots = 1 + x^2 + x^4 + x^6 + \cdots$

$$= \sum_{n=0}^{\infty} x^{2n}.$$

The interval of convergence is $(-1,1)$.

9. $f(x) = e^{-x} = 1 + (-x) + \dfrac{(-x)^2}{2!} + \dfrac{(-x)^3}{3!} + \cdots$

$$= 1 - x + \dfrac{x^2}{2!} - \dfrac{x^3}{3!} + \cdots = \sum_{n=0}^{\infty} (-1)^n \dfrac{x^n}{n!}.$$

The interval of convergence is $(-\infty, \infty)$.

11. $f(x) = xe^{-x^2} = x\left[1 - (-x^2) + \dfrac{(-x^2)^2}{2!} + \dfrac{(-x^2)^3}{3!} + \cdots\right]$

$$= x\left(1 - x^2 + \dfrac{x^4}{2!} - \dfrac{x^6}{3!} + \cdots\right) = \sum_{n=0}^{\infty} (-1)^n \dfrac{x^{2n+1}}{n!}.$$

The interval of convergence is $(-\infty, \infty)$.

13. $f(x) = \dfrac{1}{2}(e^x + e^{-x}) = \dfrac{1}{2}\left[\left(1 + x + \dfrac{x^2}{2!} + \dfrac{x^3}{3!} + \cdots\right) + \left(1 - x + \dfrac{x^2}{2!} - \dfrac{x^3}{3!} + \cdots\right)\right]$

$$= 1 + \dfrac{x^2}{2!} + \dfrac{x^4}{4!} + \dfrac{x^6}{6!} + \cdots + \dfrac{x^{2n}}{2n!} + \cdots$$

$$= \sum_{n=0}^{\infty} \frac{x^{2n}}{2n!}$$

The interval of convergence is $(-\infty, \infty)$.

15. $\ln x = (x-1) - \frac{1}{2}(x-1)^2 + \frac{1}{3}(x-1)^3 + \cdots$

Replace x by $1 + 2x$ to obtain

$$f(x) = \ln(1+2x) = 2x - \frac{1}{2}(2x)^2 + \frac{1}{3}(2x)^3 - \cdots = \sum_{n=1}^{\infty} (-1)^{n-1} \cdot \frac{2^n x^n}{n}.$$

We must have $0 < 1 + 2x \leq 2$ so $-1 < 2x \leq 1$, or $-\frac{1}{2} < x \leq \frac{1}{2}$.

So the interval of convergence is $(-\frac{1}{2}, \frac{1}{2}]$.

17. $f(x) = \ln(1+x^2) = x^2 - \frac{1}{2}x^4 + \frac{1}{3}x^6 - \cdots$

$$f(x) = x^2 - \frac{1}{2}x^4 + \frac{1}{3}x^6 + \cdots + \frac{(-1)^{n+1}}{n}x^{2n} + \cdots = \sum_{n=1}^{\infty} (-1)^{n+1} \frac{x^{2n}}{n};$$

The interval of convergence is $(-1, 1)$.

19. Replace x by $x + 1$ in the formula for $\ln x$ in Table 11.1, giving

$$\ln(x+1) = x - \frac{1}{2}x^2 + \frac{1}{3}x^3 - \cdots \qquad (-1 < x \leq 1).$$

Next, observe that $x = x - 1 + 2 = 2\left(1 + \frac{x-2}{2}\right)$. Therefore,

$$\ln x = \ln 2\left(1 + \frac{x-2}{2}\right) = \ln 2 + \ln\left(1 + \frac{x-2}{2}\right).$$

If we replace x in the expression for $\ln(x+1)$ by $\frac{x-2}{2}$, we obtain

$$\ln x = \ln 2 + \frac{x-2}{2} - \frac{1}{2}\left(\frac{x-2}{2}\right)^2 + \frac{1}{3}\left(\frac{x-2}{2}\right)^3 - \cdots$$

$$= \ln 2 + \sum_{n=1}^{\infty} (-1)^{n-1} \left(\frac{1}{n \cdot 2^n}\right)(x-2)^n.$$

Therefore, $f(x) = (x-2)\ln x = (x-2)\ln 2 + \sum_{n=1}^{\infty} (-1)^{n-1} \left(\frac{1}{n \cdot 2^n}\right)(x-2)^n.$

To find the interval of convergence, observe that x must satisfy

$$-1 < \frac{x-2}{2} < 1, \; -2 < x - 2 \leq 2, \text{ or } 0 < x \leq 4.$$

So the interval of convergence is $(0,4]$.

21. Replacing x in the formula for $\ln x$ in Table 11.1 by $1 + x$, we obtain

$$\ln(1+x) = x - \frac{1}{2}x^2 + \frac{1}{3}x^3 - \frac{1}{4}x^4 + \cdots$$

Differentiating, we obtain

$$\frac{1}{1+x} = 1 - x + x^2 - x^3 + \cdots + (-1)^n x^n + \cdots = \sum_{n=0}^{\infty} (-1)^n x^n.$$

23. $\dfrac{1}{1+x} = 1 - x + x^2 - x^3 + \cdots$

$$\int \frac{1}{1+x}\,dx = x - \frac{1}{2}x^2 + \frac{1}{3}x^3 - \frac{1}{4}x^4 + \cdots + \frac{(-1)^{n+1}}{n} + \cdots$$

Therefore,

$$f(x) = \ln(1+x) = x - \frac{1}{2}x^2 + \frac{1}{3}x^3 - \frac{1}{4}x^4 + \cdots + \frac{(-1)^{n+1}}{n}x^n + \cdots.$$

25. $\displaystyle\int_0^{0.5} \frac{1}{\sqrt{1+x^2}}\,dx = \int_0^{0.5}\left(1 - \frac{1}{2}x^2 + \frac{3}{8}x^4 - \frac{5}{16}x^6\right)dx$

$$= x - \frac{1}{6}x^3 + \frac{3}{40}x^5 - \frac{5}{112}x^7 \Big|_0^{0.5} \approx 0.4812.$$

27. $\displaystyle\int_0^1 e^{-x^2}\,dx = \int_0^1 \left(1 - x^2 + \frac{x^4}{2} - \frac{x^6}{6} + \frac{x^8}{24}\right)dx$

$$= x - \frac{1}{3}x^3 + \frac{1}{10}x^5 - \frac{1}{42}x^7 + \frac{1}{216}x^9 \Big|_0^1 \approx 0.7475.$$

29. $\displaystyle\pi = 4\int_0^1 \frac{dx}{1+x^2} = 4\int_0^1 (1 - x^2 + x^4 - x^6 + x^8)\,dx$

$$= 4\left(x - \frac{1}{3}x^3 + \frac{1}{5}x^5 - \frac{1}{7}x^7 + \frac{1}{9}x^9\right)\Big|_0^1$$

$$= 4(1 - 0.333333 + 0.2 - 0.1428571 + 0.1111111) \approx 3.34.$$

31. $P(28 \le x \le 32) = \dfrac{1}{10\sqrt{2\pi}} \displaystyle\int_{28}^{32} e^{(-1/2)[(x-30)/10]^2}\, dx$

$$= \frac{1}{10\sqrt{2\pi}} \int_{28}^{32}\left[1 - \frac{1}{2}\left(\frac{x-30}{10}\right)^2 + \frac{1}{8}\left(\frac{x-30}{10}\right)^4 - \frac{1}{48}\left(\frac{x-30}{10}\right)^6\right] dx$$

$$= \frac{1}{10\sqrt{2\pi}}\left[1 - \frac{1}{6}\left(\frac{x-30}{10}\right)^3 + \frac{1}{40}\left(\frac{x-30}{10}\right)^5 - \frac{1}{336}\left(\frac{x-30}{10}\right)^7\right]_{28}^{32}$$

$$\approx 15.85, \text{ or } 15.85 \text{ percent.}$$

EXERCISES 11.7, page 763

1. Take $f(x) = x^2 - 3$ so that $f'(x) = 2x$. We have

$$x_{n+1} = x_n - \frac{x_n^2 - 3}{2x_n} = \frac{x_n^2 + 3}{2x_n}.$$

With $x_0 = 1.5$, we find

$x_1 = 1.75,\ x_2 = 1.7321429,\ x_3 = 1.7320508$ and so $\sqrt{3} \approx 1.732051$.

3. $x_{n+1} = x_n - \dfrac{x_n^2 - 7}{2x_n} = \dfrac{x_n^2 + 7}{2x_n}$

$x_0 = 2.5,\ x_1 = 2.65,\ x_2 = 2.645755,\ x_3 = 2.645751$

Therefore, $\sqrt{7} \approx 2.64575$.

5. $x_{n+1} = x_n - \dfrac{x_n^3 - 14}{3x_n^2} = \dfrac{2x_n^3 + 14}{3x_n^2}$

$x_0 = 2.5,\ x_1 = 2.413333,\ x_2 = 2.410146,\ x_3 = 2.410142264$

Therefore, $\sqrt[3]{14} \approx 2.410142$.

7. $f(x) = x^2 - x - 3,\ f'(x) = 2x - 1$

$$x_{n+1} = x_n - \frac{x_n^2 - x_n - 3}{2x_n - 1} = \frac{2x_n^2 - x_n - x_n^2 + x_n + 3}{2x_n - 1} = \frac{x_n^2 + 3}{2x_n - 1},$$

and $x_0 = 2$, $x_1 = 2.33333$, $x_2 = 2.30303$, $x_3 = 2.30278$, and $x_4 = 2.30278$.

9. $x_{n+1} = x_n - \dfrac{x_n^3 + 2x_n^2 + x_n - 5}{3x_n^2 + 4x_n + 1} = \dfrac{2x_n^3 + 2x_n^2 + 5}{3x_n^2 + 4x_n + 1}$

$x_0 = 1$, $x_1 = 1.53333$, $x_2 = 1.19213$, $x_3 = 1.11949$, $x_4 = 1.11635$, $x_5 = 1.11634$,
$x_6 = 1.11634$.

11. $x_{n+1} = x_n - \dfrac{\sqrt{x_n + 1} - x_n}{\frac{1}{2}(x_n + 1)^{-1/2} - 1} = \dfrac{x_n + 2}{2\sqrt{x_n + 1} - 1}$ (upon simplification)

$x_0 = 1$, $x_2 = 1.640764$, $x_2 = 1.618056$, $x_3 = 1.618034$.
Therefore, the zero is approximately 1.61803.

13. $x_{n+1} = x_n - \dfrac{e^{x_n} - \dfrac{1}{x_n}}{e^{x_n} + \dfrac{1}{x_n^2}} = \dfrac{x_n^2 e^{x_n}(x_n - 1) + 2x_n}{x_n^2 e^{x_n} + 1}$ (upon simplification)

$x_0 = 0.5$, $x_1 = 0.562187$, $x_2 = 0.56712$, $x_3 = 0.567143$. Therefore, the zero is
approximatley 0.5671.

15. a. $f(0) = -2$ and $f(1) = 3$. Since $f(x)$ is a polynomial, it is continuous.
Furthermore, since f changes sign between $x = 0$ and $x = 1$, we conclude that f has
a root in the interval $(0,1)$.

b. $f(x) = 2x^3 - 9x^2 + 12x - 2$ and $f'(x) = 6x^2 - 18x + 12$.

$$x_{n+1} = x_n - \frac{2x_n^3 - 9x_n^2 + 12x_n - 2}{6x_n^2 - 18x_n + 12} = \frac{4x_n^3 - 9x_n^2 + 2}{6x_n^2 - 18x_n + 12}$$

$x_0 = 0.5$, $x_1 = \dfrac{0.25}{4.5} = 0.055556$, $x_2 = \dfrac{1.972908}{11.0185108} \approx 0.193556$.

$x_3 = \dfrac{1.734419}{8.969390} \approx 0.193371$, $x_4 = \dfrac{1.692391}{0.193371} \approx 0.179054$.

$x_5 = \dfrac{1.691830}{0.193556} \approx 0.193556$, or 0.19356.

7. a. $f(1) = -3$ and $f(2) = 1$. Since $f(x)$ has opposite signs at $x = 1$ and $x = 2$, we see that the continuous function f has at least one zero in the interval $(1,2)$.

b. $f'(x) = 3x^2 - 3$, and the required iteration formula is

$$x_{n+1} = x_n - \frac{x_n^3 - 3x_n - 1}{3x_n^2 - 3} = \frac{3x_n^3 - 3x_n - x_n^3 + 3x_n + 1}{3x_n^2 - 3} = \frac{2x_n^3 + 1}{3x_n^2 - 3}.$$

With $x_0 = 1.5$, we find

$x_1 = 2.066667,\ x_2 = 1.900876,\ x_3 = 1.879720,\ x_4 = 1.879385,\ x_5 = 1.879385,$

and the required root is 1.87939.

19. Let $F(x) = 2\sqrt{x+3} - 2x + 1$. To solve $F(x) = 0$, we use the iteration

$$x_{n+1} = x_n - \frac{2\sqrt{x_n + 3} - 2x_n + 1}{(x_n + 3)^{-1/2} - 2}$$

$x_0 = 3,\ x_1 = 2.93654,\ x_2 = 2.93649,$ and $x_3 = 2.9365$.

21. $F(x) = e^{-x} - x + 1,\ F'(x) = -e^{-x} - 1.$

Therefore, $x_{n+1} = x_n + \dfrac{e^{-x_n} - x_n + 1}{e^{-x_n} + 1}$

$x_0 = 1.2,\ x_1 = 1.2 + \dfrac{0.1011942}{1.3011942} \approx 1.2777703,\ x_2 = 1.277703 + \dfrac{0.0000955}{1.2786767} \approx 1.2784499,$

or approximately 1.2785.

$x_3 \approx 1.2784499 + \dfrac{0.0000187}{1.2784686} \approx 1.2784645,$ or approximately 1.2785.

23. To solve $F(x) = e^{-x} - \sqrt{x} = 0$, we use the iteration

$$x_{n+1} = x_n - \frac{e^{-x_n} - x_n^{1/2}}{-e^{-x_n} - \frac{1}{2}x^{-1/2}} = x_n + \frac{2\sqrt{x_n}\,e^{-x_n} - 2x_n}{2\sqrt{x_n}\,e^{-x_n} + 1}$$

$x_0 = 1,\ x_1 = 0.271649,\ x_2 = 0.411602, x_3 = 0.426303,\ x_4 = 0.426303.$

25. The daily average cost is $\overline{C}(x) = \dfrac{C(x)}{x} = 0.0002x^2 - 0.06x + 120 + \dfrac{5000}{x}.$

To find the minimum of $\overline{C}(x)$, we set

$$\overline{C}'(x) = 0.0004x - 0.06 - \frac{5000}{x^2} = 0$$

obtaining $0.0004x^3 - 0.06x^2 - 5000 = 0$, or $x^3 - 150x^2 - 12,500,000 = 0$.

Write $f(x) = x^3 - 150x^2 - 12,500,000$ and observe that $f(0) < 0$, whereas $f(500) > 0$. So the root (critical point of $\overline{C}$) lies between $x = 0$ and $x = 500$. Take $x_0 = 250$ and use the iteration

$$x_{n+1} = x_n - \frac{x_n^3 - 150x_n^2 - 12,500,000}{3x_n^2 - 300x_n} = \frac{2x_n^3 - 150x_n^2 + 12,500,000}{3x_n^2 - 300x_n}.$$

We find $x_0 = 250$, $x_1 = 305.556$, $x_2 = 294.818$, $x_3 = 294.312$, and $x_4 = 294.311$. You can show that $\overline{C}(x)$ is concave upward on $(0, \infty)$. So the level of production that minimizes $\overline{C}(x)$ is 294 units/day.

27. We solve the equation $f(t) = 0.05t^3 - 0.4t^2 - 3.8t - 15.6 = 0$.

Use the iteration $t_{n+1} = t_n - \dfrac{0.05t_n^3 - 0.4t_n^2 - 3.8t_n - 15.6}{0.15t_n^2 - 0.8t_n - 3.8} = \dfrac{0.1t_n^3 - 0.4t_n^2 + 15.6}{0.15t_n^2 - 0.8t_n - 3.8}$

with $t_0 = 14$, obtaining $t_1 = 14.6944$, $t_2 = 14.6447$, $t_3 = 14.6445$. So the temperature is $0°F$ at 8:39 P.M.

29. Here $f(x) = 120,000x^3 - 80,000x^2 - 60,000x - 40,000$
$f'(x) = 360,000x^2 - 160,000x - 60,000$.

Therefore,

$$x_{n+1} = x_n - \frac{120,000x_n^3 - 80,000x_n^2 - 60,000x_n - 40,000}{360,000x_n^2 - 160,000x_n - 60,000}$$

$$= x_n - \frac{6x_n^3 - 4x_n^2 - 3x_n - 2}{18x_n^2 - 8x_n - 3} \cdot = \frac{12x_n^3 - 4x_n^2 + 2}{18x_n^2 - 8x_n - 3}.$$

Then, $x_0 = 1.13$, $x_1 = 1.2981455$, $x_2 = 1.2692040$, $x_3 = 1.2681893$. $x_4 = 1.2681880$, $x_5 = 1.2681880$, or $r = 26.819$.

Therefore, the rate of return is approximately 26.82 percent per year.

31. Here $C = 100,000$, $R = 1053$, $N = (12)(25) = 300$. Therefore,

$$r_{n+1} = r_n - \frac{100,000r_n + 1053\left[(1+r_n)^{-300} - 1\right]}{100,000 - 315,900(1+r_n)^{-301}}.$$

With $r_0 = 0.1$, we find

$r_1 = 0.0153$, $r_2 = 0.0100043$, $r_3 = 0.00999747$, and $r_4 = 0.00999747$.

Therefore, r is approximately 0.01 and the interest rate is approximately $12(0.01) = 0.12$ or 12 percent per year.

3.	Here $C = 6000$ (75% of 8000), $N = 12(4) = 48$, and $R = 152.18$.
$$r_{n+1} = r_n - \frac{6000r_n + 152.18[(1+r_n)^{-48} - 1]}{6000 - 7304.64(1+r_n)^{-49}}.$$
With $r_0 = 0.01$, we find
$$r_1 = 0.01 - \frac{6000(0.01) + 152.18[(1.01)^{-48} - 1]}{6000 - 7304.64(1.01)^{-49}} \approx 0.00853956,$$
$r_2 \approx 0.0083388$, $r_3 \approx 0.0083346$.
So r is approximately $12(0.008335) = 0.1000$, that is, 10 percent per year.

5.	We are required to solve the equation $s(x) = d(x)$ or
$$0.1x + 20 = \frac{50}{0.01x^2 + 1}; \ 0.001x^3 + 0.2x^2 + 0.1x + 20 = 50,$$
$$0.001x^3 + 0.2x^2 + 0.1x - 30 = 0;$$

that is, the equation $F(x) = x^3 + 200x^3 + 100x - 30,000 = 0$. We use the iteration
$$x_{n+1} = x_n - \frac{x_n^3 + 200x_n^2 + 100x_n - 30,000}{3x_n^2 + 400x_n + 100} = \frac{2x_n^3 + 200x_n^2 + 30,000}{3x_n^2 + 400x_n + 100}.$$
With $x_0 = 10$, we find $x_1 = 11.8182$, $x_2 = 11.6721$, $x_3 = 11.6711$, $x_4 = 11.6711$.
Therefore, the equilibrium quantity is approximately 11,671 units and the equilibrium price is $p = 0.1(11.671) + 20$, or \$21.17/unit.

7.	a. Let $f(x) = x^n - a$. We want to solve the equation $f(x) = 0$, $f'(x) = nx^{n-1}$.
Therefore, we have the iteration
$$x_{i+1} = x_i - \frac{f(x_i)}{f'(x_i)} = x_i - \frac{x_i^n - a}{nx_i^{n-1}}$$
$$= x_i - \frac{x_i^n}{nx_i^{n-1}} + \frac{a}{nx_i^{n-1}} = x_i - \frac{1}{n}x_i + \frac{a}{nx_i^{n-1}}$$
$$= \left(1 - \frac{1}{n}\right)x_i + \frac{a}{nx_i^{n-1}} = \left(\frac{n-1}{n}\right)x_i + \frac{a}{nx_i^{n-1}}.$$
b. Use part (a) with $n = 4$ and $a = 42$ and initial guess $x_0 = 2$ (x_0 is not unique!).
We find	$$x_{i+1} = \left(\frac{3}{4}\right)x_i + \frac{42}{4x_i^3}$$
and so

$$x_1 = \left(\frac{3}{4}\right)(2) + \frac{42}{4(2^3)} = 2.8125; \quad x_2 = (.75)(2.8125) + \frac{42}{4(2.8125)^3} \approx 2.5813$$

$$x_3 \approx (.75)(2.5813) + \frac{42}{4(2.5813)^3} \approx 2.5464.$$

CHAPTER 11 REVIEW EXERCISES, page 767

1. $f(x) = \dfrac{1}{x+2} = \dfrac{1}{(x+1)+2-1} = \dfrac{1}{1+(x+1)}.$

   $f(x) = 1 - (x+1) + (x+1)^2 - (x+1)^3 + (x+1)^4.$

3. Observe that $\ln(1+x) = x - \dfrac{1}{2}x^2 + \dfrac{1}{3}x^3 - \dfrac{1}{4}x^4.$ Therefore,

   $\ln(1+x^2) = x^2 - \frac{1}{2}(x^2)^2 + \frac{1}{3}(x^2)^3 - \frac{1}{4}(x^2)^4$

   $\qquad\qquad = x^2 - \frac{1}{2}x^4.$

5. $f(x) = x^{1/3}$, $f'(x) = \frac{1}{3}x^{-2/3}$, $f''(x) = -\frac{2}{9}x^{-5/3}.$ So

   $f(8) = 2$, $f'(8) = \dfrac{1}{12}$, $f''(8) = -\dfrac{1}{144}.$ Therefore,

   $f(x) = f(8) + f'(8)(x-8) + \dfrac{f''(8)}{2!}(x-8)^2 = 2 + \dfrac{1}{12}(x-8) - \dfrac{1}{288}(x-8)^2$

   $\sqrt[3]{7.8} = f(7.8) \approx 2 + \dfrac{1}{12}(-0.2) - \dfrac{1}{288}(-0.2)^2 \approx 1.9832.$

7. $f(x) = x^{1/3}$, $f'(x) = \dfrac{1}{3}x^{-2/3}$, $f''(x) = -\dfrac{2}{9}x^{-5/3}$, $f'''(x) = \dfrac{10}{27}x^{-8/3}$

   $f(27) = 3$, $f'(27) = \dfrac{1}{27}$, $f'(27) = -\dfrac{2}{2187}.$

   $f(x) = f(27) + f'(27)(x-27) + \dfrac{f''(27)}{2!}(x-27)^2$

   $\qquad = 3 + \dfrac{1}{27}(x-27) - \dfrac{1}{2187}(x-27)^2$

   $\sqrt[3]{26.98} = f(26.98) = 3 + \dfrac{1}{27}(-0.02) - \dfrac{1}{2187}(-0.02)^2 \approx 2.9992591.$

The error is less than $\dfrac{M}{3!}|x-27|^3$ where M is a bound for $f'''(x) = \dfrac{10}{27x^{7/3}}$ on $[26.98, 26]$. Observe that $f'''(x)$ is decreasing on the interval and so

$$|f'''(x)| \le \frac{10}{27(26.98)^{8/3}} < 0.00006. \quad \text{So the error is less than}$$

$$\frac{0.0002}{6}(0.02)^3 < 8 \times 10^{-11}.$$

. $e^x = 1 + x + \dfrac{x^2}{2!} + \dfrac{x^3}{3!} + \cdots$. Therefore,

$$e^{-1} \approx 1 - 1 + \frac{1}{2} - \frac{1}{6} + \frac{1}{24} - \frac{1}{120} + \cdots \approx 0.367.$$

11. $\displaystyle\lim_{n\to\infty} \frac{2n^2+1}{3n^2-1} = \lim_{n\to\infty} \frac{2+\frac{1}{n^2}}{3-\frac{1}{n^2}} = \frac{2}{3}.$

13. $\displaystyle\lim_{n\to\infty} a_n = \lim_{n\to\infty}\left(1-\frac{1}{2^n}\right) = \lim_{n\to\infty} 1 - \lim_{n\to\infty}\frac{1}{2^n} = 1 - 0 = 1.$

5. $\displaystyle\sum_{n=1}^{\infty}\frac{2^n}{3^n} = \sum_{n=1}^{\infty}\left(\frac{2}{3}\right)^n = \frac{2}{3}\left(\frac{1}{1-\frac{2}{3}}\right) = \frac{2}{3}\cdot 3 = 2.$

17. $\displaystyle\sum_{n=1}^{\infty}(-1)^{n-1}\left(\frac{1}{\sqrt{2}}\right)^n = \frac{1}{\sqrt{2}} - \left(\frac{1}{\sqrt{2}}\right)^2 + \cdots = \dfrac{\frac{1}{\sqrt{2}}}{1-\left(-\frac{1}{\sqrt{2}}\right)} = \frac{1}{\sqrt{2}}\cdot\dfrac{1}{1+\frac{1}{\sqrt{2}}} = \frac{1}{1+\sqrt{2}}.$

9. $1.424242\ldots = 1 + \dfrac{42}{10^2} + \dfrac{42}{10^4} + \cdots$

$$= 1 + \frac{42}{100}\left[1 + \frac{1}{100} + \left(\frac{1}{100}\right)^2 + \cdots\right]$$

$$= 1 + \frac{42}{100}\left(\frac{1}{1-\frac{1}{100}}\right) = 1 + \frac{42}{100}\left(\frac{100}{99}\right) = 1 + \frac{42}{99} = \frac{141}{99}.$$

21. Let $a_n = \dfrac{n^2+1}{2n^2-1}$. Since $\lim\limits_{n\to\infty}\dfrac{n^2+1}{2n^2-1} = \lim\limits_{n\to\infty}\dfrac{1+\dfrac{1}{n^2}}{2-\dfrac{1}{n^2}} = \dfrac{1}{2} \neq 0$, the divergence test

implies that the series $\sum\limits_{n=1}^{\infty} a_n$ diverges.

23. $\sum\limits_{n=1}^{\infty}\left(\dfrac{1}{n}\right)^{1.1} = \sum\limits_{n=1}^{\infty}\dfrac{1}{n^{1.1}}$ is a convergent p-series with $p = 1.1 > 1$.

25. $R = \lim\limits_{n\to\infty}\left|\dfrac{a_n}{a_{n+1}}\right| = \lim\limits_{n\to\infty}\dfrac{\dfrac{1}{n^2+2}}{\dfrac{1}{(n+1)^2+2}} = \lim\limits_{n\to\infty}\dfrac{(n+1)^2+2}{n^2+2} = \lim\limits_{n\to\infty}\dfrac{n^2+2n+3}{n^2+2} = 1.$

The interval of convergence is $(-1,1)$.

27. $R = \lim\limits_{n\to\infty}\left|\dfrac{a_n}{a_{n+1}}\right| = \lim\limits_{n\to\infty}\dfrac{(n+1)(n+2)}{n(n+1)} = 1.$

So $R = 1$ and the interval of convergence is $(0,2)$.

29. We have $\dfrac{1}{1+x} = 1 + x + x^2 + x^3 + \cdots$ $\quad(-1 < x < 1)$

Replacing x by $2x$ in the expression, we find
$$f(x) = \dfrac{1}{2x-1} = -\dfrac{1}{1-2x} = -[1 + 2x + (2x)^2 + (2x)^3 + \cdots]$$
$$= -1 - 2x - 4x^2 - 8x^3 - \cdots - 2^n x^n - \cdots. \qquad (-\tfrac{1}{2} < x < \tfrac{1}{2}).$$

31. We know that
$$\ln(1+x) = x - \dfrac{1}{2}x^2 + \dfrac{1}{3}x^3 - \dfrac{1}{4}x^4 + \cdots + (-1)^{n+1}\dfrac{x^n}{n} \qquad (-1 < x < 1).$$
Replace x by $2x$, to obtain
$$f(x) = \ln(1+2x) = 2x - 2x^2 + \dfrac{8}{3}x^3 - \cdots + (-1)^{n+1}\dfrac{2^n x^n}{n} = \sum\limits_{n=0}^{\infty}(-1)^{n+1}\dfrac{2^n}{n}x^n.$$
The interval of convergence is $(-\tfrac{1}{2}, \tfrac{1}{2}]$..

3. $f(x) = x^3 - 12$, $f'(x) = 3x^2$

Therefore, $x_{n+1} = x_n - \dfrac{x_n^3 - 12}{3x_n^2} = \dfrac{3x_n^3 - x_n^3 + 12}{3x_n^2} = \dfrac{2x_n^3 + 12}{3x_n^2}$.

Using $x_0 = 2$, we have $x_1 = 2.3333333$, $x_1 = 2.2902491$, $x_3 = 2.2894277$, and the root is approximately 2.28943.

5. We solve the equation $F(x) = 2x - e^{-x}$. $F'(x) = 2 + e^{-x}$. So the iteration is

$$x_{n+1} = x_n - \frac{2x_n - e^{-x_n}}{2 + e^{-x_n}} = \frac{(x_n + 1)e^{-x_n}}{2 + e^{-x_n}}.$$

Taking $x_0 = 0.5$, we find $x_1 = 0.349045$, $x_2 = 0.351733$, $x_3 = 0.351733$.
So the point of intersection is approximately $(0.35173, 0.70346)$.

7. The amount required

$$A = 10,000[e^{-0.09} + e^{-0.09(2)} + \cdots] = \frac{10,000e^{-0.09}}{1 - e^{-0.09}} = 106,186.10,$$

or \$106,186.10.

9. We compute

$$P(63.5 \le x \le 65.5) = \frac{1}{2.5\sqrt{2\pi}} \int_{63.5}^{65.5} e^{-1/2[(x-64.5)/2.5]^2}.$$

Replacing x with $-\frac{1}{2}[(x - 64.5)/2.5]^2$ in the expression

$$e^x = 1 + x + \frac{1}{2!}x^2 + \frac{1}{3!}x^3,$$

we obtain

$$e^{-1/2[(x-64.5)/2.5]^2} \approx 1 - \tfrac{1}{2}[(x - 64.5)/2.5]^2 + \tfrac{1}{2!}\left\{-\tfrac{1}{2}[(x - 64.5)/2.5]^2\right\}^2$$

$$+ \tfrac{1}{3!}\left\{-\tfrac{1}{2}[(x - 64.5)/2.5]^2\right\}^3$$

$$= 1 - \tfrac{1}{12.5}(x - 64.5)^2 + \tfrac{1}{312.5}(x - 64.5)^4 - \tfrac{1}{11718.75}(x - 64.5)^6.$$

Therefore,
$P(63.5 \le x \le 65.5) \approx$

$$\frac{1}{2.5\sqrt{2\pi}} \int_{63.5}^{65.5}\left[1 - \frac{1}{12.5}(x - 64.5)^2 + \frac{1}{312.5}(x - 64.5)^4 - \frac{1}{11718.75}(x - 64.5)^6\right] dx$$

$$= \frac{1}{2.5\sqrt{2\pi}} \left[x - \frac{1}{37.5}(x-64.5)^3 + \frac{1}{1607.5}(x-64.5)^5 - \frac{1}{82031.25}(x-64.5)^7 \right]_{63.5}^{65.5}$$

$$= \frac{1}{2.5\sqrt{2\pi}} [(65.5 - 0.02667 + 0.00062 - 0.00001)$$

$$- (63.5 + 0.02667 - 0.00062 + 0.00001)]$$

$$\approx 0.3108, \text{ or approximately } 31.08\%.$$

CHAPTER 12

. $450° = \dfrac{450}{180}\pi = \dfrac{5\pi}{2}$ rad.

. $-270° = -\dfrac{270}{180}\pi = -\dfrac{3\pi}{2}$ rad.

. a. III   b. III   c. II   d. I

. $f(x) = \dfrac{\pi}{180}x$ radians, $f(75) = \dfrac{\pi}{180}(75)$ radians $= \dfrac{5\pi}{12}$ radians

. $f(x) = \dfrac{\pi}{180}x$ radians, $f(160) = \dfrac{\pi}{180}(160)$ radians $= \dfrac{8\pi}{9}$ radians

1. $f(630) = \dfrac{\pi}{180}(630)$ radians $= \dfrac{7\pi}{2}$ radians

3. $g\left(\dfrac{2\pi}{3}\right) = \left(\dfrac{180}{\pi}\right)\left(\dfrac{2\pi}{3}\right) = 120°.$ 15. $g\left(-\dfrac{3\pi}{2}\right) = \left(\dfrac{180}{\pi}\right)\left(-\dfrac{3\pi}{2}\right) = -270°.$

7. $g\left(\dfrac{22\pi}{18}\right) = \left(\dfrac{180}{\pi}\right)\left(\dfrac{22\pi}{18}\right) = 220°.$

9.

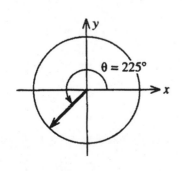

21.

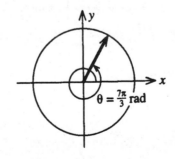

12 Trigonometric Functions

23. $\dfrac{5\pi}{6}$ rad $= 150°$;   coterminal angle is $-210°$.

25. $-\dfrac{\pi}{4}$ rad $= -45°$;   coterminal angle: $360° - 45° = 315°$

27. True. $3630 = (360)(10) + 30$ and the result is evident.

29. True. Adding $n(360)$ degrees to θ revolves the angle θ $|n|$ revolutions,

    clockwise if n is positive and counter-clockwise if n is negative.

EXERCISES 12.2, page 783

1. $\sin 3\pi = 0.$

3. $\sin\left(\dfrac{9\pi}{2}\right) = 1.$

5. $\sin\left(-\dfrac{4\pi}{3}\right) = \sin\left(\dfrac{\pi}{3}\right) = \dfrac{\sqrt{3}}{2}.$

7. $\tan\dfrac{\pi}{6} = \dfrac{\sqrt{3}}{3}.$

9. $\sec\left(-\dfrac{5\pi}{8}\right) = \sec\left(\dfrac{5\pi}{8}\right) = -2.6131.$

11. $\sin\dfrac{\pi}{2} = 1$, $\cos\dfrac{\pi}{2} = 0$, $\tan\dfrac{\pi}{2}$ is undefined, $\sec\dfrac{\pi}{2}$ is undefined,
    $\csc\dfrac{\pi}{2} = 1$, $\cot\dfrac{\pi}{2} = 0.$

13. $\sin\left(\dfrac{5\pi}{3}\right) = -\dfrac{\sqrt{3}}{2}$, $\cos\left(\dfrac{5\pi}{3}\right) = \dfrac{1}{2}$, $\tan\left(\dfrac{5\pi}{3}\right) = -\sqrt{3}$,
    $\csc\left(\dfrac{5\pi}{3}\right) = -\dfrac{2\sqrt{3}}{2}$, $\sec\left(\dfrac{5\pi}{3}\right) = 2$, $\cot\left(\dfrac{5\pi}{3}\right) = -\dfrac{\sqrt{3}}{3}.$

15. $\theta = \dfrac{7\pi}{6}$ or $\dfrac{11\pi}{6}.$

17. $\theta = \dfrac{5\pi}{6}$ or $\dfrac{11\pi}{6}$

19. $\theta = \pi$

21. $\sin\theta = \sin\left(-\dfrac{4\pi}{3}\right) = \sin\left(\dfrac{2\pi}{3}\right) = \sin\left(\dfrac{\pi}{3}\right)$, so $\theta = \dfrac{2\pi}{3}$ or $\dfrac{\pi}{3}.$

23.

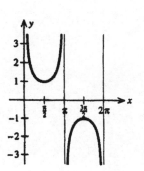

25.

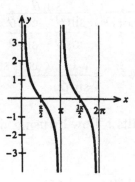

27.

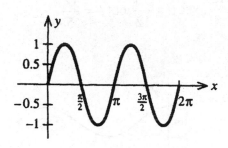

29.

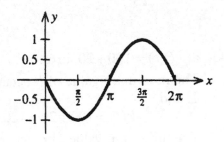

21. $\cos^2 \theta - \sin^2 \theta = \cos^2 \theta - (1 - \cos^2 \theta) = 2\cos^2 \theta - 1.$

23. $(\sec \theta + \tan \theta)(1 - \sin \theta) = \sec \theta + \tan \theta - \sin \theta \sec \theta - \tan \theta \sin \theta$
$$= \sec \theta + \tan \theta - \tan \theta - \tan \theta \sin \theta$$
$$= \frac{1}{\cos \theta} - \frac{\sin^2 \theta}{\cos \theta} = \frac{1}{\cos \theta}(1 - \sin^2 \theta)$$
$$= \frac{\cos^2 \theta}{\cos \theta} = \cos \theta.$$

25. $(1 + \cot^2 \theta)\tan^2 \theta = \csc^2 \theta \tan^2 \theta = \frac{1}{\sin^2 \theta} \cdot \frac{\sin^2 \theta}{\cos^2 \theta} = \frac{1}{\cos^2 \theta}$
$$= \sec^2 \theta.$$

12 Trigonometric Functions

37. $\dfrac{\csc\theta}{\tan\theta+\cot\theta}=\dfrac{\dfrac{1}{\sin\theta}}{\dfrac{\sin\theta}{\cos\theta}+\dfrac{\cos\theta}{\sin\theta}}=\dfrac{\dfrac{1}{\sin\theta}}{\dfrac{\sin^2\theta+\cos^2\theta}{\cos\theta\sin\theta}}$

$$=\dfrac{1}{\sin\theta}\cdot\dfrac{\cos\theta}{1}\cdot\dfrac{\sin\theta}{1}=\cos\theta.$$

39. The results follow by using similar triangles.

41. $|AB|=\sqrt{169-25}=12.$

$\sin\theta=\dfrac{5}{13},\ \cos\theta=\dfrac{12}{13},\ \tan\theta=\dfrac{5}{12},\ \csc\theta=\dfrac{13}{5},\ \sec\theta=\dfrac{13}{12},\cot\theta=\dfrac{12}{5}.$

43. a. $P(t)=100+20\sin 6t.$

The maximum value of P occurs when $\sin t=1$, and
$$P(t)=100+20=120.$$
The minimum value of P occurs when $\sin 6t=-1$ and
$$P(t)=100-20=80.$$

b. $\sin 6t=1$ implies that $6t=\frac{\pi}{2},\frac{3\pi}{2},\ldots$; that is, $t=\dfrac{\pi(4n+1)}{12}$ $\qquad(n=0,1,2,\ldots).$

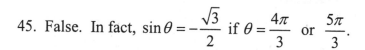

$\sin 6t=-1$ implies that $6t=\frac{3\pi}{2},\frac{7\pi}{2}$; that is, $t=\dfrac{\pi(4n+3)}{12}$ $\qquad(n=0,1,2,\ldots).$

45. False. In fact, $\sin\theta=-\dfrac{\sqrt{3}}{2}$ if $\theta=\dfrac{4\pi}{3}$ or $\dfrac{5\pi}{3}.$

47. True. $\cos 2\theta=\cos^2\theta-\sin^2\theta=\cos^2\theta-(1-\cos^2\theta)=2\cos^2\theta-1.$

EXERCISES 12.3, page 793

1. $f(x)=\cos 3x;\ f'(x)=-3\sin 3x.$

3. $f(x) = 2\cos \pi x;\ f'(x) = (-2\sin \pi x)\dfrac{d}{dx}(\pi x) = -2\pi \sin \pi x.$

5. $f(x) = \sin(x^2+1);\ f'(x) = \cos(x^2+1)\dfrac{d}{dx}(x^2+1) = 2x\cos(x^2+1).$

7. $f(x) = \tan 2x^2;\ f'(x) = (\sec^2 2x^2)\dfrac{d}{dx}(2x^2) = 4x\sec^2 2x^2.$

9. $f(x) = x\sin x\ ;\ f'(x) = \sin x + x\cos x.$

11. $f(x) = 2\sin 3x + 3\cos 2x;$
$f'(x) = (2\cos 3x)(3) - (3\sin 2x)(2) = 6(\cos 3x - \sin 2x).$

13. $f(x) = x^2\cos 2x;$
$f'(x) = 2x\cos 2x + x^2(-\sin 2x)(2) = 2x(\cos 2x - x\sin 2x).$

15. $f(x) = \sin\sqrt{x^2-1} = \sin(x^2-1)^{1/2}$
$f'(x) = \cos\sqrt{x^2-1}\,\dfrac{d}{dx}(x^2-1)^{1/2} = (\cos\sqrt{x^2-1})(\tfrac{1}{2})(x^2-1)^{-1/2}(2x)$

$\qquad = \dfrac{x\cos\sqrt{x^2-1}}{\sqrt{x^2-1}}.$

17. $f(x) = e^x\sec x$
$f'(x) = e^x\sec x + e^x(\sec x\tan x) = e^x\sec x(1+\tan x).$

19. $f(x) = x\cos\dfrac{1}{x};$

$f'(x) = \cos\dfrac{1}{x} + x(-\sin\tfrac{1}{x})\dfrac{d}{dx}\left(\dfrac{1}{x}\right) = \cos\dfrac{1}{x} - x\left(\dfrac{1}{x^2}\right)\sin\dfrac{1}{x}$

$\qquad = \dfrac{1}{x}\left(\sin\dfrac{1}{x} + x\cos\dfrac{1}{x}\right)$ or $\cos\dfrac{1}{x} + \dfrac{1}{x}\sin\dfrac{1}{x}.$

12 Trigonometric Functions

21. $f(x) = \dfrac{x - \sin x}{1 + \cos x}$

$$f'(x) = \frac{(1 + \cos x)(1 - \cos x) - (x - \sin x)(-\sin x)}{(1 + \cos x)^2}$$

$$= \frac{1 - \cos^2 x + x \sin x - \sin^2 x}{(1 + \cos x)^2} = \frac{x \sin x}{(1 + \cos x)^2}.$$

23. $f(x) = \sqrt{\tan x}; \quad f'(x) = \dfrac{1}{2}(\tan x)^{-1/2} \sec^2 x = \dfrac{\sec^2 x}{2\sqrt{\tan x}}.$

25. $f(x) = \dfrac{\sin x}{x}; \quad f'(x) = \dfrac{x \cos x - \sin x}{x^2}.$

27. $f(x) = \tan^2 x. \quad f'(x) = 2 \tan x \sec^2 x.$

29. $f(x) = e^{\cot x}; \quad f'(x) = e^{\cot x}(-\csc^2 x) = -\csc^2 x \cdot e^{\cot x}.$

31. $f(x) = \cot 2x. \quad f'(x) = -2 \csc^2 2x.$

Therefore, $f'\left(\frac{\pi}{4}\right) = -\dfrac{2}{\sin^2\left(\frac{\pi}{2}\right)} = -2.$

Then $y - 0 = -2\left(x - \frac{\pi}{4}\right)$, or $y = -2x + \frac{\pi}{2}.$

33. $f(x) = e^x \cos x.$

$f'(x) = e^x \cos x + e^x(-\sin x) = e^x(\cos x - \sin x).$

Setting $f'(x) = 0$ gives $\cos x - \sin x = 0$, or $\tan x = 1$; that is, when $x = \frac{\pi}{4}$, or $\frac{5\pi}{4}$. From the following sign diagram for f'

we see that f is increasing on $(0, \frac{\pi}{4}) \cup (\frac{5\pi}{4}, 2\pi)$ and decreasing on $(\frac{\pi}{4}, \frac{5\pi}{4}).$

35.

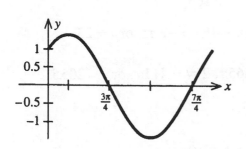

37.

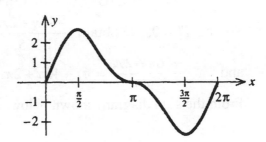

39. a. $f(x) = \sin x, \ f'(x) = \cos x, \ f''(x) = -\sin x, \ f'''(x) = -\cos x,$
$f^{(4)}(x) = \sin x, \ ... \ .$

$f(0) = 0, \ f'(0) = 1, \ f''(0) = 0, \ f'''(0) = -1, \ f^{(4)}(0) = 0, \ ...,$

Therefore, $f(x) = \sin x = x - \dfrac{x^3}{3!} + \dfrac{x^5}{5!} - \dfrac{x^7}{7!} + \cdots + (-1)^{n+1}\dfrac{x^{2n+1}}{(2n+1)!} + \cdots$

b. $\displaystyle\lim_{x \to 0}\dfrac{\sin x}{x} = \lim_{x \to 0}(1 - \dfrac{x^2}{3!} + \dfrac{x^4}{5!} - \cdots) = 1.$

41. $P(t) = 8000 + 1000\sin\left(\dfrac{\pi t}{24}\right).$

$P'(t) = [1000\cos\left(\dfrac{\pi t}{24}\right)](\dfrac{\pi}{24})$ and $P''(12) = \dfrac{1000\pi}{24}\cos\dfrac{\pi}{2} = 0;$ that is, the

wolf population is not changing during the twelfth month.

$$p(t) = 40{,}000 + 12{,}000\cos\left(\dfrac{\pi t}{24}\right)$$

$$p'(t) = [-12{,}000\sin\left(\dfrac{\pi t}{24}\right)](\dfrac{\pi}{24})$$

and $p'(12) = -500\sin\dfrac{\pi(12)}{24} = -500\pi;$ that is, the caribou population is decreasing
at the rate of 1571/month.

43. $f(t) = 3\sin\dfrac{2\pi}{365}(t - 79) + 12. \quad f'(t) = 3\cos\dfrac{2\pi}{365}(t - 79) \cdot \dfrac{2\pi}{365}$

Therefore, $f'(79) = \dfrac{6\pi}{365} \approx 0.05164.$ The number of hours of daylight is increasing
at the rate of 0.05 hrs per day on March 21.

45. $T = 62 - 18\cos\dfrac{2\pi(t-23)}{365}$. $T' = \dfrac{36\pi}{365}\sin\left(\dfrac{-46\pi + 2\pi t}{365}\right)$.

Setting $T' = 0$, we obtain $\dfrac{-46\pi + 2\pi t}{365} = 0$, $-46\pi = -2\pi t$, or $t = 23$,

and $\dfrac{-46\pi + 2\pi t}{365} = \pi$, $-46\pi + 2\pi t = 365\pi$, $2\pi t = 411\pi$, or $t = 205.5$.

From the sign diagram shown below,

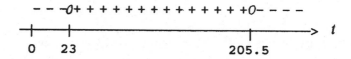

```
- - -0+ + + + + + + + + + + + + +0- - - -
+----------+----------------------------------+------------> t
0         23                                205.5
```

we see that a minimum occurs at $t = 23$ and a maximum occurs at $t = 205.5$. We conclude that the warmest day is July 25th and the coldest day is January 23rd.

47. $R(t) = 2(5 - 4\cos\frac{\pi t}{6})$. $R'(t) = \frac{4\pi}{3}\sin(\frac{\pi t}{6})$.

Setting $R' = 0$, we obtain $\frac{\pi t}{6} = 0$, and conclude that $t = 0$, $t = 6$, and $t = 12$.
From the sign diagram for R

```
0 + + + + +0- - - - - 0 + + + +
+--------------+-------------+------------> t
0              6            12
```

we conclude that a maximum occurs at $t = 6$.

49. From Problem 46, we have
$$V'(t) = \frac{6}{5\pi}\left(\sin\frac{\pi t}{2}\right)\left(\frac{\pi}{2}\right) = \frac{3}{5}\sin\frac{\pi t}{2}.$$
Then $V''(t) = \frac{3}{5}\cos\frac{\pi t}{2}\left(\frac{\pi}{2}\right) = \frac{3\pi}{10}\cos\frac{\pi t}{2}$.

Setting $V'(t) = 0$ gives $t = 1, 3, 5, 7, 9, 11, 13, 15, \dots$, as critical points. Evaluating $V''(t)$ at each of these points, we see that the rate of flow of air is at a maximum when $t = 1, 5, 9, 13, \dots$, and it is at a minimum when $t = 3, 7, 11, 15, \dots$.

51. $\tan\theta = \dfrac{y}{20}$. Therefore, $\sec^2\theta \cdot \dfrac{d\theta}{dt} = \dfrac{1}{20}\cdot\dfrac{dy}{dt}$. We want to find $\dfrac{dy}{dt}$ when $z = 30$.

But when $z = 30$, $y = \sqrt{900 - 400} = \sqrt{500} = 10\sqrt{5}$, and $\sec\theta = \dfrac{30}{20} = \dfrac{3}{2}$.

Therefore, with

$$\frac{d\theta}{dt} = \frac{\pi}{2} \text{ rad/sec, we find}$$

$$\frac{dy}{dt} = 20 \sec^2 \theta \cdot \frac{d\theta}{dt} = 20\left(\frac{3}{2}\right)^2 \cdot \frac{\pi}{2} = \frac{45\pi}{2} \approx 70.7$$

or 70.7 ft/sec.

53. The area of the cross-section is

$$A = (2)(\tfrac{1}{2})(5\cos\theta)(5\sin\theta) + 5(5\sin\theta)$$

$$= 25(\cos\theta\sin\theta + \sin\theta).$$

$$\frac{dA}{d\theta} = 25(-\sin^2\theta + \cos^2\theta + \cos\theta)$$

$$= 25(\cos^2\theta - 1 + \cos^2\theta + \cos\theta)$$

$$= 25(2\cos^2\theta + \cos\theta - 1)$$

$$= 25(2\cos\theta - 1)(\cos\theta + 1).$$

Setting $\frac{dA}{d\theta} = 0$ gives $\cos\theta = \frac{1}{2}$ or $\cos\theta = -1$, or $\theta = \frac{\pi}{3}$ or π. The absolute

maximum of A occurs at $\theta = \frac{\pi}{3}$ as can be seen from the following table:

θ	A
0	0
$\frac{\pi}{3}$	$\frac{75\sqrt{3}}{4}$
π	0

Therefore, the angle shoud be $60°$.

55. Let $f(\theta) = \theta - 0.5\sin\theta - 1$. then $f'(\theta) = 1 - 0.5\cos\theta$ and so Newton's Method
leads to the iteration.

$$\theta_{i+1} = \theta_i - \frac{f(\theta_i)}{f'(\theta_i)} = \theta_i - \frac{\theta_i - 0.5\sin\theta_i - 1}{1 - 0.5\cos\theta_i} = \frac{\theta_i - 0.5\theta_i\cos\theta_i - \theta_i + 0.5\sin\theta_i + 1}{1 - 0.5\cos\theta_i}$$

$$= \frac{1 + 0.5(\sin\theta_i - \theta_i\cos\theta_i)}{1 - 0.5\cos\theta_i}.$$

With $\theta_i = 1.5$, we find

12 Trigonometric Functions

$$\theta_1 = \frac{1 + 0.5(\sin 1.5 - 1.5\cos 1.5)}{1 - 0.5\cos 1.5} = \frac{1.4456946}{0.9646314} \approx 1.4987.$$

$$\theta_2 = \frac{1 + 0.5(\sin 1.4987 - 1.4987\cos 1.4987)}{1 - 0.5\cos 1.4987} = \frac{1.4447225}{0.9639831} \approx 1.4987.$$

57. True. Let $f(x) = \sin x$. Then $f'(0) = \lim_{x \to 0} \frac{\sin x - \sin 0}{x} = \lim_{x \to 0} \frac{\sin x}{x}$.

But $f'(0) = \cos 0 = 1$. So $\lim_{x \to 0} \frac{\sin x}{x} = 1$.

59. False. Take $x = \pi$. Then f has a relative minimum if $x = \pi$, but g does not have a relative maximum at $x = \pi$.

61. $h(x) = \csc f(x) = \dfrac{1}{\sin f(x)}$.

$$h'(x) = \frac{-\cos f(x) \cdot f'(x)}{\sin^2 f(x)} = -\big(\csc f(x) \cot f(x)\big) f'(x).$$

63. $h(x) = \cot f(x) = \dfrac{\cos f(x)}{\sin f(x)}$.

$$h'(x) = \frac{\sin f(x)[-\sin f(x)]f'(x) - (\cos f(x))[\cos f(x)]f'(x)}{\sin^2 f(x)}$$

$$= \frac{\big[-\sin^2 f(x) - \cos^2 f(x)\big]f'(x)}{\sin^2 f(x)} = \frac{-1 \cdot f'(x)}{\sin^2 f(x)}$$

$$= -\csc^2 f(x) f'(x).$$

USING TECHNOLOGY EXERCISES 12.3, page 797

1. 1.2038   3. 0.7762   5. -0.2368

7. 0.8415; -0.2172   9. 1.1271; 0.2013

1. a.

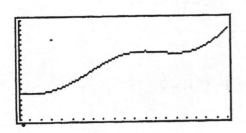

b. ≈ \$0.63   c. ≈ \$27.79

13. a.

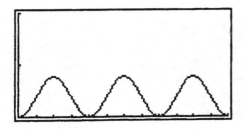

15. ≈ 0.006 ft.

EXERCISES 12.4, page 891

1. $\int \sin 3x \, dx = -\frac{1}{3}\cos 3x + C.$

3. $\int (3\sin x + 4\cos x) \, dx = -3\cos x + 4\sin x + C.$

5. Let $u = 2x$ so that $du = 2dx$. Then
$$\int \sec^2 2x \, dx = \frac{1}{2}\int \sec^2 u \, du = \frac{1}{2}\tan u + C = \frac{1}{2}\tan 2x + C.$$

7. Let $u = x^2$ so that $du = 2x \, dx$, or $x \, dx = \frac{1}{2}du$. Then
$$\int x\cos x^2 \, dx = \frac{1}{2}\int \cos u \, du = \frac{1}{2}\sin u + C = \frac{1}{2}\sin x^2 + C.$$

9. Let $u = \pi x$ so that $du = \pi \, dx$, or $dx = \frac{1}{\pi}du$. Then

$$\int \csc \pi x \cot \pi x \, dx = \frac{1}{\pi} \int \csc u \cot u \, du = -\frac{1}{\pi} \csc u + C$$

$$= -\frac{1}{\pi} \csc \pi x + C.$$

11. $\displaystyle\int_{-\pi/2}^{\pi/2} (\sin x + \cos x) dx = -\cos x + \sin x \Big|_{-\pi/2}^{\pi/2} = 1 - (-1) = 2.$

13. $\displaystyle\int_{0}^{\pi/6} \tan 2x \, dx = -\tfrac{1}{2} \ln|\cos 2x| \Big|_{0}^{\pi/6} = -\tfrac{1}{2} \ln \tfrac{1}{2}.$

15. Let $u = \sin x$ so that $du = \cos x \, dx$. Then
$$\int \sin^3 x \cos x \, dx = \int u^3 du = \tfrac{1}{4} u^4 + C = \tfrac{1}{4} \sin^4 x + C.$$

17. Let $u = \pi x$ so that $du = \pi \, dx = \tfrac{1}{\pi} du$. Then
$$\int \sec \pi x \, dx = \frac{1}{\pi} \int \sec u \, du = \frac{1}{\pi} \ln|\sec u + \tan u| + C$$
$$= \frac{1}{\pi} \ln|\sec \pi x + \tan \pi x| + C.$$

19. Let $u = 3x$ so that $du = 3dx$, or $du = \tfrac{1}{3} du$. If $x = 0$, then $u = 0$ and if $x = \frac{\pi}{12}$, then $u = \frac{\pi}{4}$. Then
$$\int_{0}^{\pi/12} \sec 3x \, dx = \tfrac{1}{3} \int_{0}^{\pi/4} \sec u \, du = \tfrac{1}{3} \ln|\sec u + \tan u| \Big|_{0}^{\pi/4}$$
$$= \tfrac{1}{3}[\ln(\sqrt{2} + 1) - \ln 1] = \tfrac{1}{3} \ln(\sqrt{2} + 1).$$

21. Let $u = \cos x$ so that $du = -\sin x \, dx$. Then
$$\int \sqrt{\cos x} \sin x \, dx = -\int \sqrt{u} \, du = -\tfrac{2}{3} u^{3/2} + C = -\tfrac{2}{3} (\cos x)^{3/2} + C.$$

23. $\displaystyle\int \cos 3x (1 - 2 \sin 3x)^{1/2} \, dx.$ Put $u = \sin 3x$, then $du = 3\cos 3x \, dx$. Therefore,
$$I = \tfrac{1}{3} \int (1 - 2u)^{1/2} \, du = -\tfrac{1}{6} \cdot \tfrac{2}{3} (1 - 2u)^{3/2} + C = -\tfrac{1}{9} (1 - 2 \sin 3x)^{3/2} + C.$$

25. $\displaystyle\int \tan^3 x \sec^2 x \, dx = \tfrac{1}{4} \tan^4 x + C.$

27. Let $u = \cot x - 1$ so that $du = -\csc^2 x\, dx$. Then

$$\int \csc^2 x(\cot x - 1)^3\, dx = -\int u^3\, du = -\tfrac{1}{4}u^4 + C = -\tfrac{1}{4}(\cot x - 1)^4 + C.$$

29. Put $u = \ln x$, $du = \dfrac{1}{x}dx$. Therefore,

$$I = \int \sin u\, du = -\cos u + C = -\cos(\ln x) + C$$

and $\displaystyle\int_1^{e^\pi} \frac{\sin(\ln x)}{x}\, dx = -\cos(\ln x)\Big|_1^{e^\pi} = -\cos\pi + \cos 0 = 2.$

31. $\displaystyle\int \sin(\ln x)\, dx$. Let $I = \displaystyle\int \sin(\ln x)\, dx$. Put $u = \sin(\ln x)$, $dv = dx$.

Then $\quad du = \dfrac{\cos(\ln x)}{x}$, $v = x$. Integrating by parts, we have

$$I = x\sin(\ln x) - \int \cos(\ln x)dx.$$

Let $J = \displaystyle\int \cos(\ln x)\, dx$ and $u = \cos(\ln x)dx$. Then $u = \cos(\ln x), dv = dx$, and

$$du = -\frac{\sin(\ln x)}{x}dx, \text{ and } v = x,$$

and $\quad J = x\cos(\ln x) + \displaystyle\int \sin(\ln x)dx = x\cos(\ln x) + I.$

So $\quad I = x[\sin(\ln x) - \cos(\ln x)] - I,$

or $\quad 2I = x[\sin(\ln x) - \cos(\ln x)].$ Therefore,

$$\int \sin(\ln x)\, dx = \tfrac{1}{2}x[\sin(\ln x) - \cos(\ln x)] + C.$$

33. Area $= \displaystyle\int_0^\pi \cos\tfrac{x}{4}\, dx = 4\sin\tfrac{x}{4}\Big|_0^\pi = 4\left(\tfrac{\sqrt{2}}{2}\right) = 2\sqrt{2}$ sq units.

35. $A = \displaystyle\int_0^{\pi/4} \tan x\, dx = -\ln|\cos x|\Big|_0^{\pi/4} = -\ln\tfrac{\sqrt{2}}{2} = -\ln\tfrac{1}{\sqrt{2}}$

$\quad = \ln\sqrt{2} = \tfrac{1}{2}\ln 2$ sq units.

37. $A = \displaystyle\int_0^\pi (x - \sin x)\, dx = \tfrac{1}{2}x^2 + \cos x\Big|_0^\pi = \tfrac{1}{2}\pi^2 - 1 - 1 = \tfrac{1}{2}\pi^2 - 2 = \tfrac{1}{2}(\pi^2 - 4)$ sq units.

39. The average is $A = \frac{1}{15}\int_0^{15}(80+3t\cos\frac{\pi t}{6})dt = 80+\frac{1}{5}\int_0^{15}t\cos\frac{\pi t}{6}\,dt.$

Integrating by parts, we find $\int t\cos\frac{\pi t}{6}\,dt = \left(\frac{6}{\pi}\right)^2\left[\cos\frac{\pi t}{6}+\frac{\pi t}{6}\sin\frac{\pi t}{6}\right].$

So $A = 80+\frac{1}{5}\left(\frac{6}{\pi}\right)^2\left[\cos\frac{\pi t}{6}+\frac{\pi t}{6}\sin\frac{\pi t}{6}\right]\Big|_0^{15} = 80+\frac{1}{5}\left(\frac{6}{\pi}\right)^2\left(\frac{15\pi}{6}-1\right)$
≈ 85, or approximately \$85 per share.

41. $R = \int_0^{12}2(5-4\cos\frac{\pi t}{6})\,dt = 10t - 8\cdot\left(\frac{6}{\pi}\right)\sin\frac{\pi t}{6}\Big|_0^{12} = 120$, or \$120,000,

43. The required volume is
$$V = \int R(t)dt = 0.6\int \sin\frac{\pi t}{2}\,dt = (0.6)\left(\frac{2}{\pi}\right)\cos\frac{\pi t}{2}+C = -\frac{1.2}{\pi}\cos\frac{\pi t}{2}+C.$$
When $t = 0$, $V = 0$, and so $0 = -\frac{1.2}{\pi}+C$, and $C = \frac{1.2}{\pi}$. Therefore,
$$V = \frac{1.2}{\pi}\left(1-\cos\frac{\pi t}{2}\right).$$

45. $\dfrac{dQ}{dt} = 0.0001(4+5\cos 2t)Q(400-Q).$

The differential equation can be written in the form
$$\frac{dQ}{Q(400-Q)} = 0.0001(4+5\cos 2t)dt \text{ or}$$

$$\frac{1}{400}\left[\frac{1}{Q}+\frac{1}{400-Q}\right]dQ = 0.0001(4+5\cos 2t)dt \text{ . So}$$

$$\int\left(\frac{1}{Q}+\frac{1}{400-Q}\right)dQ = 0.04(4+5\cos 2t)dt$$

$$\ln Q - \ln|400-Q| = 0.04(4t+2.5\sin 2t)+C_1$$

$$\ln\left|\frac{Q}{400-Q}\right| = 0.04(4t+2.5\sin 2t)+C_1$$

$$\frac{Q}{400-Q} = Ce^{0.04(4t+2.5\sin 2t)}.$$

Using the condition $Q(0) = 10$, we have $\dfrac{10}{390} = C.$

So $\dfrac{Q}{400-Q} = \dfrac{10}{390}e^{0.04(4t+25\sin 2t)}$. When $t = 20$, we have

$$\frac{Q(20)}{400-Q(20)} = \frac{10}{390}e^{0.04(80+2.5\sin 40)} \approx 0.6777.$$

$$Q(20) = 0.6777[400-Q(20)] = 271.08 - 0.6777Q(20)$$

$$1.6777Q(20) \approx 271.08, \text{ and } Q(20) \approx 161.578,$$

or approximately 162 flies.

47. $$\int_0^1 \cos t^2 \, dt = \int_0^1 \left(1 - \frac{t^4}{2} + \frac{t^8}{24} - \frac{t^{12}}{720}\right) dt = t - \frac{t^5}{10} + \frac{t^9}{216} - \frac{t^{13}}{9360}\Bigg|_0^1$$

$$= 1 - \frac{1}{10} + \frac{1}{216} - \frac{1}{9360} \approx 1 - 0.1 + 0.0046296 - 0.0001068 \approx 0.9.$$

49. True.

$$\int_0^{b+2\pi} \cos x \, dx = \sin x\Big|_a^{b+2\pi} = \sin(b+2\pi) - \sin a$$

$$= \sin b \cos 2\pi + \cos b \sin 2\pi - \sin a$$

$$= \sin b - \sin a = \int_a^b \cos x \, dx$$

51. True. $\displaystyle\int_{-\pi/2}^{\pi/2} |\sin x| \, dx = 2\int_0^{\pi/2} \sin x \, dx = -2\cos x\Big|_0^{\pi/2} = 2$

$$\int_{-\pi/2}^{\pi/2} |\cos x| \, dx = 2\int_0^{\pi/2} \cos x \, dx = -2\sin x\Big|_0^{\pi/2} = 2.$$

USING TECHNOLOGY EXERCISES 12.4, page 805

1. 0.5419   3. 0.7544   5. 0.2231

7. 0.6587   9. -0.2032   11. 0.9045

13. a.   b. 2.2687 sq units

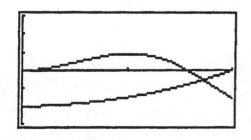

15. a.   b.   1.8239 sq units

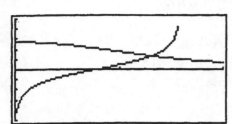

17. a.   b.   1.2484 sq units

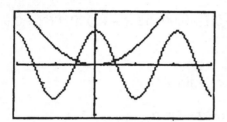

19. a.   b.   1.0983 sq units

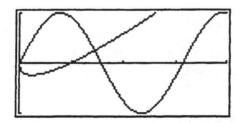

21. 7.6 ft

CHAPTER 12, REVIEW EXERCISES, page 808

1. $120° = \frac{2\pi}{3}$ rad.   3. $-225° = -\frac{225}{180}\pi = -\frac{5\pi}{4}$ rad.

5. $-\frac{5\pi}{2}$ rad $= -\frac{5}{2}(180) = -450°$.

7. $\cos\theta = \frac{1}{2}$ implies that $\theta = \frac{\pi}{3}$ or $\frac{5\pi}{3}$.

9. $f(x) = \sin 3x; \; f'(x) = 3\cos 3x.$

1. $f(x) = 2\sin x - 3\cos 2x.$

   $f'(x) = 2\cos x + 3(\sin 2x)(2) = 2(\cos x + 3\sin 2x).$

3. $f(x) = e^{-x}\tan 3x.$

   $f'(x) = -e^{-x}\tan 3x + e^{-x}(\sec^2 3x)(3) = e^{-x}(3\sec^2 3x - \tan 3x).$

15. $f(x) = 4\sin x\cos x.$

   $f'(x) = 4[\sin x(-\sin x) + (\cos x)(\cos x)] = 4(\cos^2 x - \sin^2 x) = 4\cos 2x.$

17. $f(x) = \dfrac{1 - \tan x}{1 - \cot x}.$

   $$f'(x) = \frac{(1 - \cot x)(-\sec^2 x) - (1 - \tan x)(\csc^2 x)}{(1 - \cot x)^2}$$

   $$= \frac{-\sec^2 x + \sec^2 x\cot x - \csc^2 x + \csc^2 x\tan x}{(1 - \cot x)^2}$$

   $$= \frac{(\cot x - 1)\sec^2 x - (1 - \tan x)\csc^2 x}{(1 - \cot x)^2}.$$

19. $f(x) = \sin(\sin x);\ f'(x) = \cos(\sin x)\cdot\cos x.$

21. $f(x) = \tan^2 x.\ f'(x) = 2\tan x\sec^2 x.$

   So the slope is $f'\left(\frac{\pi}{4}\right) = 2(1)(\sqrt{2})^2 = 4,$ and equation of the tangent line is

   $$y - 1 = 4(x - \tfrac{\pi}{4}),\ \text{or}\ y = 4x + 1 - \pi.$$

23. Let $u = \frac{2}{3}x$ so that $du = \frac{2}{3}dx,$ or $dx = \frac{3}{2}du.$ So

   $$\int \cos\tfrac{2}{3}x\,dx = \tfrac{3}{2}\int \cos u\,du = \tfrac{3}{2}\sin u + C = \tfrac{3}{2}\sin\tfrac{2}{3}x + C.$$

25. $\int x\csc x^2\cot x^2\,dx.$ Put $u = x^2,$ then $du = 2x\,dx.$ Then

   $$\int x\csc x^2\cot x^2\,dx = \tfrac{1}{2}\int \csc u\cot u\,du$$

   $$= -\tfrac{1}{2}\csc u + C = -\tfrac{1}{2}\csc x^2 + C.$$

27. Let $u = \sin x$ so that $du = \cos x \, dx$. Then

$$\int \sin^2 x \cos x \, dx = \int u^2 du = \tfrac{1}{3}\sin^3 x + C.$$

29. Let $u = \sin x$ so that $du = \cos x \, dx$. Then

$$\int \frac{\cos x}{\sin^2 x} dx = \int \frac{du}{u^2} = \int u^{-2} \, du = -\frac{1}{u} + C = -\frac{1}{\sin x} + C = -\csc x + C.$$

31. $\displaystyle \int_{\pi/6}^{\pi/2} \frac{\cos x}{1 - \cos^2 x} dx = \int_{\pi/6}^{\pi/2} \frac{\cos x}{\sin^2 x} dx = \int_{\pi/6}^{\pi/2} \sin^{-2} x \cos x \, dx$

$$= -\frac{1}{\sin x}\Big|_{\pi/6}^{\pi/2} = -1 + 2 = 1.$$

33. $\displaystyle A = \int_{\pi/4}^{5\pi/4} (\sin x - \cos x) \, dx = -\cos x - \sin x \Big|_{\pi/4}^{5\pi/4}$

$$= \left(\tfrac{\sqrt{2}}{2} + \tfrac{\sqrt{2}}{2}\right) - \left(-\tfrac{\sqrt{2}}{2} - \tfrac{\sqrt{2}}{2}\right) = 2\sqrt{2} \quad \text{sq units.}$$

35. $R(t) = 60 + 37 \sin^2\left(\tfrac{\pi t}{12}\right)$.

$R'(t) = 37(2)\sin\left(\tfrac{\pi t}{12}\right)\cdot \cos\left(\tfrac{\pi t}{12}\right)\left(\tfrac{\pi}{12}\right) = \tfrac{37\pi}{6}\sin\left(\tfrac{\pi t}{6}\right)$ $[\sin 2\theta = 2\sin\theta\cos\theta]$

Setting $R'(t) = 0$ gives $\tfrac{\pi t}{6} = 0,\ \pi,\ 2\pi,\ \dots$. So $t = 0, 6, 12, \dots$.
Therefore, $t = 6$ is a critical point in $(0,12)$. Now
$$R'(t) = \tfrac{37\pi}{6}\cos\left(\tfrac{\pi t}{6}\right) \cdot \tfrac{\pi}{6} = \tfrac{37\pi^2}{36}\cos\left(\tfrac{\pi t}{6}\right).$$

Since $R''(6) = -\tfrac{37\pi^2}{36} < 0$, the Second Derivative Test implies that the occupancy rate is highest when $t = 6$ (the beginning of December).Next, we set $R''(t) = 0$ giving $\tfrac{\pi t}{6} = \tfrac{\pi}{2}$, or $t = 3$. You can show that $R'''(3) < 0$ and so $R'(t)$ is maximized at $t = 3$. So the occupancy rate is increasing most rapidly at the beginning of September.